London Mathematical Society Student Texts 47

Introductory Lectures on Rings and Modules

John A. Beachy
Northern Illinois University

 CAMBRIDGE
UNIVERSITY PRESS

PUBLISHED BY THE PRESS SYNDICATE OF THE UNIVERSITY OF CAMBRIDGE
The Pitt Building, Trumpington Street, Cambridge CB2 1RP, United Kingdom

CAMBRIDGE UNIVERSITY PRESS
The Edinburgh Building, Cambridge, CB2 2RU, United Kingdom
40 West 20th Street, New York, NY 10011-4211, USA
10 Stamford Road, Oakleigh, Melbourne 3166, Australia

First published 1999

Printed in the United Kingdom at the University Press, Cambridge

A catalogue record for this book is available from the British Library

Library of Congress Cataloging-in-Publication Data

Beachy, John A.
  Introductory lectures on rings and modules / John A. Beachy
    p. cm. – (London Mathematical Society student texts; 47)
  ISBN 0 521 64340 6 (hbk.) ISBN 0 521 64407 0 (pbk.)
  1. Noncommutative rings (Algebra) 2. Modules (Algebra)
I. Title. II. Series
QA251.4.B43 1999
512'.4–dc21 9854417 CIP

40516757

ISBN 0 521 64340 6 hardback
ISBN 0 521 64407 0 paperback

Contents

PREFACE

This set of lecture notes is focused on the noncommutative aspects of the study of rings and modules. It is intended to complement the book *Steps in Commutative Algebra*, by R. Y. Sharp, which provides excellent coverage of the commutative theory. It is also intended to provide the necessary background for the book *An Introduction to Noncommutative Noetherian Rings*, by K. R. Goodearl and R. B. Warfield.

The core of the first three chapters is based on my lecture notes from the second semester of a graduate algebra sequence that I have taught at Northern Illinois University. I have added additional examples, in the hope of making the material accessible to advanced undergraduate students. To provide some variety in the examples, there is a short section on modules over the Weyl algebras. This section is marked with an asterisk, as it can be omitted without causing difficulties in the presentation. (The same is true of Section 1.5 and Section 3.4.) Chapter 4 provides an introduction to the representation theory of finite groups. Its goal is to lead the reader into an area in which there has been a very successful interaction between ring theory and group theory.

Certain books are most useful as a reference, while others are less encyclopedic in nature, but may be an easier place to learn the material for the first time. It is my hope that students will find these notes to be accessible, and a useful source from which to learn the basic material. I have included only as much material as I have felt it is reasonable to try to cover in one semester. The role of an encyclopedic text is played by any one of the standard texts by Jacobson, Hungerford, and Lang. My personal choice for a reference is *Basic Algebra* by N. Jacobson.

There are many possible directions for subsequent work. To study noncommutative rings the reader might choose one of the following books: *An Introduction to Noncommutative Noetherian Rings*, by K. R. Goodearl and R. B. Warfield, *A First Course in Noncommutative Rings*, by T. Y. Lam, and *A Course in Ring Theory*, by D. S. Passman. After finishing Chapter 4 of this text, the reader should have the background necessary to study *Representations and Characters of Finite Groups*, by M. J. Collins. Another possibility is to study *A Primer of Algebraic D-Modules*, by S. C. Coutinho.

I expect the reader to have had prior experience with algebra, either at the advanced undergraduate level, or in a graduate level course on Galois theory and the structure of groups. Virtually all of the prerequisite material can be found in undergraduate books at the level of Herstein's *Abstract Algebra*. For the sake of completeness, two definitions will be given at this point. A *group* is a nonempty set G together with a binary operation $\cdot$ on G such that the following conditions hold: (i) for all $a, b, c \in G$, we have $a \cdot (b \cdot c) = (a \cdot b) \cdot c$; (ii) there exists $1 \in G$ such that $1 \cdot a = a$ and $a \cdot 1 = a$ for all $a \in G$; (iii) for each $a \in G$ there exists an element $a^{-1} \in G$ such that $a \cdot a^{-1} = 1$ and $a^{-1} \cdot a = 1$. The group G is said to be *abelian* if $a \cdot b = b \cdot a$ for all $a, b \in G$, and in this case the symbol $\cdot$ for the operation on G is usually replaced by a $+$ symbol, and the identity element is denoted by the symbol 0 rather than by the symbol 1. The definition of an abelian group is fundamental, since the objects of study in the text (rings and modules) are constructed by endowing an abelian group with additional structure.

I sincerely hope that the reader's prior experience with algebra has included the construction of examples. Good examples provide the foundation for understanding this material. I have included a variety of them, but it is best if additional ones are constructed by the reader. A good example illustrates the key ideas of a definition or theorem, but is not so complicated as to obscure the important points. Each definition should have several associated examples that will help in understanding and remembering the conditions of the definition. It is helpful to include some that do *not* satisfy the stated conditions.

I would like to take this opportunity to thank my colleagues for many helpful conversations: Bill Blair, Harvey Blau, Harald Ellers (who made corrections in the last chapter), and George Seelinger (who class-tested the manuscript). I would also like to thank Svetlana Butler, Sonia Edghill, Lauren Grubb, Suzanne Riehl, and Adam Slagell, who gave me lists of misprints. I would like to dedicate the book to my daughter Hannah, with thanks for her patience while I was writing, and for her help in proofreading.

John A. Beachy
DeKalb, Illinois
August, 1998

Chapter 1

RINGS

The abstract definition of a ring identifies a set of axioms that underlies some familiar sets: the set $\mathbf{Z}$ of integers, the set of polynomials $\mathbf{Q}[x]$ with coefficients in the field $\mathbf{Q}$ of rational numbers, and the set $\mathrm{M}_n(\mathbf{R})$ of $n \times n$ matrices with coefficients in the field $\mathbf{R}$ of real numbers. Much of the interest in what we now call a ring had its origin in number theory. In the field $\mathbf{C}$ of complex numbers, Gauss used the subset

$$\mathbf{Z}[\mathrm{i}] = \{a + b\,\mathrm{i} \in \mathbf{C} \mid a, b \in \mathbf{Z}\}$$

to prove facts about the integers, after first showing that unique factorization into 'primes' still holds in this context. The next step was to consider subsets of the form $\mathbf{Z}[\zeta]$, where ζ is a complex root of unity. These rings were used to prove some special cases of Fermat's last theorem.

Kummer was interested in higher reciprocity laws, and found it necessary to investigate unique factorization in subrings of the field $\mathbf{C}$ of complex numbers. He realized that the analog of the prime factorization theorem in $\mathbf{Z}$ need not hold in all such subrings, but he was able to prove such a theorem in enough cases to obtain Fermat's last theorem for exponents up to 100. The modern definition of an ideal was given in 1871 by Dedekind, who proved that in certain subrings of $\mathbf{C}$ every nonzero ideal can be expressed uniquely as a product of

1

prime ideals. We will see that prime ideals play a crucial role even in the case of noncommutative rings.

The search for a way to extend the concept of unique factorization motivated much of the early work on commutative rings. An integral domain with the property that every nonzero ideal can be expressed uniquely as a product of prime ideals is now called a Dedekind domain. Lasker developed a parallel theory for polynomial rings, in which an ideal is written as an intersection of primary ideals (rather than as a product of prime ideals). Both of these theories were axiomatized by Emmy Noether, who worked with rings that satisfy the ascending chain condition for ideals. The commutative theories are beyond the scope of this text; the interested reader can consult the texts by Sharp [23], Matsumura [21], and Eisenbud [8].

The term 'number ring' was used in 1897 in a paper by Hilbert. The current definition of an abstract ring seems to have first appeared in the 1921 paper "Theory of ideals in rings" published by Emmy Noether. She played a prominent role in the early (1920–1940) development of commutative ring theory, along with Krull.

Section 1.1 and Section 1.2 introduce some of the basic definitions that will be used later in the book. For our purposes, integral domains and rings of matrices form two of the most important classes of rings. Section 1.3 and Section 1.4 provide the tools for a study of unique factorization in integral domains. Section 1.5 introduces general matrix rings, and several other noncommutative examples. This section, together with Sections 2.8 and 3.4, carries forward the noncommutative emphasis of the text, but is not crucial to the development in other sections.

1.1   Basic definitions and examples

Working with polynomials and matrices leads naturally to a study of algebraic systems with two operations, similar to ordinary addition and multiplication of integers. Although we assume that the reader has some knowledge of commutative rings, especially integral domains and fields, we will give the definition of a ring in full detail. Including the study of even the set of 2×2 matrices over the real numbers means that we cannot impose the commutative law for multiplication.

We recall that a *binary operation* on a set S is a function from the Cartesian product $S \times S$ into S. If $*$ is a binary operation on S, then for all ordered pairs $a, b \in S$, the value $a * b$ is uniquely defined and belongs to S. Thus the operation is said to satisfy the *closure* property on S.

Definition 1.1.1 *Let R be a set on which two binary operations are defined, called* addition *and* multiplication, *and denoted by* $+$ *and* $\cdot$, *respectively. Then R is called a* ring *with respect to these operations if the following properties hold.*

(i) Associative laws: *For all $a, b, c \in R$,*

$$a + (b + c) = (a + b) + c \quad and \quad a \cdot (b \cdot c) = (a \cdot b) \cdot c .$$

(ii) Commutative law (for addition): *For all $a, b \in R$,*

$$a + b = b + a .$$

(iii) Distributive laws: *For all $a, b, c \in R$,*

$$a \cdot (b + c) = a \cdot b + a \cdot c \quad and \quad (a + b) \cdot c = a \cdot c + b \cdot c .$$

(iv) Identity elements: *The set R contains elements 0 and 1 (not necessarily distinct) called, respectively, an* additive identity element *and a* multiplicative identity element, *such that for all $a \in R$,*

$$a + 0 = a \quad and \quad 0 + a = a ,$$

and

$$a \cdot 1 = a \quad and \quad 1 \cdot a = a .$$

(v) Additive inverses: *For each $a \in R$, the equations*

$$a + x = 0 \quad and \quad x + a = 0$$

have a solution x in R, called the additive inverse *of a, and denoted by $-a$.*

The definition of a ring R can be summarized by stating that R is an abelian group under addition, with a multiplication that satisfies the associative and distributive laws. Furthermore, R is assumed to have a multiplicative identity element. (Note that if $1 = 0$, then R is said to be *trivial*.) Many authors choose not to require the existence of a multiplicative identity, and there are important examples that do not satisfy the associative law, so, strictly speaking, we have given the definition of an 'associative ring with identity element.'

The distributive laws provide the only connection between the operations of addition and multiplication. In that sense, they represent the only reason for studying the operations simultaneously instead of separately. For example, in the following list of additional properties of a ring, any property using the additive identity or additive inverses in an equation involving multiplication must depend on the distributive laws.

Our observation that any ring is an abelian group under addition implies that the cancellation law holds for addition, that the additive identity element is unique, and that each element has a unique additive inverse. The remaining properties in the following list can easily be verified.

Let R be a ring, with elements $a, b, c \in R$.
(a) If $a + c = b + c$, then $a = b$.
(b) If $a + b = 0$, then $b = -a$.
(c) If $a + b = a$ for some $a \in R$, then $b = 0$.
(d) For all $a \in R$, $a \cdot 0 = 0$.
(e) For all $a \in R$, $(-1) \cdot a = -a$.
(f) For all $a \in R$, $-(-a) = a$.
(g) For all $a, b \in R$, $(-a) \cdot (-b) = a \cdot b$.

A ring can have only one multiplicative identity element. If $1 \in R$ and $e \in R$ both satisfy the definition of a multiplicative identity, then $1 \cdot e = e$ since 1 is an identity element, and $1 \cdot e = 1$ since e is an identity element. Thus $e = 1 \cdot e = 1$, showing that 1 is the unique element that satisfies the definition.

Various sets of numbers provide the most elementary examples of rings. In these sets the operation of multiplication is commutative. The set **Z** of integers should be the first example of a ring. In this ring we have the additional property that if $a \neq 0$ and $b \neq 0$, then $ab \neq 0$. The set **Q** of rational numbers, the set **R** of real numbers, and the set **C** of complex numbers also form rings, and in each of these rings every nonzero element has a multiplicative inverse.

We next review the definitions of some well-known classes of rings.

Definition 1.1.2 *Let R be a ring.*

(a) *The ring R is called a* commutative ring *if $a \cdot b = b \cdot a$ for all $a, b \in R$.*

(b) *The ring R is called an* integral domain *if R is commutative, $1 \neq 0$, and $a \cdot b = 0$ implies $a = 0$ or $b = 0$, for all $a, b \in R$.*

(c) *The ring R is called a* field *if R is an integral domain such that for each nonzero element $a \in R$ there exists an element $a^{-1} \in R$ such that $a \cdot a^{-1} = 1$.*

According to the above definition, **Z** is an integral domain, but not a field. The sets **Q**, **R**, and **C** are fields, since in each set the inverse of a nonzero element is again in the set. We assume that the reader is familiar with the ring $F[x]$ of polynomials with coefficients in a field F. This ring provides another example of an integral domain that is not a field.

If the cancellation law for multiplication holds in a commutative ring R, then for any elements $a, b \in R$, $ab = 0$ implies that $a = 0$ or $b = 0$. Conversely, if this condition holds and $ab = ac$, then $a(b - c) = 0$, so if $a \neq 0$ then $b - c = 0$ and $b = c$. Thus in a ring R with $1 \neq 0$, the cancellation law for multiplication holds if and only if R is an integral domain. It is precisely this property that is crucial in solving polynomial equations. We note that any field is an integral domain, since the existence of multiplicative inverses for nonzero elements implies that the cancellation law holds.

It is possible to consider polynomials with coefficients from any commutative ring, and it is convenient to give the general definition at this point.

Example 1.1.1 (Polynomials over a commutative ring)

Let R be any commutative ring. We let $R[x]$ denote the set of infinite tuples

$$(a_0, a_1, a_2, \ldots)$$

such that $a_i \in R$ for all i, and $a_i \neq 0$ for only finitely many terms a_i. Two elements of $R[x]$ are equal (as sequences) if and only if all corresponding entries are equal. We introduce addition and multiplication as follows:

$$(a_0, a_1, a_2, \ldots) + (b_0, b_1, b_2, \ldots) = (a_0 + b_0, a_1 + b_1, a_2 + b_2, \ldots)$$

and

$$(a_0, a_1, a_2, \ldots) \cdot (b_0, b_1, b_2, \ldots) = (c_0, c_1, c_2, \ldots),$$

where

$$c_k = \sum_{i+j=k} a_i b_j = \sum_{i=0}^{k} a_i b_{k-i}.$$

Showing that $R[x]$ is a ring under these operations is left as Exercise 2, at the end of this section.

We can identify $a \in R$ with $(a, 0, 0, \ldots) \in R[x]$, and so if R has an identity 1, then $(1, 0, 0, \ldots)$ is a multiplicative identity for $R[x]$. If we let $x = (0, 1, 0, \ldots)$, then $x^2 = (0, 0, 1, 0, \ldots)$, $x^3 = (0, 0, 0, 1, 0, \ldots)$, etc. Thus the elements of $R[x]$ can be expressed uniquely in the form

$$a_0 + a_1 x + \cdots + a_{m-1} x^{m-1} + a_m x^m,$$

allowing us to use the standard notation for the *ring of polynomials over R in the indeterminate x*. Note that two elements of $R[x]$ are equal if and only if the corresponding *coefficients* a_i are equal. We refer to R as the *ring of coefficients* of $R[x]$.

If n is the largest nonnegative integer such that $a_n \neq 0$, then we say that the polynomial has *degree n*, and a_n is called the *leading coefficient* of the polynomial.

Once we know that $R[x]$ is a ring, it is easy to work with polynomials in two indeterminates x and y. We can simply use the ring $R[x]$ as the coefficient ring, and consider all polynomials in the indeterminate y, with coefficients in $R[x]$. For example, by factoring out the appropriate terms in the following polynomial in x and y we have

$$2x - 4xy + y^2 + xy^2 + x^2y^2 - 3xy^3 + x^3y^2 + 2x^2y^3$$
$$= 2x + (-4x)y + (1 + x + x^2 + x^3)y^2 + (-3x + 2x^2)y^3 \,,$$

showing how it can be regarded as a polynomial in y with coefficients in $R[x]$. The ring of polynomials in two indeterminates with coefficients in R is usually denoted by $R[x,y]$, rather than by $(R[x])[y]$, which would be the correct notation.

Now let D be any integral domain. Then the ring $D[x]$ of all polynomials with coefficients in D is also an integral domain. To show this we note that if $f(x)$ and $g(x)$ are nonzero polynomials with leading coefficients a_m and b_n, respectively, then since D is an integral domain, the product $a_m b_n$ is nonzero. This shows that the leading coefficient of the product $f(x)g(x)$ is nonzero, and so $f(x)g(x) \neq 0$. In actuality, we have $f(x)g(x) \neq 0$ because the degree of $f(x)g(x)$ is equal to $\deg(f(x)) + \deg(g(x))$.

We note that in the definition of the polynomial ring $R[x_1, x_2, \ldots, x_n]$, the coefficients could just as easily have come from a noncommutative ring. In that case, in defining the notion of a polynomial ring we generally require that the indeterminates commute with each other and with the coefficients.

Many familiar rings fail to be integral domains. Let R be the set of all functions from the set of real numbers into the set of real numbers, with ordinary addition and multiplication of functions (not composition of functions). It is not hard to show that R is a commutative ring, since addition and multiplication are defined pointwise, and the addition and multiplication of real numbers satisfy all of the field axioms. To show that R is not an integral domain, let $f(x) = 0$ for $x < 0$ and $f(x) = x$ for $x \geq 0$, and let $g(x) = 0$ for $x \geq 0$ and $g(x) = x$ for $x < 0$. Then $f(x)g(x) = 0$ for all x, which shows that $f(x)g(x)$ is the zero function.

We next consider several noncommutative examples.

Example 1.1.2 (2×2 Matrices over a field)

We assume that the reader is familiar with the ring $M_2(F)$ of 2×2 matrices with entries in the field F. A review of the properties of matrices is provided in Section A.3 of the appendix. We recall that for matrices $\begin{bmatrix} a_{11} & a_{12} \\ a_{21} & a_{22} \end{bmatrix}$ and $\begin{bmatrix} b_{11} & b_{12} \\ b_{21} & b_{22} \end{bmatrix}$ in $M_2(F)$, their product is given by the following matrix.

$$\begin{bmatrix} a_{11} & a_{12} \\ a_{21} & a_{22} \end{bmatrix} \begin{bmatrix} b_{11} & b_{12} \\ b_{21} & b_{22} \end{bmatrix} = \begin{bmatrix} a_{11}b_{11} + a_{12}b_{21} & a_{11}b_{12} + a_{12}b_{22} \\ a_{21}b_{11} + a_{22}b_{21} & a_{21}b_{12} + a_{22}b_{22} \end{bmatrix}$$

The matrix $A = \begin{bmatrix} a_{11} & a_{12} \\ a_{21} & a_{22} \end{bmatrix}$ is invertible if and only if its determinant $\det(A) = a_{11}a_{22} - a_{12}a_{21}$ is nonzero. This follows from the next equation, found by multiplying A by its adjoint $\mathrm{adj}(A)$.

$$\begin{bmatrix} a_{11} & a_{12} \\ a_{21} & a_{22} \end{bmatrix}\begin{bmatrix} a_{22} & -a_{12} \\ -a_{21} & a_{11} \end{bmatrix} = \begin{bmatrix} a_{11}a_{22} - a_{12}a_{21} & 0 \\ 0 & a_{11}a_{22} - a_{12}a_{21} \end{bmatrix}$$

An interesting particular case is that of a lower triangular matrix with nonzero entries on the main diagonal. For a matrix of this type we have the following formula.

$$\begin{bmatrix} a & 0 \\ b & c \end{bmatrix}^{-1} = \begin{bmatrix} a^{-1} & 0 \\ -a^{-1}bc^{-1} & c^{-1} \end{bmatrix}$$

We recall the definition of a linear transformation. If V and W are vector spaces over the field F, then a *linear transformation* from V to W is a function $f : V \to W$ such that $f(c_1v_1 + c_2v_2) = c_1f(v_1) + c_1f(v_2)$ for all vectors $v_1, v_2 \in V$ and all scalars $c_1, c_2 \in F$. For a vector space V of dimension n, there is a one-to-one correspondence between $n \times n$ matrices with entries in F and linear transformations from V into V. (See Theorem A.4.1 of the appendix.)

To give the most general such example, we need to recall the definition of a homomorphism of abelian groups. If A and B are abelian groups, with the operation denoted additively, then a *group homomorphism* from A to B is a function $f : A \to B$ such that $f(x_1 + x_2) = f(x_1) + f(x_2)$, for all $x_1, x_2 \in A$. If the group homomorphism maps A into A, it is called an *endomorphism* of A. We will use the notation $\mathrm{End}(A)$ for the set of all endomorphisms of A.

We will show in Proposition 1.2.8 that rings of the form $\mathrm{End}(A)$ are the generic rings, in the same sense that permutation groups are the generic examples of groups.

Example 1.1.3 (Endomorphisms of an abelian group)

Let A be an abelian group, with its operation denoted by $+$. We define addition and multiplication of elements of $\mathrm{End}(A)$ as follows: if $f, g \in \mathrm{End}(A)$, then

$$[f + g](x) = f(x) + g(x) \quad \text{and} \quad [f \cdot g](x) = f(g(x))$$

for all $x \in A$. Thus we are using pointwise addition and composition of functions as the two operations.

Since A is an abelian group, it is easy to check that addition of functions is an associative, commutative, binary operation. The identity element is the zero function defined by $f(x) = 0$ for all

$x \in A$, where 0 is the identity of A. The additive inverse of an element $f \in \text{End}(A)$ is defined by $[-f](x) = -(f(x))$ for all $x \in A$.

Multiplication is well-defined since the composition of two group homomorphisms is again a group homomorphism. The associative law holds for composition of functions, but, in general, the commutative law does not hold. There is a multiplicative identity element, given by the identity function 1_A.

The most interesting laws to check are the two distributive laws. (We must check both since multiplication is not necessarily commutative.) If $f, g, h \in \text{End}(A)$, then $(f + g) \cdot h = (f \cdot h) + (g \cdot h)$ since

$$
\begin{aligned}
((f + g) \cdot h)(x) &= (f + g)(h(x)) \\
&= f(h(x)) + g(h(x)) \\
&= (f \cdot h)(x) + (g \cdot h)(x) \\
&= ((f \cdot h) + (g \cdot h))(x)
\end{aligned}
$$

for all $x \in A$. Furthermore, $h \cdot (f + g) = (h \cdot f) + (h \cdot g)$ since

$$
\begin{aligned}
[h \cdot (f + g)](x) &= h(f(x) + g(x)) \\
&= h(f(x)) + h(g(x)) \\
&= [(h \cdot f) + (h \cdot g)](x)
\end{aligned}
$$

for all $x \in A$. The last argument is the only place we need to use the fact that h is a group homomorphism. This completes the proof that $\text{End}(A)$ is a ring.

The list of axioms for a ring is rather exhausting to check. In many cases we will see that the set we are interested in is a subset of a known ring. If the operations on the subset are the same as those of the known ring, then only a few of the axioms need to be checked. To formalize this idea, we need the following definition.

Definition 1.1.3 *A subset S of a ring R is called a* subring *of R if S is a ring under the operations of R, and the multiplicative identity of S coincides with that of R.*

Let F and E be fields. If F is a subring of E, according to the above definition, then we usually say (more precisely) that F is a *subfield* of E, or that E is an *extension field* of F. Of course, there may be subrings of fields that are not necessarily subfields.

Any subring is a subgroup of the underlying abelian group of the larger ring, so the two rings must have the same zero element. If F is any field contained in E, then the set of nonzero elements of F is a subgroup of the multiplicative group of nonzero elements of E, and so the multiplicative identity elements of F and E must coincide. Thus F must be a subfield of E.

Suppose that D is a subring of a field F, and that $ab = 0$ for some $a, b \in D$. If a is nonzero, then the definition of a field implies that there exists an element $a^{-1} \in F$ with $a \cdot a^{-1} = 1$. Multiplying both sides of the equation $ab = 0$ by a^{-1} shows that $b = 0$. We conclude that D is an integral domain. In Section 1.3 we will show that the converse holds: if D is any integral domain, then it is possible to construct a field F in which D can be considered to be a subring. Thus integral domains can be characterized as subrings of fields.

The next proposition is useful in constructing subrings.

Proposition 1.1.4 *Let S be a ring, and let R be a nonempty subset of S. Then R is a subring of S if and only if*

   (i) *R is closed under the addition and multiplication of S,*

   (ii) *if $a \in R$, then $-a \in R$,*

  (iii) *the multiplicative identity of S belongs to R.*

Proof. If R is a subring, then the closure axioms must certainly hold. Condition (ii) holds since R is a subgroup of S under addition. Condition (iii) holds by definition of a subring.

Conversely, suppose that the given conditions hold. The first condition shows that condition (i) of Definition 1.1.1 is satisfied. Conditions (i)–(iii) of Definition 1.1.1 are inherited from S. Finally, since R is nonempty, it contains some element, say $a \in R$. Then $-a \in R$, so $0 = a + (-a) \in R$ since R is closed under addition. By assumption the multiplicative identity 1 of S belongs to R, and this serves as a multiplicative identity for R. Since identity elements are unique, this shows that 1 is the multiplicative identity for R. Thus conditions (iv) and (v) of Definition 1.1.1 are also satisfied. $\square$

The first example of a subring should be to consider $\mathbf{Z}$ as a subset of $\mathbf{Q}$. Another interesting subring is found in the field $\mathbf{C}$ of complex numbers. The set $\mathbf{Z}[\mathrm{i}]$ is called the ring of *Gaussian integers*. It is by definition the set of complex numbers of the form $m + n\mathrm{i}$, where $m, n \in \mathbf{Z}$. Since

$$(m + n\mathrm{i}) + (r + s\mathrm{i}) = (m + r) + (n + s)\mathrm{i}$$

and

$$(m + n\mathrm{i})(r + s\mathrm{i}) = (mr - ns) + (nr + ms)\mathrm{i},$$

for all $m, n, r, s \in \mathbf{Z}$, we see that $\mathbf{Z}[\mathrm{i}]$ is closed under addition and multiplication of complex numbers. Since the negative of any element in $\mathbf{Z}[\mathrm{i}]$ again has the correct form, as does $1 = 1 + 0\,i$, it follows that $\mathbf{Z}[\,i\,]$ is a commutative ring by Proposition 1.1.4. Since it is a subring of the field of complex numbers, it is an integral domain.

Example 1.1.4 (Some subrings of matrix rings)

The ring $M_2(\mathbf{C})$ of 2×2 matrices with complex entries is interesting in its own right, but in this example we consider two of its subrings.

Let R be the subring

$$M_2(\mathbf{Z}) = \left\{ \left[\begin{array}{cc} a & b \\ c & d \end{array} \right] \middle| a, b, c, d \in \mathbf{Z} \right\}$$

of $M_2(\mathbf{C})$ consisting of all matrices with integer entries. This provides an interesting example, with a much more complex structure than $M_2(\mathbf{C})$. For example, fewer matrices are invertible, since a matrix $\left[\begin{array}{cc} a & b \\ c & d \end{array} \right]$ in R has a multiplicative inverse if and only if $ad - bc = \pm 1$. Even though the entries come from an integral domain, there are many examples of nonzero matrices $A, B \in M_2(\mathbf{Z})$ with $AB = 0$, such as $A = \left[\begin{array}{cc} 1 & 0 \\ 0 & 0 \end{array} \right]$ and $B = \left[\begin{array}{cc} 0 & 0 \\ 0 & 1 \end{array} \right]$.

Let S be the subring of R consisting of all matrices $\left[\begin{array}{cc} a & b \\ c & d \end{array} \right]$ such that $b = 0$. The ring S is called the ring of *lower triangular* matrices with entries in $\mathbf{Z}$. It will provide an interesting source of examples. For instance, in S the matrix $\left[\begin{array}{cc} a & 0 \\ c & d \end{array} \right]$ is invertible if and only if $a = \pm 1$ and $d = \pm 1$.

We need to define several classes of elements.

Definition 1.1.5 *Let R be a ring, and let $a \in R$.*

(a) *If $ab = 0$ for some nonzero element $b \in R$, then a is called a* left zero divisor. *If $ba = 0$ for some nonzero element $b \in R$, then a is called a* right zero divisor.

(b) *If a is neither a left zero divisor nor a right zero divisor, then a is called a* regular *element.*

(c) *The element $a \in R$ is said to be* invertible *if there exists an element $b \in R$ such that $ab = 1$ and $ba = 1$. The element a is also called a* unit *of R, and its multiplicative inverse is usually denoted by a^{-1}.*

(d) *The set of all units of R will be denoted by $R^{\times}$.*

For any ring R, the set $R^{\times}$ of units of R is a group under the multiplication of R. To see this, if $a, b \in R^{\times}$, then a^{-1} and b^{-1} exist in R, and so $ab \in R^{\times}$ since $(ab)(b^{-1}a^{-1}) = 1$ and $(b^{-1}a^{-1})(ab) = 1$. The element $1 \in R$ is an identity element for $R^{\times}$, and $a^{-1} \in R^{\times}$ since $(a^{-1})^{-1} = a$. The associative law for multiplication holds because R is assumed to be a ring.

Now suppose that $1 \neq 0$ in R. Since $0 \cdot b = 0$ for all $b \in R$, it is impossible for 0 to have a multiplicative inverse. Furthermore, if $a \in R$ and $ab = 0$ for some nonzero element $b \in R$, then a cannot have a multiplicative inverse since multiplying both sides of the equation by the inverse of a (if it existed) would show that $b = 0$. Thus no zero divisor is a unit in R.

An element a of a ring R is called *nilpotent* if $a^n = 0$ for some positive integer n. For example, any strictly lower triangular matrix A in $M_n(F)$ is nilpotent, with $A^n = 0$. Our next observation gives an interesting connection between nilpotent elements and units. If a is nilpotent, say $a^n = 0$, then

$$(1 - a)(1 + a + a^2 + \cdots + a^{n-1}) = 1 - a^n = 1 .$$

Since $1 - a$ commutes with $1 + a + a^2 + \cdots + a^{n-1}$, this shows that $1 - a$ is a unit.

The next example provides a construction in which it is informative to consider zero divisors and units.

Example 1.1.5 (The direct sum of rings)

Let R_1, R_2, ..., R_n be rings. The set of n-tuples $(r_1, r_2, \ldots, r_n)$ such that $r_i \in R$ for all i is a group under componentwise addition. It is clear that componentwise multiplication is associative, and $(1, 1, \ldots, 1)$ serves as a multiplicative identity. It is easy to check that the distributive laws hold, since they hold in each component. This leads to the following definition.

The *direct sum* of the rings R_1, R_2, ..., R_n is defined to be the set

$$R_1 \oplus \cdots \oplus R_n = \{(r_1, r_2, \ldots, r_n) \mid r_i \in R \text{ for all } i\} ,$$

with the operations of componentwise addition and multiplication.

If R_1, R_2, ..., R_n are nontrivial rings, then zero divisors are easily found in the direct sum $R_1 \oplus \cdots \oplus R_n$. For example,

$$(1, 0, \ldots, 0)(0, 1, \ldots, 0) = (0, 0, \ldots, 0) .$$

An element $(a_1, \ldots, a_n) \in R_1 \oplus \cdots \oplus R_n$ is a unit if and only if each component is a unit. This can be shown by observing that $(a_1, \ldots, a_n)(x_1, \ldots, x_n) = (1, \ldots, 1)$ and $(x_1, \ldots, x_n)(a_1, \ldots, a_n) = (1, \ldots, 1)$ if and only if $x_i = a_i^{-1}$ for $1 \leq i \leq n$.

In the next definition we give the noncommutative analogs of integral domains and fields.

Definition 1.1.6 *Let R be a ring in which $1 \neq 0$.*

(a) *We say that R is a* noncommutative domain *if R is a noncommutative ring and $ab = 0$ implies $a = 0$ or $b = 0$, for all $a, b \in R$.*

(b) *We say that R is a* division ring *or* skew field *if every nonzero element of R is invertible.*

Strictly speaking, any field is a skew field, but we will generally use the term 'skew field' (or 'division ring') only when there is a chance that the ring is actually noncommutative. The next example is definitely noncommutative, and is probably the most familiar example of a skew field. Note that by Wedderburn's theorem (see [4], Theorem 8.5.6), it is not possible to give a finite example of a division ring that is not a field.

Example 1.1.6 (The quaternions)

The following subset of $M_2(\mathbf{C})$ is called the set of *quaternions*.

$$\mathbf{H} = \left\{ \begin{bmatrix} z & w \\ -\overline{w} & \overline{z} \end{bmatrix} \middle| z, w \in \mathbf{C} \right\}$$

If we represent the complex numbers z and w as $z = a + bi$ and $w = c + di$, then

$$\begin{bmatrix} z & w \\ -\overline{w} & \overline{z} \end{bmatrix} = a \begin{bmatrix} 1 & 0 \\ 0 & 1 \end{bmatrix} + b \begin{bmatrix} i & 0 \\ 0 & -i \end{bmatrix} + c \begin{bmatrix} 0 & 1 \\ -1 & 0 \end{bmatrix} + d \begin{bmatrix} 0 & i \\ i & 0 \end{bmatrix}.$$

If we let

$$1 = \begin{bmatrix} 1 & 0 \\ 0 & 1 \end{bmatrix}, \quad \mathbf{i} = \begin{bmatrix} i & 0 \\ 0 & -i \end{bmatrix}, \quad \mathbf{j} = \begin{bmatrix} 0 & 1 \\ -1 & 0 \end{bmatrix}, \quad \mathbf{k} = \begin{bmatrix} 0 & i \\ i & 0 \end{bmatrix},$$

then we can write

$$\mathbf{H} = \{a \cdot 1 + b\mathbf{i} + c\mathbf{j} + d\mathbf{k} \mid a, b, c, d \in \mathbf{R}\}.$$

Direct computations with the elements $\mathbf{i}, \mathbf{j}, \mathbf{k}$ show that we have the following identities:

$$\mathbf{i}^2 = \mathbf{j}^2 = \mathbf{k}^2 = -1 \, ;$$

$$\mathbf{ij} = \mathbf{k}, \quad \mathbf{jk} = \mathbf{i}, \quad \mathbf{ki} = \mathbf{j}; \qquad \mathbf{ji} = -\mathbf{k}, \quad \mathbf{kj} = -\mathbf{i}, \quad \mathbf{ik} = -\mathbf{j} \, .$$

These identities show that $\mathbf{H}$ is closed under matrix addition and multiplication, and it is easy to check that we have defined a subring of $M_2(\mathbf{C})$.

The determinant of the matrix corresponding to $a \cdot 1 + b\mathbf{i} + c\mathbf{j} + d\mathbf{k}$ is $z\overline{z} + w\overline{w} = a^2 + b^2 + c^2 + d^2$, and this observation shows that each nonzero element of $\mathbf{H}$ has a multiplicative inverse. The full name for $\mathbf{H}$ is the *division ring of real quaternions*. The notation $\mathbf{H}$ is used in honor of Hamilton, who discovered the quaternions after about ten years of trying to construct a field using 3-tuples of real numbers. He finally realized that if he would sacrifice the commutative law and extend the multiplication to 4-tuples he could construct a division ring.

The next example has its origins in analysis. It gives an example of a noncommutative domain that is not a division ring.

Example 1.1.7 (Differential operator rings)

Consider the homogeneous linear differential equation

$$a_n(z)\frac{d^n f}{dz^n} + \cdots + a_1(z)\frac{df}{dz} + a_0(z)f = 0 \,,$$

where the solution $f(z)$ is a polynomial with complex coefficients, and the terms $a_i(z)$ also belong to $\mathbf{C}[z]$. The equation can be written in compact form as

$$L(f) = 0 \,,$$

where L is the differential operator

$$L = a_n(z)\partial^n + \cdots + a_1(z)\partial + a_0(z) \,,$$

with $\partial = d/dz$. Thus the differential operator can be thought of as a polynomial in the two indeterminates z and ∂, but in this case the indeterminates do not commute, since

$$\partial(zf(z)) = f(z) + z\partial(f(z)) \,,$$

yielding the identity $\partial z = 1 + z\partial$. Repeated use of this identity makes it possible to write the composition of two differential operators in the standard form

$$a_n(z)\partial^n + \cdots + a_1(z)\partial + a_0(z) \,,$$

and we denote the resulting ring, called the *ring of differential operators*, by $\mathbf{C}[z][\partial]$ or $\mathbf{C}[z; \partial]$.

In taking the product of the terms $z^j \partial^n$ and $z^k \partial^m$, we obtain $z^{j+k} \partial^{n+m}$, together with terms having lower degree in z or ∂. This can be shown via an inductive argument using the identity $\partial z = 1 + z\partial$. In taking the product of two arbitrary elements

$$a_n(z)\partial^n + \cdots + a_1(z)\partial + a_0(z)$$

and

$$b_m(z)\partial^m + \cdots + b_1(z)\partial + b_0(z) \, ,$$

it is not difficult to show that the leading term is $a_n(z)b_m(z)\partial^{n+m}$, and so the product of two nonzero elements must be nonzero. This implies that $\mathbf{C}[z; \partial]$ is a noncommutative domain.

This construction can be made for any field F, and it can also be generalized to include polynomials in more than one indeterminate. The construction provides interesting and important noncommutative examples.

We next define an entire class of examples, in which the construction begins with a given group and a given field. This construction provides the motivation for many of the subsequent results in the text.

Example 1.1.8 (Group rings)

Let F be a field, and let G be a finite group of order n. We assume that the identity of G is denoted by 1, and that the elements of G are $g_1 = 1, g_2, \ldots, g_n$. The *group ring* FG determined by F and G is defined to be the n-dimensional vector space over F having the elements of G as a basis.

Vector addition is used as the addition in the ring. Elements of FG can be described as sums of the form

$$\textstyle\sum_{i=1}^n c_i g_i \, ,$$

where the coefficient c_i belongs to F. With this notation, the addition in FG can be thought of as componentwise addition, similar to addition of polynomials.

To define the multiplication in FG, we begin with the basis elements $\{g_i\}_{i=1}^n$, and simply use the multiplication of G. This product is extended by linearity (that is, with repeated use of the distributive law) to linear combinations of the basis elements. With this notation, the multiplication on FG is defined by

$$\textstyle(\sum_{i=1}^n a_i g_i)(\sum_{j=1}^n b_j g_j) = \sum_{k=1}^n c_k g_k \, ,$$

where $c_k = \sum_{g_i g_j = g_k} a_i b_j$.

Note that the elements of FG are sometimes simply written in the form $\sum_{g \in G} c_g g$. With this notation, the multiplication on FG is defined by

$$\left(\sum_{g \in G} a_g g\right)\left(\sum_{h \in G} b_h h\right) = \sum_{k \in G} c_k k,$$

where $c_k = \sum_{gh=k} a_g b_h$.

Note that each of the basis elements is invertible in FG, since they have inverses in G. On the other hand, zero divisors are also easy to find. If $g \in G$ has order $m > 1$, then $1, g, \ldots, g^{m-1}$ are distinct basis elements, and we have

$$(1 - g)(1 + g + \cdots + g^{m-1}) = 1 - g^m = 0 \,.$$

Thus if G is a finite nonabelian group, then FG is a noncommutative ring that is not a domain.

Let R be a commutative ring. For any $a \in R$, let

$$aR = \{x \in R \mid x = ar \text{ for some } r \in R\} \,.$$

This set is nonempty, since $a \in aR$, and it is closed under addition, subtraction, and multiplication since $ar_1 \pm ar_2 = a(r_1 \pm r_2)$ and $(ar_1)(ar_2) = a(r_1 ar_2)$, for all $r_1, r_2 \in R$. We note that $1 \in aR$ if and only if a is invertible, and that is the case if and only if $aR = R$. Thus aR is almost a subring of R, except for the fact that it does not generally contain an identity element. However, it has an important additional property: if $x \in aR$ and $r \in R$, then $xr \in aR$. This property plays a crucial role in constructing factor rings, and so in that sense the notion of an ideal of a ring (as defined below) corresponds to the notion of a normal subgroup of a group. We can give a general definition of an ideal that does not require the ring R to be commutative, but our proof that aR is an ideal definitely depends on commutativity.

Definition 1.1.7 *Let R be a ring.*

(a) *A nonempty subset I of R is called an* ideal *of R if*
    (i) *$a + b \in I$ for all $a, b \in I$ and*
    (ii) *$ra, ar \in I$ for all $a \in I$ and $r \in R$.*

(b) *If R is a commutative ring, then the ideal*

$$aR = \{x \in R \mid x = ar \text{ for some } r \in R\}$$

is called the principal ideal *generated by a. The notation (a) will also be used.*

(c) *An integral domain in which every ideal is a principal ideal is called a* principal ideal domain.

For any ring R, it is clear that the set $\{0\} = (0)$ is an ideal, which we will refer to as the *trivial* ideal. It is also evident that the set R is itself an ideal. Note that if I is an ideal, and $a \in I$, then $-a = (-1)a \in I$, and so the definition implies that I is an additive subgroup of R.

Suppose that R is a commutative ring. It is obvious from the definition of an ideal that any ideal of R that contains a must also contain aR, and so we are justified in saying that aR is the smallest ideal of R that contains a.

As an example of a nontrivial ideal in a noncommutative ring, let R be the ring $M_2(\mathbf{Z})$, as defined in Example 1.1.4. The interested reader can check that for any nonzero integer n, the set of all 2×2 matrices with entries in $n\mathbf{Z}$ is a nontrivial ideal of R.

The next example shows that fields can be characterized in terms of their ideals.

Example 1.1.9 (A characterization of fields)

Let R be a nonzero commutative ring. Then R is a field if and only if it has no proper nontrivial ideals.

To show this, first assume that R is a field, and let I be any ideal of R. Either $I = (0)$, or else there exists $a \in I$ such that $a \neq 0$. In the second case, since R is a field, there exists an inverse a^{-1} for a, and then for any $r \in R$ we have $r = r \cdot 1 = r(a^{-1}a) = (ra^{-1})a$, so by the definition of an ideal we have $r \in I$. We have shown that either $I = (0)$ or $I = R$.

Conversely, assume that R has no proper nontrivial ideals, and let a be a nonzero element of R. Then aR is a nonzero ideal of R, so by assumption $aR = R$. Since $1 \in R$, we have $1 = ar$ for some $r \in R$. This implies that a is invertible, completing the proof.

Example 1.1.10 ($F[x]$ is a principal ideal domain)

If F is any field, then the division algorithm holds in the ring $F[x]$ of polynomials over F. That is, for any polynomial $f(x)$ in $F[x]$, and any nonzero polynomial $g(x)$ in $F[x]$, there exist unique polynomials $q(x)$ and $r(x)$ in $F[x]$ such that $f(x) = q(x)g(x) + r(x)$, where either $\deg(r(x)) < \deg(g(x))$ or $r(x) = 0$.

An easy application of the division algorithm shows that the ring $F[x]$ of polynomials over F is a principal ideal domain. If I is any nonzero ideal of $F[x]$, then $f(x)$ is a generator for I if and only if $f(x)$ has minimal degree among the nonzero elements of I. Since a generator of I is a factor of every element of I, it is easy to see that there is only one generator for I that is a monic polynomial.

We have also considered the ring of polynomials $R[x]$ over any ring R. However, if the coefficients do not come from a field, then the proof of the division algorithm is no longer valid, and so we should not expect $R[x]$ to be a principal ideal domain. In fact, the ring $\mathbf{Z}[x]$ of polynomials with integer coefficients is a domain, but not every ideal is principal. (See Exercise 15.)

It is possible to develop an 'arithmetic' of ideals, using the operations on ideals that are defined in the next proposition. For example, if $m\mathbf{Z}$ and $n\mathbf{Z}$ are ideals of the ring of integers, then the ideals $m\mathbf{Z} \cap n\mathbf{Z}$, $m\mathbf{Z} + n\mathbf{Z}$, and $(m\mathbf{Z})(n\mathbf{Z})$ correspond, respectively, to the ideals determined by the least common multiple, the greatest common divisor, and the product of m and n.

Proposition 1.1.8 *Let R be a ring, and let I, J be ideals of R. The following subsets of R are ideals of R.*

(a) $I \cap J$,

(b) $I + J = \{x \in R \mid x = a + b \text{ for some } a \in I, b \in J\}$,

(c) $IJ = \{\sum_{i=1}^{n} a_i b_i \mid a_i \in I, b_i \in J, n \in \mathbf{Z}^+\}$.

Proof. (a) Since $0 \in I$ and $0 \in J$, we have $0 \in I \cap J$ and $I \cap J \neq \emptyset$. If $a, b \in I \cap J$, then $a, b \in I$ and $a, b \in J$. Hence $a + b \in I$ and $a + b \in J$, so $a + b \in I \cap J$. If $a \in I \cap J$ and $r \in R$, then $a \in I$ and $a \in J$ and so $ra \in I$ and $ra \in J$. Hence $ra \in I \cap J$. Similarly, $ar \in I \cap J$, and so $I \cap J$ is an ideal of R.

(b) Since $0 = 0 + 0 \in I + J$, we have $I + J \neq \emptyset$. If $x_1, x_2 \in I + J$, then $x_1 = a_1 + b_1$ and $x_2 = a_2 + b_2$, where $a_1, a_2 \in I$ and $b_1, b_2 \in J$. Since $a_1 + a_2 \in I$ and $b_1 + b_2 \in J$, we have

$$x_1 + x_2 = (a_1 + b_1) + (a_2 + b_2) = (a_1 + a_2) + (b_1 + b_2) \in I + J \, .$$

If $x \in I + J$ and $r \in R$, then $x = a + b$, where $a \in I$ and $b \in J$. Then $ra \in I$ and $rb \in J$, and so $rx = r(a + b) = ra + rb \in I + J$. Similarly, $xr \in I + J$, and so $I + J$ is an ideal of R.

(c) Since $0 = 0 \cdot 0 \in IJ$ we have $IJ \neq \emptyset$. If $x, y \in IJ$, then $x = \sum_{i=1}^{n} a_i b_i$ where $a_i \in I$, $b_i \in J$ for $i = 1, \ldots, n$ and $y = \sum_{j=1}^{m} c_j d_j$ where $c_j \in I$, $d_j \in J$ for $j = 1, \ldots, m$. We have

$$x + y = a_1 b_1 + \cdots + a_n b_n + c_1 d_1 + \cdots + c_m d_m \in IJ$$

since $a_i, c_j \in I$ and $b_i, d_j \in J$ for all i, j. If $r \in R$, then $rx = r\left(\sum_{i=1}^{n} a_i b_i\right) = \sum_{i=1}^{n} (ra_i) b_i \in IJ$ since $ra_i \in I$ and $b_i \in J$ for $i = 1, \ldots, n$. A similar argument shows that $xr \in IJ$, and thus IJ is an ideal of R. $\square$

In any commutative ring R it is possible to develop a general theory of divisibility. For $a, b \in R$, we say that b is a *divisor* of a if $a = bq$ for some $q \in R$. We also say that a is a *multiple* of b, and we will use the standard notation $b|a$ when b is a divisor of a.

Divisibility can also be expressed in terms of the principal ideals generated by a and b. Since aR is the smallest ideal that contains a, we have $aR \subseteq bR$ if and only if $a \in bR$, and this occurs if and only if $a = bq$ for some $q \in R$. Thus we have shown that $aR \subseteq bR$ if and only if $b|a$.

Let a and b be elements of a commutative ring R with 1. Then a is called an *associate* of b if $a = bu$ for some unit $u \in R$. Of course, if $a = bu$, then $b = au^{-1}$, and so b is also an associate of a.

In the ring $\mathbf{Z}$, two integers are associates if they are equal or differ in sign. In the ring $F[x]$, since the only units are the nonzero constant polynomials, two polynomials are associates if and only if one is a nonzero constant multiple of the other. Part (d) of the following result shows that for nonzero elements a and b of any integral domain, $aR = bR$ if and only if a and b are associates.

The following properties can easily be verified for any elements a, b, c of a commutative ring R.

(a) If $c|b$ and $b|a$, then $c|a$.

(b) If $c|a$, then $c|ab$.

(c) If $c|a$ and $c|b$, then $c|(ax + by)$, for any $x, y \in R$.

(d) If R is an integral domain, a is nonzero, and both $b|a$ and $a|b$ hold, then a and b are associates.

EXERCISES: SECTION 1.1

1. Let R be a ring. Verify that the following properties hold in R.

    (a) For all $a \in R$, $a \cdot 0 = 0$.

    (b) For all $a \in R$, $(-1) \cdot a = -a$.

    (c) For all $a \in R$, $-(-a) = a$.

    (d) For all $a, b \in R$, $(-a) \cdot (-b) = a \cdot b$.

2. Let R be any commutative ring. Verify that the set $R[s]$ of polynomials as defined in Example 1.1.1 is a ring.

3. Verify that the set $\mathbf{C}[z; \partial]$ of differential operators defined in Example 1.1.7 is a ring.

4. Let F be a field, and let G be a finite group. Verify that the set FG defined in Example 1.1.8 is a ring.

5. Let R be the set $M_2(\mathbf{R})$ of all 2×2 matrices with entries in the field of real numbers. Show that each of the following subsets is a subring of R.

(a) $\left\{ \begin{bmatrix} a & b \\ c & d \end{bmatrix} \in R \mid a = d \text{ and } b = 0 \right\}$

(b) $\left\{ \begin{bmatrix} a & b \\ c & d \end{bmatrix} \in R \mid a \in \mathbf{Q} \text{ and } b = 0 \right\}$

(c) $\left\{ \begin{bmatrix} a & b \\ c & d \end{bmatrix} \in R \mid a \in \mathbf{Z}, b = 0, \text{ and } c, d \in \mathbf{Q} \right\}$

6. Show that if $m\mathbf{Z}$ and $n\mathbf{Z}$ are ideals of the ring of integers, then the ideals $m\mathbf{Z} \cap n\mathbf{Z}$, $m\mathbf{Z} + n\mathbf{Z}$, and $(m\mathbf{Z})(n\mathbf{Z})$ correspond, respectively, to the ideals determined by the least common multiple, the greatest common divisor, and the product of m and n.

7. Let $R = \{m + n\sqrt{2} \mid m, n \in \mathbf{Z}\}$. Show that R is a subring of the field $\mathbf{R}$ of real numbers, and that $m + n\sqrt{2}$ is a unit in R if and only if $m^2 - 2n^2 = \pm 1$.

8. Prove that any finite integral domain is a field.

9. Let R be a commutative ring, let a be a nilpotent element of R with $a^n = 0$, and let u be any unit of R. Prove that $u - a$ is a unit of R.

10. Let $\mathcal{C}(\mathbf{R})$ be the set of all continuous functions from the $\mathbf{R}$ into $\mathbf{R}$.

    (a) Show that the set $\mathcal{C}(\mathbf{R})$ is a commutative ring under the following operations: $[f+g](x) = f(x)+g(x)$ and $[f \cdot g](x) = f(x)g(x)$ for all $x \in \mathbf{R}$ and all $f, g \in \mathcal{C}(\mathbf{R})$. Is $\mathcal{C}(\mathbf{R})$ an integral domain?

    (b) Which properties in the definition of a commutative ring fail if the product of two functions is defined to be $[f \circ g](x) = f(g(x))$, for all $x \in \mathbf{R}$?

11. A ring R is called a *Boolean ring* if $x^2 = x$ for all $x \in R$.

    (a) Show that if R is a Boolean ring, then $a + a = 0$ for all $a \in R$.

    (b) Show that any Boolean ring is commutative.

    (c) Let X be any set and let R be the collection of all subsets of X. Define addition and multiplication of subsets $A, B \subseteq X$ as follows:

    $$A + B = (A \cup B) \setminus (A \cap B) \quad \text{and} \quad A \cdot B = A \cap B .$$

    (Here $S \setminus T$ denotes the difference $\{s \in S \mid s \notin T\}$ of sets S and T.) Show that R is a Boolean ring under this addition and multiplication.

12. Let R be a set that satisfies all of the ring axioms except the existence of a multiplicative identity element. Set $R_1 = \{(r, n) \mid r \in R, n \in \mathbf{Z}\}$. Define $+$ on R_1 by $(r, n) + (s, m) = (r + s, n + m)$ and define $\cdot$ on R_1 by $(r, n) \cdot (s, m) = (rs + ns + mr, nm)$. Show that R_1 is a ring with multiplicative identity $(0, 1)$.

13. Let R be a ring, and let $Z(R) = \{x \in R \mid xr = rx \text{ for all } r \in R\}$. This subset is called the *center* of R.

    (a) Prove that $Z(R)$ is a subring of R.

    (b) Prove that the center of a division ring is a field.

    (c) Find the center $Z(\mathbf{H})$ of the division ring of quaternions.

14. Let F be a field, let G be a finite group, and let C be a conjugacy class of G. Show that $\sum_{g \in C} g$ is in the center $Z(FG)$ of the group ring FG.

15. Let I be the smallest ideal of $\mathbf{Z}[x]$ that contains both 2 and x. Show that I is not a principal ideal.

16. Let R be a commutative ring. Define the ring $R[[x]]$ of all *formal power series* over R as the set of all sequences $a = (a_0, a_1, a_2, \ldots)$, with componentwise addition, and multiplication defined as follows:

$$(a_0, a_1, a_2, \ldots) \cdot (b_0, b_1, b_2, \ldots) = (c_0, c_1, c_2, \ldots) \text{, where } c_i = \sum_{i=j+k} a_j b_k.$$

Check that $R[[x]]$ is a ring, that R can be identified with elements of the form $(a_0, 0, 0, \ldots)$, and that the polynomial ring $R[x]$ can be identified with the subset of all sequences with finitely many nonzero terms.

1.2   Ring homomorphisms

After defining the concept of a ring, we need to study the appropriate functions that can be used to compare rings. These functions should preserve the essential algebraic structure. We also require (although this is not a universal standard) that the functions should preserve the multiplicative identity element. We note that this does not necessarily follow from parts (i) and (ii) of the definition.

Definition 1.2.1 *Let R and S be rings.*

(a) *A function $\phi : R \to S$ is called a* ring homomorphism *if*
   (i) $\phi(a + b) = \phi(a) + \phi(b)$, *for all $a, b \in R$,*
   (ii) $\phi(ab) = \phi(a)\phi(b)$, *for all $a, b \in R$, and*
   (iii) $\phi(1) - 1$.

(b) *A ring homomorphism that is one-to-one and onto is called an* isomorphism. *If there is an isomorphism from R onto S, we say that R is* isomorphic *to S, and write $R \cong S$. An isomorphism from the ring R onto itself is called an* automorphism *of R.*

The statement that a ring homomorphism $\phi : R \to S$ preserves addition is equivalent to the statement that ϕ is a group homomorphism (of the underlying additive groups of the respective rings). This means that we have at our disposal all of the results that hold for group homomorphisms. In particular, $\phi(0) = 0$, $\phi(-a) = -\phi(a)$ for all $a \in R$, and $\phi(R)$ must be an additive subgroup of S.

It follows from the next proposition that 'is isomorphic to' is reflexive, symmetric, and transitive. Recall that a function is one-to-one and onto if and only if it has an inverse. Thus a ring isomorphism always has an inverse, but it is not evident that this inverse respects addition and multiplication.

Proposition 1.2.2

(a) *The inverse of a ring isomorphism is a ring isomorphism.*

(b) *The composition of two ring isomorphisms is a ring isomorphism.*

Proof. (a) Let $\phi : R \to S$ be an isomorphism of rings. It follows from elementary group theory that ϕ^{-1} is an isomorphism of the underlying additive groups. To show that ϕ^{-1} is a ring homomorphism, let $s_1, s_2 \in S$. Since ϕ is onto, there exist $r_1, r_2 \in R$ such that $\phi(r_1) = s_1$ and $\phi(r_2) = s_2$. Then $\phi^{-1}(s_1 s_2)$ must be the unique element $r \in R$ for which $\phi(r) = s_1 s_2$. Since ϕ preserves multiplication,

$$\phi^{-1}(s_1 s_2) = \phi^{-1}(\phi(r_1)\phi(r_2)) = \phi^{-1}(\phi(r_1 r_2)) = r_1 r_2 = \phi^{-1}(s_1)\phi^{-1}(s_2) .$$

Finally, since $\phi(1) = 1$, we have $\phi^{-1}(1) = 1$.

(b) If $\phi : R \to S$ and $\theta : S \to T$ are ring homomorphisms, then

$$\theta\phi(ab) = \theta(\phi(ab)) = \theta(\phi(a)\phi(b)) = \theta(\phi(a)) \cdot \theta(\phi(b)) = \theta\phi(a) \cdot \theta\phi(b) .$$

Since $\theta\phi$ preserves sums (this follows from results for group homomorphisms), and $\theta\phi(1) = 1$, the desired result follows from the fact that the composition of two one-to-one and onto functions is again one-to-one and onto. $\square$

We now give some examples of ring homomorphisms. First, the inclusion mapping from a subring into a ring is always a ring homomorphism. Next, the projection mapping $\pi : R \oplus S \to R$ from the direct sum $R \oplus S$ of two rings onto the component R given by $\pi(r, s) = r$ is a ring homomorphism. Finally, on the ring of Gaussian integers, the function $\phi : \mathbf{Z}[i] \to \mathbf{Z}[i]$ given by $\phi(m + ni) = m - ni$ is an automorphism.

Example 1.2.1 (Homomorphisms defined on a polynomial ring)

Let R and S be commutative rings, let $\theta : R \to S$ be a ring homomorphism, and let $\eta : \{x_1, x_2, \ldots, x_n\} \to S$ be any mapping into S. Then there exists a unique ring homomorphism

$$\widehat{\theta} : R[x_1, x_2, \ldots, x_n] \to S$$

such that $\widehat{\theta}(r) = \theta(r)$ for all $r \in R$ and $\widehat{\theta}(x_i) = \eta(x_i)$, for $i = 1, 2, \ldots, n$.

By induction, it is sufficient to prove the result in the case of a single indeterminate x. We assume that $\theta : R \to S$ and $\eta : \{x\} \to S$ are given, with $\eta(x) = s$. We will first show the uniqueness of

$\widehat{\theta} : R[x] \to S$ with $\widehat{\theta}(x) = s$. If $\phi : R[x] \to S$ is any homomorphism with the required properties, then for any polynomial

$$f(x) = a_0 + a_1 x + \cdots + a_m x^m$$

in $R[x]$ we must have

$$
\begin{aligned}
\phi(f(x)) &= \phi(a_0 + a_1 x + \cdots + a_m x^m) \\
&= \phi(a_0) + \phi(a_1 x) + \cdots + \phi(a_m x^m) \\
&= \phi(a_0) + \phi(a_1)\phi(x) + \ldots + \phi(a_m)\phi(x)^m \\
&= \phi(a_0) + \phi(a_1)s + \cdots + \phi(a_m)s^m \ .
\end{aligned}
$$

This shows that the only possible way to define $\widehat{\theta}$ is the following:

$$\widehat{\theta}(a_0 + a_1 x + \cdots + a_m x^m) = \theta(a_0) + \theta(a_1)s + \cdots + \theta(a_m)s^m \ .$$

Given this definition, we must show that $\widehat{\theta}$ is a homomorphism. Since addition of polynomials is defined componentwise, and θ preserves sums, it can be checked that $\widehat{\theta}$ preserves sums of polynomials. If

$$g(x) = b_0 + b_1 x + \cdots + b_n x^n \ ,$$

then the coefficient c_k of the product $h(x) = f(x)g(x)$ is given by the formula

$$c_k = \sum_{i+j=k} a_i b_j \ .$$

Applying θ to both sides gives

$$\theta(c_k) = \theta(\sum_{i+j=k} a_i b_j) = \sum_{i+j=k} \theta(a_i)\theta(b_j) \ ,$$

since θ preserves both sums and products. The above formula is the tool needed to check that

$$\widehat{\theta}(f(x)g(x)) = \widehat{\theta}(h(x)) = \widehat{\theta}(f(x))\widehat{\theta}(g(x)) \ .$$

This completes the proof that $\widehat{\theta}$ is a ring homomorphism.

A *matrix representation* of a finite group G is a group homomorphism from G into the *general linear group* $\mathrm{GL}_n(F)$ of $n \times n$ invertible matrices over a field F. The next example shows that using the group ring FG we can convert a representation of G into a ring homomorphism from FG into the ring $\mathrm{M}_n(F)$.

Example 1.2.2 (Homomorphisms defined on a group ring)

Let F be a field, and let G be a finite group, with elements $g_1 = 1, g_2, \ldots, g_n$. We will show that the group ring FG converts certain group homomorphisms to ring homomorphisms.

Let R be any ring, and suppose that θ is a group homomorphism from G into the group $R^\times$ of units of R. We can define a ring homomorphism $\widehat{\theta} : FG \to R$ as follows:

$$\widehat{\theta}\left(\sum_{i=1}^n c_i g_i\right) = \sum_{i=1}^n c_i \theta(g_i) \, .$$

It is clear that $\widehat{\theta}$ respects sums. It can also be shown that it respects products, since

$$\widehat{\theta}(g_i)\widehat{\theta}(g_j) = \theta(g_i)\theta(g_j) = \theta(g_i g_j) = \widehat{\theta}(g_i g_j)$$

for all $g_i, g_j \in G$, and the general result follows from the distributive law. Finally, $\widehat{\theta}$ maps the identity element of G to the identity of R since θ is a group homomorphism. Thus $\widehat{\theta}$ is a ring homomorphism.

If $\rho : G \to \mathrm{GL}_n(F)$ is any representation of G, then we define $\widehat{\rho} : FG \to \mathrm{M}_n(F)$ by

$$\widehat{\rho}\left(\sum_{i=1}^n c_i g_i\right) = \sum_{i=1}^n c_i \rho(g_i) \, .$$

Since $\mathrm{GL}_n(F)$ is the group of units of $\mathrm{M}_n(F)$, it follows that $\widehat{\rho}$ is a ring homomorphism.

Given a ring R, the *opposite ring* R^{op} is constructed by using the set R and the existing addition on R. A new multiplication $\cdot$ is defined on R by letting $a \cdot b = ba$, for all $a, b \in R$. Of course, if R is a commutative ring, the construction yields nothing new. On the other hand, if R is noncommutative, then the opposite ring interchanges properties on the left and right sides.

Example 1.2.3 (Symmetry of group rings)

If F is a field and G is a finite group, then there is a natural group homomorphism ι from G into the group of units of $(FG)^{op}$ defined by $\iota(g) = g^{-1}$. If $\cdot$ denotes the multiplication in $(FG)^{op}$, then ι is a group homomorphism since

$$\iota(g_1 g_2) = (g_1 g_2)^{-1} = g_2^{-1} g_1^{-1} = \iota(g_2)\iota(g_1) = \iota(g_1) \cdot \iota(g_2) \, .$$

It follows from Example 1.2.2 that ι can be extended by linearity to a ring homomorphism $\widehat{\iota} : FG \to (FG)^{op}$, with

$$\widehat{\iota}\left(\sum_{g \in G} c_g g\right) = \sum_{g \in G} c_g g^{-1} \,.$$

It is easily seen that $\widehat{\iota}$ is an isomorphism. We conclude that the group ring FG has left–right symmetry.

Before giving some additional examples, we need to give a definition for ring homomorphisms analogous to the definition of the kernel of a group homomorphism. In fact, the kernel of a ring homomorphism will be defined to be just the kernel of the mapping when viewed as a group homomorphism of the underlying additive groups of the respective rings.

Definition 1.2.3 *Let* $\phi : R \to S$ *be a ring homomorphism. The set*

$$\{a \in R \mid \phi(a) = 0\}$$

is called the kernel *of* ϕ*, denoted by* $\ker(\phi)$*.*

Proposition 1.2.4 *Let* $\phi : R \to S$ *be a ring homomorphism.*

(a) *The subset* $\ker(\phi)$ *is an ideal of* R*.*

(b) *The homomorphism* ϕ *is an isomorphism if and only if* $\ker(\phi) = (0)$ *and* $\phi(R) = S$*.*

Proof.   (a) If $a, b \subset \ker(\phi)$, then

$$\phi(a \pm b) = \phi(a) \pm \phi(b) = 0 \pm 0 = 0 \,,$$

and so $a + b \in \ker(\phi)$. If $r \in R$, then

$$\phi(ra) = \phi(r) \cdot \phi(a) = \phi(r) \cdot 0 = 0 \,,$$

showing that $ra \in \ker(\phi)$. Similarly, $ar \in \ker(\phi)$.

(b) This part follows from the fact that ϕ is a group homomorphism, since ϕ is one-to-one if and only if $\ker(\phi) = (0)$ and onto if and only if $\phi(R) = S$. $\square$

Note that $\ker(\phi) = \phi^{-1}\left((0)\right)$, and so the fact that $\ker(\phi)$ is an ideal could be deduced from the more general result that if $\phi : R \to S$ is a ring homomorphism then the inverse image of any ideal of S is an ideal of R. It is also true that if ϕ maps R onto S, then the image of an ideal of R is an ideal of S.

Example 1.2.4 (Extension of group homomorphisms)

Let G and H be finite groups, and let F be a field. If $\theta : G \to H$ is a group homomorphism, we can extend the mapping θ to a ring homomorphism $\widehat{\theta} : FG \to FH$ as follows:

$$\widehat{\theta}\left(\sum_{g \in G} c_g g\right) = \sum_{g \in G} c_g \theta(g) .$$

The proof that $\widehat{\theta}$ is a ring homomorphism is left as an exercise.

If H is the trivial group, then FH is just the field F, and so as a special case of the above ring homomorphism we have the *augmentation map* $\epsilon : FG \to F$ defined by

$$\epsilon\left(\sum_{g \in G} c_g g\right) = \sum_{g \in G} c_g .$$

This mapping plays an important role in the study of group rings; its kernel is called the *augmentation ideal* of FG.

Let $\phi : R \to S$ be a ring homomorphism. The fundamental homomorphism theorem for groups implies that the abelian group $R/\ker(\phi)$ is isomorphic to the abelian group $\phi(R)$, which is a subgroup of S. In order to obtain a homomorphism theorem for rings we need to consider the cosets of $\ker(\phi)$. Intuitively, the situation may be easiest to understand if we consider the cosets to be defined by the equivalence relation $\sim$ given by $a \sim b$ if $\phi(a) = \phi(b)$, for all $a, b \in R$. The sum of equivalence classes $[a]$ and $[b]$ in $R/\ker(\phi)$ is well-defined, using the formula $[a] + [b] = [a + b]$, for all $a, b \in R$. The product of equivalence classes $[a]$ and $[b]$ in $R/\ker(\phi)$ is defined by the expected formula $[a] \cdot [b] = [ab]$, for all $a, b \in R$. To show that this multiplication is well-defined, we note that if $a \sim c$ and $b \sim d$, then $ab \sim cd$ since

$$\phi(ab) = \phi(a)\phi(b) = \phi(c)\phi(d) = \phi(cd) .$$

Basic results from group theory imply that $R/\ker(\phi)$ is an abelian group. To show that $R/\ker(\phi)$ is a ring we only need to verify the distributive laws, the associative law for multiplication, and show that $R/\ker(\phi)$ has a multiplicative identity. We have

$$[a]([b] + [c]) = [a][b + c] = [a(b + c)] = [ab + ac] = [ab] + [ac] = [a][b] + [a][c] ,$$

showing that the distributive laws follows directly from the definitions of addition and multiplication for equivalence classes, and the corresponding distributive law in R. The proof that the associative law holds is similar. Finally, the coset $[1]$ serves as a multiplicative identity.

If I is any ideal of the ring R, then I is a subgroup of the underlying additive group of R, so the cosets of I in R determine the abelian group R/I. The cosets of I are usually denoted additively, in the form $a + I$, for all elements $a \in R$. It is important to remember that these are additive cosets, not multiplicative cosets. We now show that R/I has a ring structure.

Theorem 1.2.5 *Let I be an ideal of the ring R.*

(a) *The abelian group R/I has a natural ring structure.*

(b) *The natural projection mapping $\pi : R \to R/I$ defined by $\pi(a) = a + I$ for all $a \in R$ is a ring homomorphism, and* $\ker(\pi) = I$.

Proof. (a) It follows from standard results in group theory that the set of cosets R/I is a group under the addition. (See Theorem 2.6.1 of [11] or Proposition 3.8.4 of [4].) To show that multiplication of cosets is well-defined, suppose that $a + I = c + I$ and $b + I = d + I$, for elements $a, b, c, d \in R$. Then $a - c \in I$ and $b - d \in I$, so $(a - c)b \in I$ and $c(b - d) \in I$, and therefore $ab - cd = (ab - cb) + (cb - cd) \in I$, showing that $ab + I = cd + I$.

To show that multiplication is associative, let $a, b, c \in R$. Then

$$
\begin{aligned}
((a + I)(b + I))(c + I) &= (ab + I)(c + I) = (ab)c + I \\
&= a(bc) + I = (a + I)(bc + I) \\
&= (a + I)((b + I)(c + I)) \, .
\end{aligned}
$$

Similarly, the argument that the distributive laws hold in R/I depends directly on the fact that these laws hold in R. The coset $1 + I$ serves as the multiplicative identity, completing the proof that R/I is a ring.

(b) The projection mapping preserves addition and multiplication as a direct consequence of the definition of addition and multiplication of congruence classes. $\square$

Definition 1.2.6 *Let I be an ideal of the ring R. With the following addition and multiplication for all $a, b \in R$, the set of cosets $\{a + I \mid a \in R\}$ is denoted by R/I, and is called the* factor ring *of R modulo I.*

$$ (a + I) + (b + I) = (a + b) + I \, , \qquad\qquad (a + I)(b + I) = ab + I \, . $$

The most familiar factor ring is $\mathbf{Z}/n\mathbf{Z}$, for which we will use the notation $\mathbf{Z}_n$. In this ring, multiplication can be viewed as repeated addition, and it is easy to show that any subgroup is an ideal. Since every subgroup of $\mathbf{Z}_n$ is cyclic, the lattice of ideals of $\mathbf{Z}_n$ corresponds to the lattice of divisors of n.

Having introduced the necessary notation, we can now prove the analog of the fundamental homomorphism theorem for groups. (See Theorem 2.7.1 of [11] or Proposition 3.8.8 of [4].)

Theorem 1.2.7 (The fundamental homomorphism theorem) *Let $\phi : R \to S$ be a ring homomorphism. Then $\phi(R)$ is a subring of S, and $\phi(R)$ is isomorphic to the factor ring $R/\ker(\phi)$.*

Proof. Let $K = \ker(\phi)$. Since we have already shown that R/K is a ring, we can apply the fundamental homomorphism theorem for groups to show that the mapping $\overline{\phi}$ given by $\overline{\phi}(a + K) = \phi(a)$, for all $a \in R$, defines an isomorphism of the abelian groups R/K and $\phi(R)$. That $\overline{\phi}$ preserves multiplication follows from the computation

$$\overline{\phi}((a + K)(b + K)) = \overline{\phi}(ab + K) = \phi(ab) = \phi(a)\phi(b) = \overline{\phi}(a + K)\overline{\phi}(b + K) .$$

Finally, $\overline{\phi}(1 + K) = 1$, and this completes the proof. $\square$

Example 1.2.5 *(FG, when G is cyclic)*

As an interesting application of the fundamental homomorphism theorem, let F be a field, and let G be a cyclic group of order n, with generator g. By Example 1.2.1 there is a unique ring homomorphism $\phi : F[x] \to FG$ with $\phi(x) = g$. Then ϕ is onto, and $\ker(\phi)$ is easily shown to be $(x^n - 1)$ since g has order n, so it follows from the fundamental homomorphism theorem that the group ring FG is isomorphic to $F[x]/(x^n - 1)$.

We recall Cayley's theorem, from elementary group theory: any group of order n is isomorphic to a subgroup of the symmetric group $\mathcal{S}_n$. This theorem shows that the symmetric groups in some sense are the most basic finite groups. The next proposition shows that in the study of rings, the corresponding role is played by endomorphism rings of abelian groups. (See Example 1.1.3.)

Proposition 1.2.8 *For any ring R there exists an abelian group A such that R is isomorphic to a subring of the endomorphism ring $\mathrm{End}(A)$.*

Proof. Let R be any ring. As the required abelian group we choose the underlying additive group of R. For each element $a \in R$ we define $\lambda_a : R \to R$ by $\lambda_a(x) = ax$. Then $\lambda_a \in \mathrm{End}(R)$ since the distributive law holds, and so we can define $\Phi : R \to \mathrm{End}(R)$ by $\Phi(a) = \lambda_a$. It is easy to check that the following formulas hold. For $a, b \in R$, the formula $\lambda_a + \lambda_b = \lambda_{a+b}$ follows from the distributive law, and the formula $\lambda_a \circ \lambda_b = \lambda_{ab}$ follows from the associative law.

It is clear that λ_1 is the identity function on R, and these are the facts needed to show that Φ is a ring homomorphism. If $\lambda_a = \lambda_b$, then $a = \lambda_a(1) = \lambda_b(1) = b$, and so Φ is one-to-one. Finally, it can be checked that the image $\Phi(R)$ is a subring of $\text{End}(R)$. $\square$

The next examples provide additional applications of the fundamental homomorphism theorem.

Example 1.2.6 $(F[x,y]/(y) \cong F[x])$

Let F be a field, let $R = F[x,y]$, the ring of polynomials in two indeterminates with coefficients in F, and let $I = (y)$. That is, I is the set of all polynomials that have y as a factor. In forming R/I we make the elements of I congruent to 0, and so in some sense we should be left with just polynomials in x. This is made precise in the following way: define $\phi : F[x,y] \to F[x]$ by $\phi(f(x,y)) = f(x,0)$. It is necessary to check that ϕ is a ring homomorphism. Then it is clear that $\ker(\phi) = (y)$, and so we can conclude from the fundamental homomorphism theorem for rings that $R/I \cong F[x]$.

Example 1.2.7 (Correspondence of factor rings)

Let $\phi : R \to S$ be a ring homomorphism from R onto S. Since $S \cong R/\ker(\phi)$, there is a one-to-one correspondence between the ideals of S and the ideals of R that contain $\ker(\phi)$. We will show that the correspondence extends to factor rings.

Let I be any ideal of R with $I \supseteq \ker(\phi)$, and let $J = \phi(I)$. We will show that R/I is isomorphic to S/J. To do this, let π be the natural projection from S onto S/J, and consider $\theta = \pi\phi$. Then θ is onto since both π and ϕ are onto, and

$$\ker(\theta) = \{r \in R \mid \phi(r) \in J\} = I \, ,$$

so it follows from the fundamental homomorphism theorem for rings that R/I is isomorphic to S/J.

Example 1.2.8 (The characteristic of a ring)

As an application of the results in this section we define the characteristic of a ring. Let R be a ring. The smallest positive integer n such that $n \cdot 1 = 0$ is called the *characteristic* of R, denoted by $\text{char}(R)$. If no such positive integer exists, then R is said to have *characteristic zero*.

The characteristic of a ring R is just the order of 1 in the underlying additive group. If $\mathrm{char}(R) = n$, then it follows from the distributive law that $n \cdot a = (n \cdot 1) \cdot a = 0 \cdot a = 0$, and so the exponent of the underlying abelian group is n.

A more sophisticated way to view the characteristic is to define a ring homomorphism $\phi : \mathbf{Z} \to R$ by $\phi(n) = n \cdot 1$. Then either 1 has infinite order as an element of the additive group of R, and $\ker(\phi) = (0)$, or 1 has order $k > 0$, in which case $\ker(\phi) = k\mathbf{Z}$. Thus the characteristic of R is determined by $\ker(\phi)$.

Recall that in $\mathbf{Z}$ we have $m|k$ if and only if $m\mathbf{Z} \supseteq k\mathbf{Z}$. Therefore the lattice of ideals of $\mathbf{Z}_n$ corresponds to the lattice of all ideals of $\mathbf{Z}$ that contain $n\mathbf{Z}$. As shown by the next proposition, the analogous result holds in any factor ring.

Proposition 1.2.9 *Let I be an ideal of the ring R.*

(a) *There is a one-to-one correspondence between the ideals of R/I and the ideals of R that contain I.*

(b) *If K is an ideal of R with $I \subseteq K \subseteq R$, then $(R/I)/(K/I) \cong R/K$.*

Proof. (a) From elementary group theory there is a one-to-one correspondence between subgroups of R/I and subgroups of R that contain I. (See Theorem 2.7.2 of [11] or Proposition 3.8.6 of [4].) We must show that this correspondence preserves ideals. If J is an ideal of R that contains I, then it corresponds to the additive subgroup

$$\pi(J) = \{a + I \mid a \in J\}.$$

For any elements $r + I \in R/I$ and $a + I \in \pi(J)$, we have $(r + I)(a + I) = ra + I$, and then $ra + I \in \pi(J)$ since $ra \in J$. On the other hand, if J is an ideal of R/I, then it corresponds to the subgroup

$$\pi^{-1}(J) = \{a \in R \mid a + I \in J\}.$$

For any $r \in R$ and $a \in \pi^{-1}(J)$, we have $ra + I = (r + I)(a + I) \in J$. Thus $ra \in \pi^{-1}(J)$, and a similar argument shows that $ar \in \pi^{-1}(J)$.

(b) Define $\theta : R/I \to R/K$ by $\theta(r + I) = r + K$. Then θ is an onto ring homomorphism, with $\ker(\theta) = K/I$. The fundamental homomorphism theorem for rings shows that $(R/I)/(K/I)$ and R/K are isomorphic rings. $\square$

Example 1.2.9 $(\mathbf{Q}[x]/\left(x^2 - 2x + 1\right))$

Let $R = \mathbf{Q}[x]$ and let $I = \left(x^2 - 2x + 1\right)$. Using the division algorithm, it is possible to show that the cosets modulo I correspond

to the possible remainders upon division by $x^2 - 2x + 1$, so that we
only need to consider cosets of the form $a + bx + I$, for all $a, b \in \mathbf{Q}$.
Since $x^2 - 2x + 1 \in I$, we can use the formula $x^2 + I = 2x - 1 + I$
to simplify products. This gives us the following formulas:

$$(a + bx + I) + (c + dx + I) = (a + c) + (b + d)x + I$$

and

$$(a + bx + I) \cdot (c + dx + I) = (ac - bd) + (bc + ad + 2bd)x + I \ .$$

By Proposition 1.2.9, the ideals of R/I correspond to the ideals of R
that contain I. Since $\mathbf{Q}[x]$ is a principal ideal domain, these ideals
are determined by the divisors of $x^2 - 2x + 1$, showing that there is
only one proper nontrivial ideal in R/I, corresponding to the ideal
generated by $x - 1$.

Definition 1.2.10 *Let I be a proper ideal of the ring R.*

(a) *The ideal I is said to be a* maximal ideal *of R if for all ideals J of R
such that $I \subseteq J \subseteq R$, either $J = I$ or $J = R$.*

(b) *The ring R is said to be a* simple ring *if the zero ideal (0) is a maximal
ideal of R.*

Proposition 1.2.11 *Let I be a proper ideal of the ring R.*

(a) *The ideal I is a maximal ideal of R if and only if the factor ring R/I
is a simple ring.*

(b) *If R is a commutative ring, then the ideal I is a maximal ideal of R if
and only if the factor ring R/I is a field.*

Proof. (a) By definition, I is a maximal ideal if and only if there are no
ideals properly contained between I and R. By Proposition 1.2.9 (a), this is
equivalent to the statement that there are no proper nonzero ideals in R/I, and
this is equivalent to the statement that R/I is a simple ring.

(b) If R is commutative, then by Example 1.1.9, R/I is a field if and only
if it has no proper nontrivial ideals. Thus R/I is a field if and only if it is a
simple ring. $\square$

The ideals $I_1, \ldots, I_n$ of the ring R are called *comaximal* if $I_j + I_k = R$, for
all $j \neq k$.

Theorem 1.2.12 (Chinese remainder theorem)   *Let $I_1, \ldots, I_n$ be co-maximal ideals of the ring R.*

(a) *The factor ring $R/\left(\bigcap_{j=1}^{n} I_j\right)$ is isomorphic to $(R/I_1) \oplus \cdots \oplus (R/I_n)$.*

(b) *If R is commutative, then $\bigcap_{j=1}^{n} I_j = \prod_{j=1}^{n} I_j$.*

Proof. We define the function $\pi : R \to (R/I_1) \oplus \cdots \oplus (R/I_n)$ by setting $\pi(r) = (r + I_1, \ldots, r + I_n)$. The jth component of π is the canonical projection from R onto R/I_j. The fact that each component is a ring homomorphism can be used to show that π is a ring homomorphism. The kernel of π is $\bigcap_{j=1}^{n} I_j$, so the desired result will follow from the fundamental homomorphism theorem, provided we show that π is an onto function.

Let e_j be the element of $(R/I_1) \oplus \cdots \oplus (R/I_n)$ with 1 in the jth component, and 0 in every other component. It suffices to prove that there exist $x_j \in R$ with $\pi(x_j) = e_j$, for each $1 \leq j \leq n$. To show this, let $r = (r_1 + I_1, \ldots, r_n + I_n)$ be any element of $(R/I_1) \oplus \cdots \oplus (R/I_n)$, and let $x = \sum_{i=1}^{n} r_i x_i$. Then indeed

$$\pi(x) = \pi\left(\sum_{i=1}^{n} r_i x_i\right) = \sum_{i=1}^{n} \pi(r_i)\pi(x_i) = \sum_{i=1}^{n} \pi(r_i)e_i = r \ .$$

We next show how to solve the equation $\pi(x_j) = e_j$. In general, suppose that I, J, K are ideals of R with $I + J = R$ and $I + K = R$. Then there exist $a, b \in I$, $y \in J$, and $z \in K$ with $a + y = 1$ and $b + z = 1$. In the product $1 = (ab + yb + az) + yz$ we have $ab + yb + az \in I$ and $yz \in JK$, so it follows that $I + JK = R$. An easy extension of this result shows that $I_j + \prod_{k \neq j} I_k = R$. Thus there exist elements $a_j \in I_j$ and $x_j \in \prod_{k \neq j} I_k$ with $a_j + x_j = 1$. Since $x_j \in I_k$ for $k \neq j$ and $x_j + I_j = 1 + I_j$, we have $\pi(x_j) = e_j$.

(b) Suppose that I, J are ideals of R with $I + J = R$. Then

$$I \cap J = (I \cap J)R = (I \cap J)(I + J) \subseteq JI + IJ \subseteq I \cap J \ .$$

It follows that if R is commutative, then $I \cap J = IJ$.

Now it is clear that $I_1 \cap I_2 = I_1 I_2$. The proof of part (a) shows that $I_3 + I_1 I_2 = R$, and therefore

$$\bigcap_{j=1}^{3} I_j \subseteq I_3 \cap (I_1 I_2) = \prod_{j=1}^{3} I_j \ .$$

An induction argument can be given to show that $\bigcap_{j=1}^{n} I_j = \prod_{j=1}^{n} I_j$. $\quad\square$

Example 1.2.10 ($\mathbf{Z}[x]/\left(x^4 - 1\right)$)

Consider the ring $R = \mathbf{Z}[x]/\left(x^4 - 1\right)$. Since $x^4 - 1$ factors as $x^4 - 1 = (x - 1)(x + 1)(x^2 + 1)$, we have $I = I_1 \cap I_2 \cap I_3$ for $I = \left(x^4 - 1\right)$, $I_1 = (x - 1)$, $I_2 = (x + 1)$, and $I_3 = \left(x^2 + 1\right)$. Furthermore,

$$1 = \frac{1}{2}(1 - x) + \frac{1}{2}(1 + x) \quad \text{and} \quad 1 = \frac{1}{2}(1 - x)(1 + x) + \frac{1}{2}(1 + x^2) \ ,$$

so the ideals I_1, I_2, I_3 are comaximal.

The proof of the Chinese remainder theorem shows that for the isomorphism $\widehat{\pi} : R/I \to R/I_1 \oplus R/I_2 \oplus R/I_3$, the preimage of $(1, 0, 0)$ is the coset of

$$\frac{1}{2}(1 + x) \cdot \frac{1}{2}(1 + x^2) = \frac{1}{4}(1 + x + x^2 + x^3) \, .$$

Similarly, the preimage of $(0, 1, 0)$ is represented by

$$\frac{1}{2}(1 - x) \cdot \frac{1}{2}(1 + x^2) = \frac{1}{4}(1 - x + x^2 - x^3) \, ,$$

and the preimage of $(0, 0, 1)$ is represented by

$$\frac{1}{2}(1 - x)(1 + x) \cdot \frac{1}{2}(1 - x)(1 + x) = \frac{1}{4}(1 - x^2)^2 \, ,$$

which is equivalent modulo I to $\frac{1}{2}(1 - x^2)$.

Example 1.2.11 $(\mathbf{C}[x]/\left(x^3 - 1\right))$

Consider the ring $R = \mathbf{C}[x]/\left(x^3 - 1\right)$. Over $\mathbf{C}$, the polynomial $x^3 - 1$ factors as $x^3 - 1 = (x - 1)(x - \omega)(x - \omega^2)$, where ω is a primitive cube root of unity. We have $I = I_1 \cap I_2 \cap I_3$ for $I = \left(x^3 - 1\right)$, $I_1 = (x - 1)$, $I_2 = (x - \omega)$, and $I_3 = (x - \omega^2)$. The ideals I_1, I_2, I_3 are maximal, so they are certainly comaximal. In fact, we have

$$1 = \frac{\omega + 2}{3}(1 - x) - \frac{\omega + 2}{3}(\omega - x) \, ,$$

$$1 = \frac{1 - \omega}{3}(1 - x) + \frac{\omega - 1}{3}(\omega^2 - x) \, ,$$

and

$$1 = \frac{-1 - 2\omega}{3}(\omega - x) + \frac{1 + 2\omega}{3}(\omega^2 - x) \, .$$

The proof of the Chinese remainder theorem shows that in R/I the following elements correspond, respectively, to $(1, 0, 0)$, $(0, 1, 0)$, and $(0, 0, 1)$ in $R/I_1 \oplus R/I_2 \oplus R/I_3$:

$$\frac{1}{3}(1 + x + x^2) = -\frac{\omega + 2}{3}(\omega - x) \cdot \frac{\omega - 1}{3}(\omega^2 - x) \, ,$$

$$\frac{1}{3}(1 + \omega^2 x + \omega x^2) = \frac{\omega + 2}{3}(1 - x) \cdot \frac{1 + 2\omega}{3}(\omega^2 - x) \, ,$$

$$\frac{1}{3}(1 + \omega x + \omega^2 x^2) = \frac{1 - \omega}{3}(1 - x) \cdot \frac{1 + 2\omega}{3}(\omega^2 - x) \, .$$

EXERCISES: SECTION 1.2

1. (a) Show that the only ring homomorphism from the ring $\mathbf{Z}$ into itself is the identity mapping.

   (b) Find all ring homomorphisms from $\mathbf{Z} \oplus \mathbf{Z}$ into $\mathbf{Z} \oplus \mathbf{Z}$.

   Hint: An element e of a ring R is called *idempotent* if $e^2 = e$. Use the fact that any ring homomorphism must map idempotents to idempotents.

2. Show that the ring $\operatorname{End}(\mathbf{Z}_n)$ is isomorphic to the ring $\mathbf{Z}_n$ itself.

3. Show that the ring $\operatorname{End}(\mathbf{Z} \oplus \mathbf{Z})$ is isomorphic to the subring $\mathrm{M}_2(\mathbf{Z})$ of $\mathrm{M}_2(\mathbf{Q})$.

4. Let X be a set and let R be the ring of all subsets of X, with the operations defined as in Exercise 11 (c) of Section 1.1. Show that if X has n elements, then R is isomorphic to the direct sum of n copies of $\mathbf{Z}_2$.

5. Let R be the set of all matrices $\begin{bmatrix} a & b \\ c & d \end{bmatrix}$ over $\mathbf{Q}$ such that $b = 0$ and $a = d$. Let I be the set of all such matrices for which $a = d = 0$. Show that I is an ideal of R, and use the fundamental homomorphism theorem for rings to show that $R/I \cong \mathbf{Q}$.

6. Given a ring R, the *opposite ring* R^{op} is constructed by using the set R and the existing addition on R. Prove that R^{op} is a ring, using the multiplication $\cdot$ defined on R by setting $a \cdot b = ba$, for all $a, b \in R$.

7. Let R be the set of all matrices $\begin{bmatrix} a & b \\ c & d \end{bmatrix}$ over $\mathbf{Q}$ such that $b = 0$. Determine the lattice of all ideals of R.

8. Let R and S be rings.

   (a) Show that if I is an ideal of R and J is an ideal of S, then $I \oplus J$ is an ideal of $R \oplus S$.

   (b) Show that any ideal of $R \oplus S$ must have the form of the ideals described in part (a).

   (c) Show that any maximal ideal of $R \oplus S$ must have the form $I \oplus S$ for a maximal ideal I of R or $R \oplus I$ for a maximal ideal I of S.

9. Let p be a prime number, and let R be the set of all rational numbers m/n such that p is not a factor of n.

   (a) Show that R is a subring of $\mathbf{Q}$.

   (b) For any positive integer k, let $p^k R = \{m/n \in R \mid m \text{ is a multiple of } p^k\}$. Show that $p^k R$ is an ideal of R.

   (c) Show that $R/p^k R$ is isomorphic to $\mathbf{Z}_{p^k}$.

10. Let F be a field, let R be the polynomial ring $F[x]$, and let I be any maximal ideal of R. Prove that $I = (p(x))$ for some irreducible polynomial $p(x) \in F[x]$.

11. Let $\mathcal{C}(\mathbf{R})$ be the set of all continuous functions from $\mathbf{R}$ into $\mathbf{R}$. Exercise 10 of Section 1.1 shows that $\mathcal{C}(\mathbf{R})$ is a commutative ring under the operations $[f + g](x) = f(x) + g(x)$ and $[f \cdot g](x) = f(x)g(x)$, for all $x \in \mathbf{R}$. Let I be the set of all functions $f(x) \in \mathcal{C}(\mathbf{R})$ such that $f(1) = 0$. Show that I is a maximal ideal of $\mathcal{C}(\mathbf{R})$.

12. Let R_1, R_2, ..., R_n be commutative rings, and let $R = R_1 \oplus \cdots \oplus R_n$.

(a) Show that the respective groups of units $R^\times$ and $R_1^\times \times R_2^\times \times \cdots \times R_n^\times$ are isomorphic.

(b) Let n be a positive integer with prime decomposition $n = p_1^{\alpha_1} p_2^{\alpha_2} \cdots p_m^{\alpha_m}$. Prove that $\mathbf{Z}_n^\times \cong \mathbf{Z}_{p_1^{\alpha_1}}^\times \times \mathbf{Z}_{p_2^{\alpha_2}}^\times \times \cdots \times \mathbf{Z}_{p_m^{\alpha_m}}^\times$.

(c) Recall the definition of the Euler φ-function: for any positive integer n, the number of positive integers less than or equal to n that are relatively prime to n is denoted by $\varphi(n)$. Show that $\varphi(n) = n \left(1 - \frac{1}{p_1}\right) \left(1 - \frac{1}{p_2}\right) \cdots \left(1 - \frac{1}{p_m}\right)$.

1.3   Localization of integral domains

In this section (and the next) we will restrict the discussion to commutative rings. We first introduce and discuss the notion of a prime ideal. The next step is to construct the quotient field of an integral domain. Finally, we consider the construction of rings of fractions in which the 'denominators' are taken from the complement of a prime ideal of an integral domain.

To motivate the definition of a prime ideal, consider the ring of integers $\mathbf{Z}$. We know that the proper nontrivial ideals of $\mathbf{Z}$ correspond to the positive integers, with $n\mathbf{Z} \subseteq m\mathbf{Z}$ if and only if $m|n$. Euclid's lemma states that an integer $p > 1$ is prime if and only if it satisfies the following condition:

if $p|ab$ for integers a, b, then either $p|a$ or $p|b$.

In the language of ideals, this says that p is prime if and only if $ab \in p\mathbf{Z}$ implies $a \in p\mathbf{Z}$ or $b \in p\mathbf{Z}$, for all integers a, b.

Similarly, over a field F a polynomial $p(x) \in F[x]$ is irreducible if and only if it satisfies the following condition:

$p(x) \mid f(x)g(x)$ implies $p(x) \mid f(x)$ or $p(x) \mid g(x)$.

When formulated in terms of the principal ideal $(p(x))$, this shows that $p(x)$ is irreducible over F if and only if $f(x)g(x) \in (p(x))$ implies $f(x) \in (p(x))$ or $g(x) \in (p(x))$ Thus irreducible polynomials play the same role in $F[x]$ as do prime numbers in $\mathbf{Z}$.

These two examples probably would not provide sufficient motivation to introduce a general definition of a 'prime' ideal. In both $\mathbf{Z}$ and $F[x]$, where F is a field, we have shown that each element can be expressed as a product of primes and irreducibles, respectively. This is not true in general for all commutative rings. In fact, certain subrings of $\mathbf{C}$ that are important in number theory do not have this property. One of the original motivations for introducing the notion of an ideal (or 'ideal number') was to be able to salvage at least the property that every ideal could be expressed as a product of prime ideals.

Definition 1.3.1 *Let I be a proper ideal of the commutative ring R. Then I is said to be a* prime ideal *of R if $ab \in I$ implies $a \in I$ or $b \in I$, for all $a, b \in R$.*

The definition of a prime ideal can be reformulated in a useful way in terms of ideals. We need to recall the construction of the product of ideals A and B, from Proposition 1.1.8 (c):

$$AB = \left\{ \sum_{j=1}^{n} a_j b_j \mid a_j \in A,\ b_j \in B \right\}.$$

Prime ideals can also be characterized by conditions on the corresponding factor ring.

Proposition 1.3.2 *Let I be a proper ideal of the commutative ring R.*

(a) *The ideal I is a prime ideal of R if and only if $AB \subseteq I$ implies $A \subseteq I$ or $B \subseteq I$, for all ideals A, B of R.*

(b) *The ideal I is a prime ideal of R if and only if the factor ring R/I is an integral domain.*

(c) *If I is a maximal ideal of R, then it is a prime ideal of R.*

Proof. (a) First assume that I is a prime ideal, and that $AB \subseteq I$ for ideals A, B of R. If neither A nor B is contained in I, then there exist $a \in A \setminus I$ and $b \in B \setminus I$. Then $ab \in I$ since $AB \subseteq I$, which contradicts the assumption that I is prime.

Conversely, suppose that I satisfies the given condition. If $ab \in I$ for elements $a, b \in R$, then $(Ra)(Rb) \subseteq I$ since R is commutative, so by assumption either $Ra \subseteq I$ or $Rb \subseteq I$, showing that $a \in I$ or $b \in I$, and therefore I is a prime ideal.

(b) Assume that R/I is an integral domain, and let $a, b \in R$ with $ab \in I$. Remember that the zero element of R/I is the coset consisting of all elements of I. Thus in R/I we have a product $(a + I)(b + I)$ of congruence classes that is equal to the zero coset, and so by assumption this implies that either $a + I$ or $b + I$ is the zero coset. This implies that either $a \in I$ or $b \in I$, and so I must be a prime ideal.

Conversely, assume that I is a prime ideal. If $a, b \in R$ with $(a + I)(b + I) = 0 + I$ in R/I, then we have $ab \in I$, and so by assumption either $a \in I$ or $b \in I$. This shows that either $a + I = 0 + I$ or $b + I = 0 + I$, and so R/I is an integral domain.

(c) Since every field is an integral domain, this follows from part (b) and Proposition 1.2.11 (b), which states that I is maximal if and only if R/I is a field. $\square$

If R is a commutative ring, then R is an integral domain if and only if the zero ideal of R is a prime ideal. This observation provides an example, namely, the zero ideal of $\mathbf{Z}$, of a prime ideal that is not maximal. The ideal $(0) \oplus \mathbf{Z}$ of the direct sum $\mathbf{Z} \oplus \mathbf{Z}$ is a nonzero prime ideal that is not maximal. On the other hand, Proposition 1.3.3 shows that in a commutative ring any maximal ideal is also a prime ideal.

Example 1.3.1 (Some prime ideals of $R[x]$)

Let R be a commutative ring, and let P be a prime ideal of R. We will show that the set $P[x]$ of all polynomials with coefficients in P is a prime ideal of $R[x]$.

Let $\overline{R}$ be the factor ring R/P, which is by assumption an integral domain, and let $\theta : R \to \overline{R}[x]$ be defined by $\theta(r) = r + P$, for all $r \in R$. As in Example 1.2.1, the ring homomorphism θ extends to $\widehat{\theta} : R[x] \to \overline{R}[x]$. It is easy to check that the kernel of $\widehat{\theta}$ is $P[x]$, which shows that $P[x]$ is an ideal of $R[x]$. Furthermore, since $\widehat{\theta}$ is an onto mapping and $\overline{R}[x]$ is an integral domain, it follows that $P[x]$ is a prime ideal of $R[x]$.

Example 1.3.2 (Isomorphisms preserve prime and maximal ideals)

Let $\phi : R \to S$ be an isomorphism of commutative rings. The isomorphism gives a one-to-one correspondence between ideals of the respective rings, and it is not hard to show directly that prime (or maximal) ideals of R correspond to prime (or maximal) ideals of S. This can also be proved as an application of Proposition 1.3.2, since in Example 1.2.7 we observed that if I is any ideal of R, then $\phi(I)$ is an ideal of S with $R/I \cong S/\phi(I)$. It then follows immediately from Proposition 1.3.2 that I is prime (or maximal) in R if and only if $\phi(I)$ is prime (or maximal) in S.

The prime numbers in $\mathbf{Z}$ can be characterized in another way. Since a prime p has no divisors except $\pm p$ and ± 1, there cannot be any ideals properly contained between $p\mathbf{Z}$ and $\mathbf{Z}$, and so in $\mathbf{Z}$, every nonzero prime ideal is maximal. One reason behind this fact is that any finite integral domain is a field. It is also true that if F is a field, then irreducible polynomials in $F[x]$ determine maximal ideals. The next proposition gives a general reason applicable in both of these special cases.

Proposition 1.3.3 *Every nonzero prime ideal of a principal ideal domain is maximal.*

Proof. Let P be a nonzero prime ideal of a principal ideal domain R, and let J be any ideal with $P \subseteq J \subseteq R$. Since R is a principal ideal domain, we can assume that $P = Ra$ and $J = Rb$ for some elements $a, b \in R$. Since $a \in P$, we have $a \in J$, and so there exists $r \in R$ such that $a = rb$. This implies that $rb \in P$, and so either $b \in P$ or $r \in P$. In the first case, $b \in P$ implies that $J = P$. In the second case, $r \in P$ implies that $r = sa$ for some $s \in R$, since P is generated by a. This gives $a = sab$, and using the assumption that R is an integral domain allows us to cancel a to get $1 = sb$. This shows that $1 \in J$, and so $J = R$. $\square$

Example 1.3.3 (Ideals of $F[x]$)

We can now summarize the information we have regarding polynomials over a field, using a ring theoretic point of view. Let F be any field. The nonzero ideals of $F[x]$ are all principal, of the form $(f(x))$, where $f(x)$ is the unique monic polynomial of minimal degree in the ideal. The ideal is prime (and hence maximal) if and only if $f(x)$ is irreducible. If $p(x)$ is irreducible, then the factor ring $F[x]/(p(x))$ is a field.

Example 1.3.4 (The evaluation mapping)

Let F be a subfield of E, and for any element $\alpha \in E$ define the evaluation mapping $\phi_\alpha : F[x] \to E$ by $\phi_\alpha(f(x)) = f(\alpha)$, for all $f(x) \in F[x]$. It is easily checked that ϕ_α defines a ring homomorphism. Since $\phi_\alpha(F[x])$ is a subring of the field E, it must be an integral domain. By the fundamental homomorphism theorem for rings this image is isomorphic to $F[x]/\ker(\phi_\alpha)$, and so by Proposition 1.3.2, the kernel of ϕ_α must be a prime ideal. If $\ker(\phi_\alpha)$ is nonzero, then it follows from Proposition 1.3.3 that it is a maximal ideal. By Proposition 1.3.2, we know that $F[x]/\ker(\phi_\alpha)$ is a field, so it follows from the fundamental homomorphism theorem for rings that the image of ϕ_α is in fact a subfield of E, which is denoted by $F(\alpha)$.

Example 1.3.5 (The characteristic of an integral domain)

We can easily show that an integral domain has characteristic 0 or p, for some prime number p.

Let D be an integral domain, and consider the mapping $\phi :$ $\mathbf{Z} \to D$ defined by $\phi(n) = n \cdot 1$. The fundamental homomorphism theorem for rings shows that $\mathbf{Z}/\ker(\phi)$ is isomorphic to the subring $\phi(\mathbf{Z})$. Since any subring of an integral domain inherits the property that it has no nontrivial divisors of zero, this shows that $\mathbf{Z}/\ker(\phi)$ must be an integral domain. Thus either $\ker(\phi) = (0)$, in which case $\operatorname{char}(D) = 0$, or $\ker(\phi) = k\mathbf{Z}$ for positive integer k. Since $\mathbf{Z}/k\mathbf{Z} = \mathbf{Z}_k$ is an integral domain if and only if k is prime, we conclude that in this case $\operatorname{char}(D) = p$, for some prime p.

We are now ready to construct the quotient field of an integral domain. We start with any integral domain D. (You may want to keep in mind the ring $\mathbf{R}[x]$ of polynomials with real coefficients.) To formally construct fractions with numerator and denominator in D we will consider ordered pairs (a, b), where a represents the numerator and b represents the denominator, assuming that $b \neq 0$. In $\mathbf{Q}$, two fractions m/n and r/s may be equal even though their corresponding numerators and denominators are not equal. We can express the fact that $m/n = r/s$ by writing $ms = nr$. This shows that we must make certain identifications within the set of ordered pairs (a, b), and the appropriate way to do this is to introduce an equivalence relation.

To show this, given $(a, b) \in W$ we have $(a, b) \sim (a, b)$ since $ab = ba$. If $(c, d) \in W$ with $(a, b) \sim (c, d)$, then $ad = bc$, and so $cb = da$, showing that $(c, d) \sim (a, b)$. If $(a, b) \sim (c, d)$ and $(c, d) \sim (u, v)$ in W, then $ad = bc$ and $cv = du$, and $adv = bcv$ and $bcv = bdu$. Thus $adv = bdu$, and since d is a nonzero element of an integral domain, we obtain $av = bu$, so $(a, b) \sim (u, v)$.

Definition 1.3.4 *Let D be an integral domain. The equivalence classes of the set*

$$\{(a, b) \mid a, b \in D \text{ and } b \neq 0\}$$

under the equivalence relation defined by $(a, b) \sim (c, d)$ if $ad = bc$ will be denoted by $[a, b]$. The set of all such equivalence classes will be denoted by $Q(D)$.

Theorem 1.3.5 *Let D be an integral domain. Then $Q(D)$ is a field that contains a subring isomorphic to D.*

Proof. We will show that the following operations are well-defined on $Q(D)$. For $[a, b], [c, d] \in Q(D)$,

$$[a, b] + [c, d] = [ad + bc, bd] \quad \text{and} \quad [a, b] \cdot [c, d] = [ac, bd] .$$

Let $[a, b] = [a', b']$ and $[c, d] = [c', d']$ be elements of $Q(D)$. To show that addition is well-defined, we must show that $[a, b] + [c, d] = [a', b'] + [c', d']$, or, equivalently, that $[ad + bc, bd] = [a'd' + b'c', b'd']$. In terms of the equivalence relation on $Q(D)$, we must show that $(ad+bc)b'd' = bd(a'd'+b'c')$. We are given that $ab' = ba'$ and $cd' = dc'$. Multiplying the first of these two equations by dd' and the second by bb' and then adding terms gives us the desired equation.

In order to show that multiplication is well-defined, we must show that $[ac, bd] = [a'c', b'd']$. This is left as Exercise 3.

It is relatively straightforward to verify the commutative and associative laws for addition. Let 1 be the identity element of D. For all $[a, b] \in Q(D)$, we have $[0, 1] + [a, b] = [a, b]$, and so $[0, 1]$ serves as an additive identity element. Furthermore, $[-a, b] + [a, b] = [-ab + ba, b^2] = [0, b^2]$, which shows that $[-a, b]$ is the additive inverse of $[a, b]$ since $[0, b^2] = [0, 1]$.

Again, verifying the commutative and associative laws for multiplication is straightforward. The equivalence class $[1, 1]$ clearly acts as a multiplicative identity element. Note that we have $[1, 1] = [d, d]$ for any nonzero $d \in D$. If $[a, b]$ is a nonzero element of $Q(D)$, that is, if $[a, b] \neq [0, 1]$, then we must have $a \neq 0$. Therefore $[b, a]$ is an element of $Q(D)$, and since $[b, a] \cdot [a, b] = [ab, ab]$, we have $[b, a] = [a, b]^{-1}$. Thus every nonzero element of $Q(D)$ is invertible, and so to complete the proof that $Q(D)$ is a field we only need to check that the distributive law holds. Let $[a, b]$, $[c, d]$, and $[u, v]$ be elements of $Q(D)$. We have

$$([a, b] + [c, d]) \cdot [u, v] = [(ad + bc)u, (bd)v]$$

and

$$[a, b] \cdot [u, v] + [c, d] \cdot [u, v] = [(au)(dv) + (bv)(cu), (bv)(dv)] .$$

We can factor $[v, v]$ out of the second expression, showing equality.

Finally, consider the mapping $\phi : D \to Q(D)$ defined by $\phi(d) = [d, 1]$, for all $d \in D$. It is easy to show that ϕ preserves sums and products, and so it is a ring homomorphism. If $[d, 1] = [0, 1]$, then we must have $d = 0$, which shows that $\ker(\phi) = (0)$. By the fundamental homomorphism theorem for rings, $\phi(D)$ is a subring of $Q(D)$ that is isomorphic to D. $\square$

Definition 1.3.6 *Let D be an integral domain. The field $Q(D)$ defined in Definition 1.3.4 is called the* quotient field *or* field of fractions *of D. The equivalence classes $[a, b]$ of $Q(D)$ will usually be denoted by a/b or ab^{-1}.*

If we use the notation a/b, then we identify an element $d \in D$ with the fraction $d/1$, and this allows us to assume that D is a subring of $Q(D)$. If $b \in D$ is nonzero, then $1/b \in Q(D)$, and $(1/b) \cdot (b/1) = b/b = 1$ shows that $1/b = b^{-1}$ (where we have identified b and $b/1$). Thus we can also write $a/b = (a/1) \cdot (1/b) = ab^{-1}$, for $a, b \in D$, with $b \neq 0$.

Theorem 1.3.7 *Let D be an integral domain, and let $\theta : D \to F$ be a one-to-one ring homomorphism from D into a field F. Then there exists a unique extension $\widehat{\theta} : Q(D) \to F$ that is one-to-one and satisfies $\widehat{\theta}(d) = \theta(d)$ for all $d \in D$.*

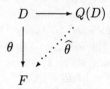

Proof. For $ab^{-1} \in Q(D)$, define $\widehat{\theta}(ab^{-1}) = \theta(a)\theta(b)^{-1}$. Since $b \neq 0$ and θ is one-to-one, $\theta(b)^{-1}$ exists in F, and the definition makes sense. We must show that $\widehat{\theta}$ is well-defined. If $ab^{-1} = cd^{-1}$, then $ad = bc$, and applying θ to both sides of this equation gives $\theta(a)\theta(d) = \theta(b)\theta(c)$ since θ is a ring homomorphism. Both $\theta(b)^{-1}$ and $\theta(d)^{-1}$ exist, so we must have $\theta(a)\theta(b)^{-1} = \theta(c)\theta(d)^{-1}$.

The proof that $\widehat{\theta}$ is a homomorphism is left as an exercise. To show that $\widehat{\theta}$ is one-to-one, suppose that $\widehat{\theta}(ab^{-1}) = \widehat{\theta}(cd^{-1})$, for $a, b, c, d \in D$, with b and d nonzero. Then $\theta(a)\theta(b)^{-1} = \theta(c)\theta(d)^{-1}$, so $\theta(ad) = \theta(bc)$, and then $ad = bc$ since θ is one-to-one. Using the equivalence relation on the classes of $Q(D)$, we see that $ab^{-1} = cd^{-1}$.

To prove the uniqueness of $\widehat{\theta}$, suppose that $\phi : Q(D) \to F$ with $\phi(d) = \theta(d)$ for all $d \in D$. Then for any element $ab^{-1} \in Q(D)$ we have $\phi(a) = \phi(ab^{-1}b) = \phi(ab^{-1})\phi(b)$, and so $\theta(a) = \phi(ab^{-1})\theta(b)$ or $\phi(ab^{-1}) = \theta(a)\theta(b)^{-1}$. $\square$

Corollary 1.3.8 *Let D be an integral domain that is a subring of a field F. If each element of F has the form ab^{-1} for some $a, b \in D$, then F is isomorphic to the quotient field $Q(D)$ of D.*

Proof. By Theorem 1.3.7, the inclusion mapping $\theta : D \to F$ can be extended to a one-to-one mapping $\widehat{\theta} : Q(D) \to F$. The given condition is precisely the one necessary to guarantee that $\widehat{\theta}$ is onto. $\square$

Example 1.3.6 (The quotient field of $\mathbf{Z}_{(p)}$)

Let $\mathbf{Z}_{(p)}$ be the integral domain consisting of all rational numbers $m/n \in \mathbf{Q}$ such that n is relatively prime to p. (Refer to Exercise 9 in Section 1.2.) If a/b is any element of $\mathbf{Q}$ such that $\gcd(a, b) = 1$, then either b is relatively prime to p, in which case $a/b \in D$, or b is divisible by p, in which case a is relatively prime to p and $a/b = 1 \cdot (b/a)^{-1}$, with $b/a \in D$. Applying Corollary 1.3.8 shows that $\mathbf{Q}$ is isomorphic to the quotient field $Q(\mathbf{Z}_{(p)})$ of $\mathbf{Z}_p$.

Example 1.3.7 (The ring $F(x)$ of rational functions)

If F is any field, then we know that the ring of polynomials $F[x]$ is an integral domain. Applying Theorem 1.3.5 shows that we can construct a field that contains $F[x]$ by considering all fractions of the form $f(x)/g(x)$, where $f(x), g(x)$ are polynomials with $g(x) \neq 0$. This field is called the *field of rational functions* in x, and is denoted by $F(x)$.

Example 1.3.8 (The prime subfield of a field)

Any field F contains a subfield isomorphic to $\mathbf{Q}$ or $\mathbf{Z}_p$, for some prime number p. This is called the *prime subfield* of F.

To see this, let F be any field, and let ϕ be the homomorphism from $\mathbf{Z}$ into F defined by $\phi(n) = n \cdot 1$. If $\ker(\phi) \neq (0)$, then we have $\ker(\phi) = p\mathbf{Z}$ for some prime p, and so the image of ϕ is a subfield isomorphic to $\mathbf{Z}_p$. If ϕ is one-to-one, then by Theorem 1.3.7 it extends to an embedding of $\mathbf{Q}$ (the quotient field of $\mathbf{Z}$) into F, and so the image of this extension is a subfield isomorphic to $\mathbf{Q}$.

Proposition 1.3.9 *Let D be an integral domain with quotient field $Q(D)$, and let P be a prime ideal of D. Then*

$$D_P = \{ab^{-1} \in Q(D) \mid b \notin P\}$$

is an integral domain with $D \subseteq D_P \subseteq Q(D)$.

Proof. Let ab^{-1} and cd^{-1} belong to D_P. Then $b, d \in D \setminus P$, and so $bd \in D \setminus P$ since P is a prime ideal. It follows that

$$ab^{-1} + cd^{-1} = (ad + bc)(bd)^{-1} \in D_P$$

and

$$ab^{-1}cd^{-1} = (ac)(bd)^{-1} \in D_P \,,$$

and so D_P is closed under addition and multiplication. Since $1 \notin P$, the given set includes D. Finally, since $Q(D)$ is a field, the subring D_P is an integral domain. $\square$

Definition 1.3.10 *Let D be an integral domain with quotient field $Q(D)$, and let P be a prime ideal of D.*

(a) The ring $D_P = \{ab^{-1} \in Q(D) \mid b \notin P\}$ is called the localization *of D at P.*

(b) If I is any ideal of D, then $I_P = \{ab^{-1} \in Q(D) \mid a \in I$ and $b \notin P\}$ is called the extension *of I to D_P.*

Theorem 1.3.11 *Let D be an integral domain with quotient field $Q(D)$, and let P be a prime ideal of D.*

(a) *If J is any proper ideal of D_P, then there exists an ideal I of D with $I \subseteq P$ such that $J = I_P$. If J is a prime ideal, then so is I.*

(b) *If J is any prime ideal of D with $J \subseteq P$, then J_P is a prime ideal of D_P with $D \cap J_P = J$.*

(c) *There is a one-to-one correspondence between prime ideals of D_P and prime ideals of D that are contained in P. Furthermore, P_P is the unique maximal ideal of D_P.*

Proof. (a) Let J be an ideal of D_P, and let $I = J \cap D$. If I contains an element a not in P, then J contains an invertible element, and must equal D_P. Thus $J \cap D \subseteq P$ if J is a proper ideal.

If $x = ab^{-1} \in J$, then $a = xb \in I$, and so $x \in I_P$. On the other hand, if $x = ab^{-1} \in I_P$, then $a \in I \subseteq J$, and so $ab^{-1} \in J$, showing that $J = I_P$.

If J is a prime ideal of D_P, then the fact that D/I is isomorphic to a subring of D_P/J shows that I is a prime ideal of D.

(b) Let J be a prime ideal of D with $J \subseteq P$. It is easy to show that J_P is closed under addition and multiplication by any element of D_P, so it is an ideal of D_P. It is clear that $J \subseteq J_P \cap D$, and if $a \in J_P \cap D$, then $a = cd^{-1}$ for some $c \in J$ and $d \notin P$. Since $J \subseteq P$, we have $d \notin J$, and then $ad = c \in J$ implies $a \in J$ because J is a prime ideal. Thus $J = J_P \cap D$.

If $ab^{-1}cd^{-1} \in J_P$, for $ab^{-1}, cd^{-1} \in D_P$, then $ac \in D \cap J_P = J$, so either $a \in J$ or $c \in J$. Thus either $ab^{-1} \in J_P$ or $cd^{-1} \in J_P$, showing that J_P is a prime ideal.

(c) This part of the proof is clear, since parts (a) and (b) establish the desired one-to-one correspondence. □

EXERCISES: SECTION 1.3

1. Let R and S be commutative rings. Show that any prime ideal of $R \oplus S$ must have the form $P \oplus S$ for a prime ideal P of R or $R \oplus P$ for a prime ideal P of S.

2. Let P be a prime ideal of the commutative ring R. Prove that if I and J are ideals of R and $I \cap J \subseteq P$, then either $I \subseteq P$ or $J \subseteq P$.

3. Complete the proof of Theorem 1.3.5, by showing that multiplication of equivalence classes in $Q(D)$ is well-defined, where D is an integral domain.

4. Verify that the function $\widehat{\theta}$ defined in Theorem 1.3.7 is a ring homomorphism.

5. For the ring $D = \{m + n\sqrt{2} \mid m, n \in \mathbf{Z}\}$, determine the quotient field $Q(D)$.

6. Let p be a prime number, and let $D = \{m/n \in \mathbf{Q} \mid m, n \in \mathbf{Z} \text{ and } p \nmid n\}$. Determine the quotient field $Q(D)$.

7. For the ring $D = \{m + ni \mid m, n \in \mathbf{Z}\} \subseteq \mathbf{C}$, determine the quotient field $Q(D)$.

8. Let D be a principal ideal domain. Prove that if P is any prime ideal of D, then the localization D_P is a principal ideal domain.

9. Let R be a commutative ring. A subset S of R is called a *multiplicative set* if $0 \notin S$ and $a, b \in S$ implies $ab \in S$.

    (a) Let R be a commutative ring, and let S be a multiplicative set in R. Prove that the relation defined on $R \times S$ by $(a, c) \sim (b, d)$ if $s(ad - bc) = 0$ for some $s \in S$ is an equivalence relation.

    (b) Prove that if the equivalence classes of $\sim$ are denoted by $[a, c]$, for $a \in R$, $c \in S$, then the following addition and multiplication are well-defined.

$$[a, c] + [b, d] = [ad + bc, cd] \qquad \text{and} \qquad [a, c] \cdot [b, d] = [ab, cd]$$

10. Let R be a commutative ring, and let S be a multiplicative set in R, with a fixed element $s_1 \in S$. The set of equivalence classes of $R \times S$, with the addition and multiplication defined in Exercise 9, will be denoted by R_S.

    (a) Prove that R_S is a commutative ring, and the mapping $\phi_S : R \to R_S$ defined by $\phi_S(a) = [as_1, s_1]$ is a ring homomorphism such that $\phi_S(c)$ is invertible in R_S, for all $c \in S$.

    (b) Prove that if $\theta : R \to R'$ is any ring homomorphism such that $\theta(c)$ is invertible for all elements $c \in S$, then there exists a unique ring homomorphism $\widehat{\theta} : R_S \to R'$ such that $\widehat{\theta}\phi_S = \theta$.

11. Let R be a commutative ring, and let S be a multiplicative set in R. Show that if I is an ideal of R, then $I_S = \{[a, c] \in R_S \mid a \in I \text{ and } c \in S\}$ is an ideal of R_S.

12. Let R be a commutative ring, let P be a prime ideal of R, and let S be the multiplicative set $R \setminus P$. Then the ring R_S is called the *localization of R at P*, and is denoted by R_P.

    (a) Show that there is a one-to-one correspondence between the prime ideals of R_P and the prime ideals of R that are contained in P.

    (a) Show that the ideal P_P is the unique maximal ideal of R_P.

1.4 Unique factorization

In this section we present a theory of unique factorization for certain integral domains, beginning with principal ideal domains. We start with the general notion of a greatest common divisor. Recall that by definition $b|a$ in the ring R if $a = bq$ for some $q \in R$, and so $aR \subseteq bR$ if and only if $b|a$.

Definition 1.4.1 *Let R be a commutative ring.*

(a) *For elements* $a, b \in R$, *an element d of R is called a* greatest common divisor *of a and b if*

    (i) $d|a$ *and* $d|b$ *and*

    (ii) *if* $c|a$ *and* $c|b$, *for* $c \in R$, *then* $c|d$.

(b) *We say that a and b are* relatively prime *if* $aR + bR = R$.

(c) *A nonzero element p of R is said to be* irreducible *if*

    (i) *p is not a unit of R and*

    (ii) *if* $p = ab$ *for* $a, b \in R$, *then either a or b is a unit of R.*

In the ring $\mathbf{Z}$ the only units are ± 1, so we can define a unique greatest common divisor by choosing the positive number among the two possibilities. For a field F, in the ring $F[x]$ we can choose a monic greatest common divisor, again producing a unique answer. In the general setting we no longer have uniqueness.

If R is any commutative ring, with $a, b, d \in R$, then d is a greatest common divisor of a and b if $aR + bR = dR$. Thus if D is a principal ideal domain, and a and b are nonzero elements of D, then we can choose a generator d for the ideal $aR + bR$, and so D contains a greatest common divisor of a and b, of the form $d = as + bt$ for $s, t \in D$. If d' is another greatest common divisor, then $d'|d$ and $d|d'$, and so d and d' are associates.

The first result on irreducible elements generalizes a familiar property of integers and polynomials.

Proposition 1.4.2 *Let D be a principal ideal domain, and let p be a nonzero element of D.*

(a) *If p is an irreducible element of D, and* $a, b \in D$ *with* $p|ab$, *then either* $p|a$ *or* $p|b$.

(b) *The element p is irreducible in D if and only if the ideal pD is a prime ideal of D.*

Proof. (a) If $p|ab$ but $p \nmid a$, then the only common divisors of p and a are units, so the greatest common divisor of p and a is 1. Therefore there exist $q_1, q_2 \in D$ with $1 = pq_1 + aq_2$, and multiplying by b gives $b = p(bq_1) + (ab)q_2$. Then $p|b$ since $p|ab$.

(b) First suppose that p is irreducible in D. If $ab \in pD$ for $a, b \in D$, then part (a) shows that either $a \in pD$ or $b \in pD$, and therefore pD is a prime ideal.

Conversely, suppose that pD is a prime ideal of D, and $p = ab$ for some nonunits a, b of D. Then $ab \in pD$ implies $a \in pD$ or $b \in pD$, so either $p \mid a$ or $p \mid b$. If $p \mid a$, then $a = pc$ for some c, and $p = pcb$. Since D is an integral domain we can cancel p, so $1 = cb$, contradicting the assumption that b is not a unit. The case $p \mid b$ leads to a similar contradiction, showing that p is irreducible. $\square$

Definition 1.4.3 *Let D be an integral domain. Then D is called a* unique factorization domain *if*

(i) *each nonzero element a of D that is not a unit can be expressed as a finite product of irreducible elements of D, and*

(ii) *in any two such factorizations $a = p_1 p_2 \cdots p_n = q_1 q_2 \cdots q_m$ the integers n and m are equal and it is possible to rearrange the factors so that q_i is an associate of p_i, for $1 \le i \le n$.*

Lemma 1.4.4 *Let D be a principal ideal domain. For any chain of ideals $I_1 \subseteq I_2 \subseteq I_3 \subseteq \cdots$ in D, there exists a positive integer m such that $I_n = I_m$ for all $n > m$.*

Proof. If $I_1 \subseteq I_2 \subseteq I_3 \subseteq \cdots$ is an ascending chain of ideals, let $I = \bigcup_{n=1}^{\infty} I_n$. Since the ideals form a chain, it is easy to check that I is an ideal of D. Then $I = aD$ for some $a \in D$, since D is a principal ideal domain, and since a must belong to I_m for some m, we have $I \subseteq I_m$, so $I_n = I_m$ for all $n > m$. $\square$

For sets S and T, we have been using the notation $S \subseteq T$ or $T \supseteq S$ to indicate that S is a subset of T. It is convenient to use the notation $S \subset T$ or $T \supset S$ when S is a *proper* subset of T, that is, when $S \subseteq T$ but $S \ne T$.

Theorem 1.4.5 *Any principal ideal domain is a unique factorization domain.*

Proof. Let D be a principal ideal domain, and let d be a nonzero element of D that is not a unit. Suppose that d cannot be written as a product of irreducible elements. Then since d itself cannot be irreducible, there exist nonunits $a_1, b_1 \in D$ with $d = a_1 b_1$, such that either a_1 or b_1 is not a product of irreducible elements. Assume that a_1 cannot be written as a product of irreducible elements. Now b_1 is not a unit, so $dD \subset a_1 D$. Repeating this argument, we obtain a factor a_2 of a_1 that cannot be written as a product of irreducible elements, with $a_1 D \subset a_2 D$. Thus the assumption that d cannot be written as a product of irreducible elements leads to a strictly ascending chain of ideals $dD \subset a_1 D \subset a_2 D \subset a_3 D \subset \cdots$. By Lemma 1.4.4, this contradicts the assumption that D is a principal ideal domain.

Now suppose that d can be written in two ways as a product of irreducible elements. If $d = p_1 p_2 \cdots p_n = q_1 q_2 \cdots q_m$, where $n \le m$ and p_i, q_j are irreducible for all i and m, we will show that $n = m$. The proof is by induction on n. Since p_1 is irreducible and a divisor of $q_1 q_2 \cdots q_m$, it follows that p_1 is a divisor of q_i, for some i, and we can assume that $i = 1$. Since both p_1 and q_1 are irreducible and $p_1|q_1$, it follows that they are associates. Canceling gives $p_2 \cdots p_n = uq_2 \cdots q_m$, where u is a unit, and so by induction we have $n - 1 = m - 1$. The elements q_i can be reordered in such a way that q_i and p_i are associates. □

If F is a field, then the polynomial ring $F[x]$ is a principal ideal domain, and so it follows from Theorem 1.4.5 that $F[x]$ is a unique factorization domain. This argument fails for the ring $\mathbf{Z}[x]$ of polynomials with integer coefficients, since Exercise 15 of Section 1.1 shows that $\mathbf{Z}[x]$ is not a principal ideal domain. We will show more generally that the ring $D[x]$ still has unique factorization, provided that the coefficients come from a unique factorization domain. This will be proved by working in the ring $Q(D)[x]$, which *is* a principal ideal domain, and therefore has unique factorization.

We begin the development with a generalization of Euclid's lemma.

Lemma 1.4.6 *Let D be a unique factorization domain, and let p be an irreducible element of D. If $a, b \in D$ and $p|ab$, then $p|a$ or $p|b$.*

Proof. If p is irreducible and $p|ab$, then $ab = pd$ for some $d \in D$, and we can assume that $ab \neq 0$. Let $a = q_1 q_2 \cdots q_n$, $b = r_1 r_2 \cdots r_m$, and $d = p_1 p_2 \cdots p_k$ be the factorizations of a, b, and d as products of irreducible elements. Then ab has the two factorizations $ab = q_1 q_2 \cdots q_n r_1 r_2 \cdots r_m = p p_1 p_2 \cdots p_k$, and it follows from the definition of a unique factorization domain that p is an associate of q_i or r_j, for some i or j. Thus $p|a$ or $p|b$. □

Definition 1.4.7 *Let D be a unique factorization domain. A nonconstant polynomial $f(x) = a_n x^n + a_{n-1} x^{n-1} + \cdots + a_1 x + a_0$ in $D[x]$ is called* primitive *if there is no irreducible element $p \in D$ such that $p|a_i$ for all i.*

Lemma 1.4.8 (Gauss) *The product of two primitive polynomials is again primitive.*

Proof. Let p be an irreducible element of the integral domain D. Then pD is a prime ideal of D, so D/pD is an integral domain, and therefore $(D/pD)[x]$ is an integral domain. As in Example 1.3.1, we have $D[x]/pD[x] \cong (D/pD)[x]$, showing that $pD[x]$ is a prime ideal of $D[x]$. If p is a divisor of each coefficient of a polynomial $f(x) = g(x)h(x)$ in $D[x]$, then $f(x) \in pD[x]$. Since $pD[x]$ is a prime ideal, either $g(x) \in pD[x]$ or $h(x) \in pD[x]$. It follows that if $g(x)$ and $h(x)$ are primitive, then $f(x)$ must be primitive. □

If $f(x) = a_n x^n + a_{n-1} x^{n-1} + \cdots + a_1 x + a_0 \in D[x]$ and p is an irreducible element of D such that $p|a_i$ for all i, then we can factor p out of each coefficient of $f(x)$. This allows us to write $f(x) = pf_1(x)$, for an appropriate polynomial $f_1(x)$. Since D is a unique factorization domain, each of the finite number of coefficients has only finitely many nonassociated irreducible factors. After only a finite number of steps we can write $f(x) = df^*(x)$, where $f^*(x)$ is primitive. We can obtain a stronger result, stated for polynomials over the quotient field of D. It is interesting to note that if Q is the quotient field of D, then every polynomial in $Q[x]$ has the form $\frac{1}{c}f(x)$, where $f(x) \in D[x]$. Therefore $Q[x]$ can be obtained by localizing $D[x]$ at the multiplicative set of all nonzero elements of D.

Lemma 1.4.9 *Let Q be the quotient field of the integral domain D, and let $f(x) \in Q[x]$.*

(a) *The polynomial $f(x)$ can be written in the form $f(x) = (a/b)f^*(x)$, where $f^*(x)$ is a primitive element of $D[x]$, $a, b \in D$, and a and b have no common irreducible divisors. This expression is unique, up to units of D.*

(b) *If D is a unique factorization domain, and $f(x)$ is a primitive polynomial in $D[x]$, then $f(x)$ is irreducible in $D[x]$ if and only if $f(x)$ is irreducible in $Q[x]$.*

Proof. (a) Let Q denote the quotient field of D. The required factorization of $f(x)$ can be found as follows: let

$$f(x) = \frac{a_n}{b_n}x^n + \cdots + \frac{a_1}{b_1}x + \frac{a_0}{b_0} \, ,$$

where $a_i, b_i \in D$ for $0 \le i \le n$. There exists a common nonzero multiple b of the denominators $b_0, b_1, \ldots, b_n$, so $f(x) = (1/b) \cdot bf(x)$, where $bf(x)$ has coefficients in D. Next we can write $bf(x) = af^*(x)$, where $f^*(x)$ is primitive in $D[x]$. By factoring out common irreducible divisors of a and b we obtain the required form $f(x) = (a/b)f^*(x)$.

To show uniqueness, suppose that $(a/b)f^*(x) = (c/d)g^*(x)$, where $g^*(x)$ is also primitive. Then $adf^*(x) = bcg^*(x)$, and so the irreducible factors of ad and bc must be the same. Since a and b are relatively prime and D is a unique factorization domain, this shows that a and c are associates and that b and d are also associates. This in turn implies that $f^*(x)$ and $g^*(x)$ are associates.

(b) If $f(x)$ is irreducible in $Q[x]$, then any factorization of $f(x)$ in $D[x]$ can be regarded as a factorization in $Q[x]$. Therefore $f(x)$ cannot have a proper factorization into polynomials of lower degree.

On the other hand, suppose that $f(x)$ is irreducible in $D[x]$, but has a factorization $f(x) = g(x)h(x)$ in $Q[x]$, where $g(x)$ and $h(x)$ both have lower degree than $f(x)$. Then $g(x) = (a/b)g^*(x)$ and $h(x) = (c/d)h^*(x)$, where $g^*(x)$ and $h^*(x)$ are primitive polynomials in $D[x]$.

Without loss of generality we can assume that $tf(x) = sg^*(x)h^*(x) \in D[x]$, where $s, t \in D$ and s and t have no common irreducible divisors. If p is any irreducible factor of t, then p is a divisor of every coefficient of $sg^*(x)h^*(x)$. Since p is not a divisor of s, by our choice of s and t, it follows that p is a divisor of every coefficient of $g^*(x)h^*(x)$. By Lemma 1.4.8, the product of primitive polynomials in $D[x]$ is again primitive, so this is a contradiction.

Thus t is a unit of D, so there is a factorization $f(x) = tg^*(x)h^*(x)$ in $D[x]$, in which the degrees of the factors are the same. This contradicts the assumption that $f(x)$ is irreducible in $D[x]$. □

Theorem 1.4.10 *If D is a unique factorization domain, then the polynomial ring $D[x]$ is also a unique factorization domain.*

Proof. Assume that the nonzero element $f(x) \in D[x]$ is not a unit. We can write $f(x) = df^*(x)$, where $d \in D$ and $f^*(x)$ is primitive. We will show by induction on the degree of $f(x)$ that it can be written as a product of irreducible elements.

First, if $f(x)$ has degree 0, then d is not a unit. Since D is a unique factorization domain, d can be written as a product of irreducible elements of D, and these elements are still irreducible in $D[x]$. Next, suppose that $f(x)$ has degree n, and that the result holds for all polynomials of degree less than n. If $f^*(x)$ is not irreducible, then it can be expressed as a product of polynomials of lower degree, which must all be primitive, and the induction hypothesis can be applied to each factor. If d is a unit we are done, and if not we can factor d in D. The combined factorizations give the required factorization of $df^*(x)$.

If $f(x)$ is primitive, then it can be factored into a product of irreducible primitive polynomials, which are irreducible over $Q[x]$. Since $Q[x]$ is a unique factorization domain, in any two factorizations the factors will be associates in $Q[x]$. Lemma 1.4.9 implies that they will be associates in $D[x]$. If $f(x)$ is not primitive, we can write $f(x) = df^*(x)$, where $f^*(x)$ is primitive. If we also have $f(x) = cg^*(x)$, then by Lemma 1.4.9, c and d must be associates, and $f^*(x)$ and $g^*(x)$ must be associates. It follows that the factorization of $f(x)$ is unique up to associates, completing the proof. □

Corollary 1.4.11 *For any field F, the ring of polynomials $F[x_1, x_2, \ldots, x_n]$ in n indeterminates is a unique factorization domain.*

Proof. Since $F[x_1, x_2] = (F[x_1])[x_2]$, Theorem 1.4.10 implies that $F[x_1, x_2]$ is a unique factorization domain. This argument can be extended inductively to $F[x_1, x_2, \ldots, x_n]$. □

The next example shows that unique factorization can fail, even in a subring of the field of complex numbers.

Example 1.4.1 ($\mathbf{Z}[\sqrt{-5}]$ is not a unique factorization domain)

Let $\zeta = \sqrt{5}\,i$, and consider the subring $\mathbf{Z}[\zeta] = \{m + n\zeta \mid m, n \in \mathbf{Z}\}$ of $\mathbf{C}$. The units of $\mathbf{Z}[\zeta]$ are ± 1, since if $a = m + n\zeta$ is a unit, then $|a|^2 = 1$, and the only integer solutions of $m^2 + 5n^2 = 1$ are $m = \pm 1$, $n = 0$.

We claim that 3, 7, $1 + 2\zeta$, and $1 - 2\zeta$ are irreducible in $\mathbf{Z}[\zeta]$. If $3 = ab$ in $\mathbf{Z}[\zeta]$, where a, b are not units, then $|a|^2|b|^2 = 9$ implies that $|a|^2 = 3$ and $|b|^2 = 3$. This is impossible since there are no integer solutions to the equation $m^2 + 5n^2 = 3$. Similarly, there are no integer solutions to the equation $m^2 + 5n^2 = 7$, and so 7 is also irreducible. If $1 + 2\zeta = ab$, where $a, b \in \mathbf{Z}[\zeta]$ are not units, then $|a|^2|b|^2 = |1 + 2\zeta|^2 = 21$. This implies that $|a|^2 = 3$ or $|a|^2 = 7$, a contradiction. The same argument implies that $1 - 2\zeta$ is irreducible in $\mathbf{Z}[\zeta]$. The two factorizations $21 = 3 \cdot 7$ and $21 = (1 + 2\zeta)(1 - 2\zeta)$ show that $\mathbf{Z}[\zeta]$ is not a unique factorization domain.

EXERCISES: SECTION 1.4

1. Let R be a commutative ring. An element $a \in R$ is said to be *nilpotent* if $a^n = 0$ for some positive integer n. Let N be the set of all nilpotent elements of R.

   (a) Show that N is an ideal of R.

   (b) Show that R/N has no nonzero nilpotent elements.

   (c) Show that $N \subseteq P$ for each prime ideal P of R.

2. Define $\pi : \mathbf{Z} \to \mathbf{Z}_n$ by $\pi(x) = [x]_n$, for all $x \in \mathbf{Z}$. Then π is a ring homomorphism with $\ker(\pi) = n\mathbf{Z}$. Now consider the polynomial rings $\mathbf{Z}[x]$ and $\mathbf{Z}_n[x]$. Define $\widehat{\pi} : \mathbf{Z}[x] \to \mathbf{Z}_n[x]$ as follows: for any polynomial $f(x) = a_0 + a_1 x + \cdots + a_m x^m$ set $\widehat{\pi}(f(x)) = \pi(a_0) + \pi(a_1)x + \cdots + \pi(a_m)x^m$.

   (a) Prove that $\widehat{\pi}$ is a ring homomorphism whose kernel is the set of all polynomials for which each coefficient is divisible by n.

   (b) Show that if $f(x) \in \mathbf{Z}[x]$ has a nontrivial factorization $f(x) = g(x)h(x)$ in $\mathbf{Z}[x]$, then there is a nontrivial factorization of $\widehat{\pi}(f(x))$ in $\mathbf{Z}_n[x]$.

   (c) Let $f(x) = a_n x^n + a_{n-1}x^{n-1} + \cdots + a_0$ be a polynomial with integer coefficients. Prove Eisenstein's criterion, which is stated as follows. If there exists a prime number p such that $a_{n-1} \equiv a_{n-2} \equiv \cdots \equiv a_0 \equiv 0 \pmod{p}$ but $a_n \not\equiv 0 \pmod{p}$ and $a_0 \not\equiv 0 \pmod{p^2}$, then $f(x)$ is irreducible over the field of rational numbers.

3. Let D be a unique factorization domain, and let P be any prime ideal of D. Prove that the localization D_P is a unique factorization domain.

4. An integral domain D is called a *Euclidean domain* if for each nonzero element $x \in D$ there is assigned a nonnegative integer $\delta(x)$ such that (i) $\delta(ab) \geq \delta(b)$ for all nonzero $a, b \in D$, and (ii) for any nonzero elements $a, b \in D$ there exist $q, r \in D$ such that $a = bq + r$, where either $r = 0$ or $\delta(r) < \delta(b)$.

   Prove that any Euclidean domain is a principal ideal domain.

5. This exercise outlines the proof that the ring $\mathbf{Z}[i]$ of Gaussian integers is a Euclidean domain.

   (a) For each nonzero element $a = m + ni$ of $\mathbf{Z}[i]$, define $\delta(a) = m^2 + n^2$. Show that if $b = s + ti$ is nonzero, then $\delta(ab) \geq \delta(b)$.

   (b) For nonzero elements a and b, consider the quotient a/b as a complex number. Letting $a = m + ni$ and $b = s + ti$, for $m, n, s, t \in \mathbf{Z}$, choose $u, v \in \mathbf{Z}$ closest to $c = (ms + nt)/(s^2 + t^2)$ and $d = (ns - mt)/(s^2 + t^2)$, respectively. Show that $|c - u| \leq \frac{1}{2}$ and $|d - v| \leq \frac{1}{2}$.

   (c) Continuing the above notation, let $q = u + vi$ and $r = a - bq$. Show that $\delta(r) \leq \delta(b)/2$, and then finish the argument that $\mathbf{Z}[i]$ is a Euclidean domain.

6. A commutative ring R is said to be a *Noetherian* ring if every ideal of R has a finite set of generators. Prove that if R is a commutative ring (with identity 1), then R is Noetherian if and only if for any ascending chain of ideals $I_1 \subseteq I_2 \subseteq \cdots$ there exists a positive integer n such that $I_k = I_n$ for all $k > n$.

7. Let R be a Noetherian integral domain, and let S be any multiplicative subset of R. Prove that the ring of fractions R_S is Noetherian.

8. Let R be a commutative Noetherian ring. This exercise provides an outline of the steps in a proof of the Hilbert basis theorem, which states that the polynomial ring $R[x]$ is a Noetherian ring.

   (a) Let I be any ideal of $R[x]$, and let I_k be the set of all $r \in R$ such that $r = 0$ or r occurs as the leading coefficient of a polynomial of degree k in I. Prove that I_k is an ideal of R.

   (b) For the ideals I_k in part (a), prove that there exists an integer n such that $I_n = I_{n+1} = \cdots$.

   (c) By assumption, each left ideal I_k is finitely generated (for $k \leq n$), and we can assume that it has $m(k)$ generators. Each generator of I_k is the leading coefficient of a polynomial of degree k, so we let $\{p_{jk}(x)\}_{j=1}^{m(k)}$ be the corresponding polynomials. Prove that $\mathcal{B} = \bigcup_{k=1}^{n} \{p_{jk}(x)\}_{j=1}^{m(k)}$ is a set of generators for I.

   Hint: If not, then among the polynomials that cannot be expressed as linear combinations of polynomials in $\mathcal{B}$ there exists one of minimal degree.

1.5 *Additional noncommutative examples

In this section we will consider a number of noncommutative examples. These will include the ring of endomorphisms of an abelian group, the ring of $n \times n$ matrices over an arbitrary ring R, the ring of skew polynomials over a ring R, and the ring of differential polynomials over a ring R. As a special case of the last example, we will introduce the Weyl algebras.

We have assumed that the reader is familiar with the ring of $n \times n$ matrices whose entries come from a field. We now consider matrices with entries from an arbitrary ring R. An $n \times n$ matrix over R is an array of elements of R, of the form

$$[a_{ij}] = \begin{bmatrix} a_{11} & a_{12} & \cdots & a_{1n} \\ a_{21} & a_{22} & \cdots & a_{2n} \\ \vdots & \vdots & & \vdots \\ a_{n1} & a_{n2} & \cdots & a_{nn} \end{bmatrix}.$$

Two matrices $A = [a_{ij}]$ and $B = [b_{ij}]$ are equal if $a_{ij} = b_{ij}$ for all i, j.

Definition 1.5.1 *Let R be a ring. The set of all $n \times n$ matrices with entries in R will be denoted by $\mathrm{M}_n(R)$.*

For $[a_{ij}]$ and $[b_{ij}]$ in $\mathrm{M}_n(R)$, we define a sum, given by the formula

$$[a_{ij}] + [b_{ij}] = [a_{ij} + b_{ij}],$$

and a product, given by

$$[a_{ij}][b_{ij}] = [c_{ij}],$$

where $[c_{ij}]$ is the matrix whose (i, j)-entry is

$$c_{ij} = \sum_{k=1}^{n} a_{ik} b_{kj}.$$

Proposition 1.5.2 *Matrix multiplication is associative.*

Proof. Let $[a_{ij}], [b_{ij}], [c_{ij}] \in \mathrm{M}_n(R)$. Then the (i, j)-entry of the matrix $[a_{ij}]([b_{ij}][c_{ij}])$ is

$$\begin{aligned} \sum_{k=1}^{n} a_{ik} \left(\sum_{m=1}^{n} b_{km} c_{mj} \right) &= \sum_{k=1}^{n} \sum_{m=1}^{n} a_{ik} (b_{km} c_{mj}) \\ &= \sum_{m=1}^{n} \sum_{k=1}^{n} (a_{ik} b_{km}) c_{mj} \\ &= \sum_{m=1}^{n} \left(\sum_{k=1}^{n} a_{ik} b_{km} \right) c_{mj}, \end{aligned}$$

and so we have $[a_{ij}]([b_{ij}][c_{ij}]) = ([a_{ij}][b_{ij}])[c_{ij}]$. $\square$

Theorem 1.5.3 *For any ring R, the set $\mathrm{M}_n(R)$ of all $n \times n$ matrices with entries in R is a ring under matrix addition and multiplication.*

Proof. Since R is an abelian group under addition, it is easily checked that $M_n(R)$ is an abelian group under matrix addition. The zero matrix, in which each entry is zero, serves as an identity element for addition, and the additive inverse of a matrix is found by taking the additive inverse of each of its entries.

The associative law for multiplication was verified in the preceding proposition. For $[a_{ij}], [b_{ij}], [c_{ij}] \in M_n(R)$, the (i, j)-entries of $[a_{ij}]([b_{ij}] + [c_{ij}])$ and $[a_{ij}][b_{ij}] + [a_{ij}][c_{ij}]$ are $\sum_{k=1}^{n} a_{ik}(b_{kj} + c_{kj})$ and $\sum_{k=1}^{n}(a_{ik}b_{kj} + a_{ik}c_{kj})$ respectively. These entries are equal since the distributive law holds in R. The second distributive law is verified is the same way. The $n \times n$ identity matrix I_n, with 1 in each entry on the main diagonal and zeros elsewhere, is an identity element for multiplication. □

Let R be a ring, and let $M_n(R)$ be the ring of matrices over R. If I is an ideal of R, we let

$$M_n(I) = \{[a_{ij}] \in M_n(R) \mid a_{ij} \in I \text{ for all } i, j\}$$

denote the set of all matrices with entries in I. It is easy to check that $M_n(I)$ is a subgroup. If $A \in M_n(R)$ and $X \in M_n(I)$, then since I is an ideal of R it is closed under addition and closed under multiplication by elements of R. It follows immediately that the entries of AX and XA belong to I, so $AX \in M_n(I)$ and $XA \in M_n(I)$, showing that $M_n(I)$ is an ideal of $M_n(R)$.

The next proposition shows that every ideal of $M_n(R)$ is of this type. To give the proof, we need to introduce some notation. Let e_{ij} be the matrix with 1 in the (i, j)-entry and 0 elsewhere. A matrix of this type is called a *matrix unit*. Note that $e_{ij}e_{jk} = e_{ik}$, and for any matrix $A = [a_{ij}]$, we have the identity $e_{mi}Ae_{jk} = a_{ij}e_{mk}$.

Proposition 1.5.4 *Let R be a ring. Then every ideal of $M_n(R)$ has the form $M_n(I)$, for some ideal I of R.*

Proof. Given an ideal J of $M_n(R)$, let $I = \{r \in R \mid re_{11} \in J\}$. Since J is an ideal, it is easily checked that I is an ideal of R. For any matrix $A = [a_{ij}]$ in J, we have $a_{ij}e_{11} = e_{1i}Ae_{j1}$, and since J is an ideal, this shows that $a_{ij} \in I$, so $J \subseteq M_n(I)$. Furthermore, if $r \in I$, then $e_{i1}(re_{11})e_{1j} = re_{ij}$. It follows that $M_n(I) \subseteq J$, and so we have $J = M_n(I)$. □

Proposition 1.5.4 shows that if F is a field, then the matrix ring $M_n(F)$ has no proper nontrivial ideals. Even in the matrix ring $M_2(\mathbf{Q})$, it becomes evident that we need a definition of a 'left' ideal, to study more subsets of the ring.

Definition 1.5.5 *Let R be a ring.*

(a) *A nonempty subset I of R is called a* left ideal *of R if*
   (i) *$a \pm b \in I$ for all $a, b \in I$ and*
   (ii) *$ra \in I$ for all $a \in I$ and $r \in R$.*

(b) *If $a \in R$, then the left ideal*

$$Ra = \{x \in R \mid x = ra \text{ for some } r \in R\}$$

is called the principal left ideal *generated by a.*

It is easily shown that if $a \in R$, then Ra is a left ideal. The matrix ring $M_2(\mathbf{Q})$ contains many nontrivial left ideals. For example, the left ideal generated by the matrix unit $e_{11} = \begin{bmatrix} 1 & 0 \\ 0 & 0 \end{bmatrix}$ is the set of matrices with zeros in the second column. This principal left ideal is also generated by e_{21}. Left ideals (together with right ideals, defined analogously) will play a major role in our study of noncommutative rings.

For the next example we will consider the set of all linear transformations from a vector space V into itself. Such linear transformations are referred to as F-endomorphisms of V. It is easy to check that the sum of two linear transformations is again a linear transformation, and that the composition of two linear transformations is again a linear transformation. Since V is an abelian group under addition, and every linear transformation from V to V must preserve the addition, it follows that the set of F-endomorphisms of V is a subring of $\text{End}(V)$. Recall that if V is n-dimensional, then each choice of a basis for V determines an isomorphism between $M_n(F)$ and $\text{End}_F(V)$.

Definition 1.5.6 *Let V be a vector space over the field F. The ring of all F-endomorphisms of V (that is, the ring of all F-linear transformations from V into V) will be denoted by $\text{End}_F(V)$.*

Example 1.5.1 (Endomorphism rings in infinite dimensions)

Let V be a vector space over the field F, and suppose that the dimension of V is countably infinite. We will show that the set of all F-endomorphisms of finite rank is the only proper nontrivial ideal of $\text{End}_F(V)$. (In fact, the number of proper nontrivial ideals depends on the cardinality of the basis. See page 93 of [14] for further details.)

Let $R = \text{End}_F(V)$, and let I be the set of all F-endomorphisms f of V such that the image $f(V)$ has finite dimension. It is clear that

I is closed under addition. If $f \in I$, and $g \in R$, then $fg(V) \subseteq f(V)$, and so fg also has finite rank, and thus $fg \in I$. Furthermore, since $f(V)$ is finite dimensional, it follows that $g(f(V))$ must also be finite dimensional, and thus gf has finite rank, so $gf \in R$. Therefore I is an ideal of R.

First suppose that J is an ideal of R that contains an endomorphism f of infinite rank. By Theorem A.2.6 of the appendix, which states that any subspace of V has a complement, we can find subspaces V' and V'', with $V = V' \oplus \ker(f)$ and $V = V'' \oplus f(V)$. Then f is one-to-one on V', so $f(V)$ and V' both have countably infinite dimension. The dimension of V is the same, so we can find a one-to-one linear transformation h such that h maps V onto V', and an endomorphism g which maps V'' to zero and maps $f(V)$ onto V. Then the composition gfh is one-to-one, and maps V onto V, so it has an inverse k, and therefore $(kg)fh$ is the identity mapping. Since $kgfh \in J$, it follows that $J = R$.

Next suppose that J is a nonzero ideal of R that consists entirely of endomorphisms of finite rank. We can choose a basis for V, and observe that there is a least one basis element v that is mapped to a nonzero element by an endomorphism in J. It follows easily that the endomorphism which maps v to itself and all other basis vectors to zero must be in the ideal J. Then each of the 'unit' endomorphisms f_w, which map v to another basis vector w and all others to zero, must belong to J. The argument is essentially the same as that in Proposition 1.5.4, and it follows that $J = I$. This completes the proof that I is the only proper nontrivial ideal of $\mathrm{End}_F(V)$.

Vector spaces which have an additional ring structure play an important role in many examples.

Definition 1.5.7 *Let R be a ring, and let F be a field. Then R is called an* algebra *over F if*

(i) *R is a vector space over F,*

(ii) *$c(rs) = (cr)s = r(cs)$ for all $c \in F$ and all $r, s \in R$.*

If R has finite dimension as a vector space over F, then it is called a finite dimensional *algebra over F.*

For a given field F, we have already introduced a number of important algebras over F. For examples of finite dimensional algebras we have the matrix rings $\mathrm{M}_n(F)$, for any n, and the group rings FG, for any finite group G. If G is an infinite group, then FG is no longer finite dimensional. Similarly, if V

is an infinite dimensional vector space over F, then $\mathrm{End}_F(V)$ is an infinite dimensional algebra over F.

Theorem 1.5.8 *Let R be a finite dimensional algebra over the field F. Then for some positive integer n, the algebra R is isomorphic to a subring of $\mathrm{M}_n(F)$.*

Proof. The proof is similar to that of Proposition 1.2.8. We only need to observe that if $a \in R$, then $\lambda_a : R \to R$ defined by $\lambda_a(x) = ax$, for all $x \in R$, is an F-endomorphism of R. This produces an embedding into $\mathrm{End}_F(R)$, which is in turn isomorphic to $\mathrm{M}_n(F)$, where $n = \dim(R)$. $\square$

After matrix rings, the next class of rings to focus on is polynomial rings. In Example 1.1.1 we considered the ring of polynomials over a commutative ring. We now consider polynomials with coefficients from an arbitrary ring R, and no longer require that the coefficients commute with the indeterminates.

Let R be any ring, and let $\sigma : R \to R$ be any ring homomorphism. We let $R[x; \sigma]$ denote the set of infinite tuples $(a_0, a_1, a_2, \ldots)$ such that $a_i \in R$ for all i, and $a_i \neq 0$ for only finitely many terms a_i. Two sequences are equal if and only if all corresponding terms are equal. We use the standard addition of polynomials, defined as follows:

$$(a_0, a_1, a_2, \ldots) + (b_0, b_1, b_2, \ldots) = (a_0 + b_0, a_1 + b_1, a_2 + b_2, \ldots) .$$

If $a \in R$, then we have a 'scalar multiplication' defined by

$$a \cdot (b_0, b_1, b_2, \ldots) = (ab_0, ab_1, ab_2, \ldots) .$$

We can identify an element $a \in R$ with $(a, 0, 0, \ldots) \in R[x]$. If we let $x = (0, 1, 0, \ldots)$, then the elements of $R[x]$ can be expressed in the form

$$a_0 + a_1 x + \cdots + a_{m-1} x^{m-1} + a_m x^m ,$$

allowing us to use the standard notation for 'left' polynomials.

The endomorphism σ of R is used to introduce a 'twisted' multiplication. For $r \in R$ we define $xr = \sigma(r)x$. For $a, b \in R$ we have $ax^i \cdot bx^j = a\sigma^i(b)x^{i+j}$. This multiplication is extended to polynomials by repeatedly using the distributive laws.

For example, let R be the field $\mathbf{C}$ of complex numbers, and let σ be the automorphism given by complex conjugation. In $\mathbf{C}[x; \sigma]$ we have $xc = \overline{c}x$, for any $c \in \mathbf{C}$. For instance, for the complex number i we have

$$(\mathrm{i}x)^2 = (\mathrm{i}x)(\mathrm{i}x) = \mathrm{i}(x\mathrm{i})x = \mathrm{i}(-\mathrm{i}x)x = -\mathrm{i}^2 x^2 = x^2 .$$

We will usually require that σ should be an automorphism, but the general definition allows the construction of interesting counterexamples. If σ is not one-to-one, say $\sigma(a) = 0$ for some nonzero $a \in R$, we note that $ax \neq 0$ while $xa = \sigma(a)x = 0$.

If n is the largest nonnegative integer such that $a_n \neq 0$, then as usual we say that the polynomial has *degree* n, and a_n is called the *leading coefficient* of the polynomial.

Definition 1.5.9 *Let R be a ring, and let σ be an endomorphism of R. The skew polynomial ring $R[x; \sigma]$ is defined to be the set of all left polynomials of the form $a_0 + a_1 x + \cdots + a_n x^n$ with coefficients $a_0, \ldots, a_n$ in R. Addition is defined as usual, and multiplication is defined by using the relation*

$$xa = \sigma(a)x \, ,$$

for all $a \in R$.

Proposition 1.5.10 *Let R be a ring, and let σ be an endomorphism of R. The set $R[x; \sigma]$ of skew polynomials over R is a ring.*

Proof. It is clear that $R[x; \sigma]$ is a group under addition. The associative law holds for multiplication of monomials, as shown by the following computations:

$$
\begin{aligned}
(ax^i \cdot bx^j) \cdot cx^k &= (a\sigma^i(b)x^{i+j}) \cdot cx^k = (a\sigma^i(b))(x^{i+j} \cdot cx^k) \\
&= (a\sigma^i(b))(\sigma^{i+j}(c)x^{i+j+k}) = (a\sigma^i(b))(\sigma^{i+j}(c))x^{i+j+k} \, ;
\end{aligned}
$$

$$
\begin{aligned}
ax^i \cdot (bx^j \cdot cx^k) &= ax^i \cdot (b\sigma^j(c)x^{j+k}) = a(x^i \cdot b\sigma^j(c))x^{j+k} \\
&= a(\sigma^i(b\sigma^j(c))x^i)x^{j+k} = a(\sigma^i(b)\sigma^i(\sigma^j(c)))x^{i+j+k} \\
&= a(\sigma^i(b)\sigma^{i+j}(c))x^{i+j+k} = (a\sigma^i(b))\sigma^{i+j}(c)x^{i+j+k} \, .
\end{aligned}
$$

We extend the definition of multiplication to all polynomials in $R[x; \sigma]$ by repeatedly using the distributive laws.

Since σ is a ring homomorphism, we have $\sigma(1) = 1$, and so the constant polynomial 1 serves as a multiplicative identity element. $\square$

We recall Definition 1.1.6 (a): the ring R is a *domain* if $ab \neq 0$ for all nonzero elements $a, b \in R$. If $f(x) = a_0 + \cdots + a_m x^m$ and $g(x) = b_0 + \cdots + b_n x^n$ belong to $R[x; \sigma]$ and have degrees m and n respectively, then the coefficient of x^{m+n} in the product $f(x)g(x)$ is $a_m\sigma(b_n)$. If R is a domain and σ is one-to-one, then this coefficient is nonzero, and so $\deg(f(x)g(x)) = \deg(f(x)) + \deg(g(x))$. This shows that if R is a domain and σ is one-to-one, then $R[x; \sigma]$ is a domain.

Proposition 1.5.11 *Let K be a division ring, and let σ be a nontrivial endomorphism of K. Then the skew polynomial ring $K[x; \sigma]$ is a noncommutative domain in which every left ideal is a principal left ideal.*

Proof. Let I be a nonzero left ideal of $K[x; \sigma]$. Among the nonzero elements of I we can choose one of minimal degree m, say $p(x) = a_0 + \cdots + a_m x^m$. Since K is a division ring, we can assume without loss of generality that $a_m = 1$. (If not, consider the polynomial $a_m^{-1} p(x)$, which is also in I and still has degree m.)

We claim that I is the left ideal generated by $p(x)$, so that $I = K[x; \sigma] \cdot p(x)$. The proof is by induction on the degree of the nonzero elements of I. Let $f(x) = b_0 + \cdots + b_n x^n$ belong to I, with $\deg(f(x)) = n$. Since $n \geq m$, consider the polynomial $g(x) = f(x) - b_n x^{n-m} p(x)$. Because the leading coefficient of $p(x)$ is 1, the automorphism σ has no effect on the product $x^{n-m} \cdot x^m$, and so the degree of $g(x)$ is strictly less than the degree of $f(x)$. If $n = m$, then we conclude from the choice of $p(x)$ that $g(x) = 0$, and so $f(x) \in K[x; \sigma] \cdot p(x)$. Now assume that the induction hypothesis holds for all elements of I with degree $\leq k$, and that $n = k + 1$. Then it follows that $g(x)$ belongs to I, and so $f(x) = g(x) + b_n x^{n-m} p(x)$ belongs to I. $\square$

Our next class of examples involves a slightly different twist to the multiplication, introduced in order to extend the definition of a ring of differential operators. In Example 1.1.7 we considered differential operators acting on the ring $\mathbf{C}[z]$. Rather than considering coefficients from a polynomial ring, we can allow coefficients from any ring R, assuming that we have an operator δ defined on R that acts like differentiation. We require that δ is an additive endomorphism of R, and that $\delta(ab) = a\delta(b) + \delta(a)b$, for all $a, b \in R$. Such a function is called a *derivation* on the ring R.

To construct a polynomial ring analogous to $\mathbf{C}[z; \partial]$, we assume that δ is a derivation of R and consider left polynomials with coefficients from R, using the standard addition of polynomials. We use the multiplication determined by the relation $xa = ax + \delta(a)$, for all $a \in R$. We need the defining properties of δ to verify even the simplest cases of the distributive and associative laws, as shown by the following computations:

$$x(a+b) = (a+b)x + \delta(a+b) = ax + bx + \delta(a) + \delta(b) = xa + xb$$

and

$$
\begin{aligned}
x(ab) &= (ab)x + \delta(ab) = a(bx) + a\delta(b) + \delta(a)b \\
&= a((bx) + \delta(b)) + \delta(a)b = a(xb) + \delta(a)b \\
&= (ax)b + \delta(a)b = (ax + \delta(a))b = (xa)b \,,
\end{aligned}
$$

for all $a, b \in R$.

It follows from the relation $xa = ax + \delta(a)$ that

$$
\begin{aligned}
x^2 a &= x(ax + \delta(a)) = (xa)x + x\delta(a) \\
&= ax^2 + \delta(a)x + \delta(a)x + \delta^2(a) \\
&= ax^2 + 2\delta(a)x + \delta^2(a) \ .
\end{aligned}
$$

The general formula

$$
x^n a = \sum_{i=0}^{n} \binom{n}{i} \delta^{n-i}(a) x^i
$$

can be proved by induction. This formula is used to verify the associative law for multiplication, so checking these details is quite tedious.

Definition 1.5.12 *Let R be a ring, and let δ be a derivation of R. The differential polynomial ring $R[x; \delta]$ is defined to be the set of all left polynomials of the form $a_0 + a_1 x + \cdots + a_n x^n$ with coefficients $a_0, \ldots, a_n$ in R. Addition is defined as usual, and multiplication is defined by using the relation*

$$
xa = ax + \delta(a) \ ,
$$

for all $a \in R$.

Proposition 1.5.13 *Let R be a ring, and let δ be a derivation of R.*

(a) *The set $R[x; \delta]$ of differential polynomials over R is a ring.*

(b) *If the coefficient ring R is a domain, then so is $R[x; \delta]$.*

Proof. (a) Let y be an indeterminate, and consider the endomorphism ring of $R[y]$, viewed as an abelian group. We let $E = \operatorname{End}(R[y])$, and our plan is to show that there is an additive isomorphism between the additive group of $R[x; \delta]$ and a subring of E. The multiplication on the subring can be transferred to $R[x; \delta]$, so that we can avoid many of the details necessary to prove directly that $R[x; \delta]$ is a ring. For any element $a \in R$, define $\lambda_a : R[y] \to R[y]$ by $\lambda_a(f(y)) = af(y)$. Since λ_a defines an element of the endomorphism ring E, the mapping $\phi : R \to E$ defined by $\phi(a) = \lambda_a$ is a one-to-one ring homomorphism. We will identify R with the subring $\phi(R)$ of E.

We can extend δ to a derivation of $R[y]$, by defining

$$
\delta(a_0 + a_1 y + \cdots + a_n y^n) = \delta(a_0) + \delta(a_1)y + \cdots + \delta(a_n)y^n \ .
$$

It can be checked that $\delta(f(y)g(y)) = f(y)\delta(g(y)) + \delta(f(y))g(y)$, since $\delta(a+b) = \delta(a) + \delta(b)$ and $\delta(ab) = \delta(a)b + a\delta(b)$, for all $a, b \in R$.

We next define an element of E, which we denote by z, by letting $z(f(y)) = f(y)y + \delta(f(y))$. For any $a \in R$, we have

$$
\begin{aligned}
[za](f(y)) &= z(af(y)) = (af(y))y + \delta(af(y)) \\
&= a(f(y)y) + \delta(a)f(y) + a\delta(f(y)) \\
&= a(f(y)y + \delta(f(y))) + \delta(a)f(y) \\
&= a(zf(y)) + \delta(a)f(y) = [az + \delta(a)](f(y)) .
\end{aligned}
$$

Thus, as functions in E, the actions of za and $az + \delta(a)$ are equal.

Let S be the subset $\sum_{i=0}^{\infty} Rz^i$ of E. We claim that S is a subring E isomorphic to $R[x; \delta]$. It is clear that S is a subgroup under addition, and from the relation $zR \subseteq Rz + R$ it follows by induction that $z^i R \subseteq Rz^i + Rz^{i-1} + \cdots + Rz + R$, for all i. This implies, in turn, that S is closed under multiplication. Now define $\Phi : R[x; \delta] \to E$ by letting $\Phi(a_0 + \cdots + a_n x^n) = \phi(a_0) + \cdots + \phi(a_n)z^n$, for all $a_0 + \cdots + a_n x^n \in R[x; \delta]$. Because of the identification we have made, we can write this as $\Phi(a_0 + \cdots + a_n x^n) = a_0 + \cdots + a_n z^n$. If $\Phi(a_0 + \cdots + a_n x^n) = 0$, then the endomorphism $a_0 + \cdots + a_n z^n$ must be identically zero in E. The action of z on the constant polynomial 1 is given by $z(1) = 1 \cdot y + \delta(1) = y$, since $\delta(1) = 0$. Thus the action of $a_0 + \cdots + a_n z^n$ on 1 yields $a_0 + \cdots + a_n y^n$, and if this is zero, then $a_0 = \cdots = a_n = 0$. This shows that Φ is one-to-one, and since the image of Φ is S, we can use Φ^{-1} to transfer the multiplication of S back to $R[x; \delta]$. Finally, we have $\Phi(x) = z$, and since $za = az + \delta(a)$ in S, the corresponding identity holds in $R[x; \delta]$.

(b) If $f(x) = a_0 + \cdots + a_m x^m$ and $g(x) = b_0 + \cdots + b_n x^n$ belong to $R[x; \delta]$ and have degrees m and n respectively, consider the coefficient of x^{m+n} in the product $f(x)g(x)$. We have $a_m(x^m b_n)x^n = a_m b_n x^{m+n} + h(x)$, where $h(x)$ contains terms of strictly lower degree. Thus if R is a domain we have $\deg(f(x)g(x)) = \deg(f(x)) + \deg(g(x))$, and $R[x; \delta]$ is a domain. $\square$

If K is a division ring, then the proof of Proposition 1.5.11, regarding left ideals of the skew polynomial ring $K[x; \sigma]$, avoids the endomorphism σ, and so it carries over directly to the differential polynomial ring $K[x; \delta]$. We conclude that if K is a division ring, then the ring $K[x; \delta]$ of differential polynomials over K is a domain in which every left ideal is a principal left ideal.

Just as for ordinary polynomials, the construction of differential polynomial rings can be repeated, to obtain $(R[x_1; \delta_1])[x_2; \delta_2]$, provided δ_2 is a derivation of $R[x_1; \delta_1]$. This construction can be done if we begin with derivations δ_1 and δ_2 of R for which $\delta_1 \delta_2 = \delta_2 \delta_1$. Define $\widehat{\delta_2}$ on $R[x_1; \delta_1]$ by setting $\widehat{\delta_2}(\sum a_i x_1^i) = \sum \delta_2(a_i)x_1^i$. It can be checked that $\widehat{\delta_2}$ is a derivation of $R[x_1; \delta_1]$. The ring $(R[x_1; \delta_1])[x_2; \widehat{\delta_2}]$ is simply denoted by $R[x_1, x_2; \delta_1, \delta_2]$.

Let F be a field, and let R be the ring $F[x_1, x_2, \ldots, x_n]$ of polynomials over F. The partial derivatives $\dfrac{\partial}{\partial x_i}$ commute with each other, and so we can form the iterated differential operator ring.

Definition 1.5.14 *Let F be a field, and let $R = F[x_1, x_2, \ldots, x_n]$. The iterated differential operator ring*

$$R[y_1, \ldots, y_n; \frac{\partial}{\partial x_1}, \ldots, \frac{\partial}{\partial x_n}]$$

is called the nth Weyl algebra over F, *and is denoted by* $\mathrm{A}_n(F)$.

We conclude this section with a construction that provides many counterexamples for noncommutative rings. We have already seen some interesting examples defined by subrings of lower triangular matrices in $\mathrm{M}_2(\mathbf{C})$, as in Example 1.1.4, Exercise 5 of Section 1.1, and Exercises 5 and 7 of Section 1.2. The subring of $\mathrm{M}_2(\mathbf{C})$ defined by

$$R = \left\{ \left[\begin{array}{cc} a & b \\ c & d \end{array} \right] \middle| a \in \mathbf{Z}, c, d \in \mathbf{Q} \text{ and } b = 0 \right\}$$

has interesting properties. The notation $\left[\begin{array}{cc} \mathbf{Z} & 0 \\ \mathbf{Q} & \mathbf{Q} \end{array} \right]$ gives a convenient way to describe the ring R. It is also convenient to use matrix notation for a more general construction. If V is a vector space over the field F, we can use the notation $\left[\begin{array}{cc} F & 0 \\ V & F \end{array} \right]$ for the set of ordered triples $\{(a, v, b) \mid a, b \in F, v \in V\}$. We use componentwise addition, but the multiplication is defined using matrix multiplication as a model:

$$(a_1, v_1, b_1) \cdot (a_2, v_2, b_2) = (a_1 a_2, v_1 b_2 + a_1 v_2, b_1 b_2) .$$

In this notation, we need to be able to write the scalar product on either side of a vector. The vector space axioms can be used to show that the operations define a ring structure on the set of triples.

More generally, we can start with two rings S and T, not necessarily commutative, and an abelian group V. Our goal is to construct a ring of 'formal triangular matrices' that will be denoted by

$$\left[\begin{array}{cc} T & 0 \\ V & S \end{array} \right] = \left\{ \left[\begin{array}{cc} t & 0 \\ v & s \end{array} \right] \middle| t \in T, v \in V, s \in S \right\} .$$

In order to define a multiplication of elements the following product must make sense for all $t_1, t_2 \in T$, $v_1, v_2 \in V$, and $s_1, s_2 \in S$.

$$\left[\begin{array}{cc} t_1 & 0 \\ v_1 & s_1 \end{array} \right] \left[\begin{array}{cc} t_2 & 0 \\ v_2 & s_2 \end{array} \right] = \left[\begin{array}{cc} t_1 t_2 & 0 \\ v_1 t_2 + s_1 v_2 & s_1 s_2 \end{array} \right]$$

We need to define 'scalar' multiplications $\mu : S \times V \to V$ and $\eta : V \times T \to V$ that satisfy a number of axioms. It is left as an exercise for the reader to show that we need

$$
\begin{array}{rclcrcl}
1v & = & v, & \qquad s(vt) = (sv)t, & \qquad v1 & = & v, \\
s(v_1 + v_2) & = & sv_1 + sv_2, & & (v_1 + v_2)t & = & v_1t + v_2t, \\
(s_1 + s_2)v & = & s_1v + s_2v, & & v(t_1 + t_2) & = & vt_1 + vt_1, \\
s_1(s_2v) & = & (s_1s_2)v, & & (vt_1)t_2 & = & v(t_1t_2)
\end{array}
$$

for all $s, s_1, s_2 \in S$, $v, v_1, v_2 \in V$, and $t, t_1, t_2 \in T$. The abelian group V is called an S-T-*bimodule* if it satisfies the above axioms. The corresponding matrix operations are called *formal matrix addition* and *formal matrix multiplication*. The verification that these operations define a ring is left to the reader in Exercise 8.

Definition 1.5.15 *Let S and T be rings, and let V be an S-T-bimodule. The set* $\begin{bmatrix} T & 0 \\ V & S \end{bmatrix}$, *under the operations of formal matrix addition and multiplication, is called a* formal triangular matrix ring.

In the formal triangular matrix ring $R = \begin{bmatrix} T & 0 \\ V & S \end{bmatrix}$, there are several classes of ideals. For example, it can be shown that if A is an ideal of T, B is an ideal of S, and W is a subgroup of V that is an S-T-bimodule under the given operations, then $I = \begin{bmatrix} A & 0 \\ V & B \end{bmatrix}$ is an ideal of R if and only if $BV \subseteq W$ and $VA \subseteq W$. In particular, $\begin{bmatrix} T & 0 \\ V & 0 \end{bmatrix}$ and $\begin{bmatrix} 0 & 0 \\ V & S \end{bmatrix}$ are ideals of R.

EXERCISES: SECTION 1.5

1. Let I be an ideal of the ring R. For any matrix $[a_{ij}] \in M_n(R)$, define $\phi([a_{ij}]) = [a_{ij} + I]$. Show that ϕ defines a ring homomorphism from $M_n(R)$ onto $M_n(R/I)$, and that $\ker(\phi) = M_n(I)$.

2. For any ring R, compute the center of $M_n(R)$.

3. Show that the characteristic of a simple ring is either 0 or a prime number.

4. Let V be a countably infinite dimensional vector space over the field F. Identify $\mathrm{End}_F(V)$ with the set R of all countably infinite matrices over F in which each column has only finitely many nonzero entries. Let S be the set of matrices $[a_{ij}]$ in R such that there exist a positive integer N and an element $d \in F$ satisfying the following conditions: (i) $a_{ij} = d$ if $i > N$ and $i = j$; (ii) $a_{ij} = 0$ if $i > N$ and $i > j$; (iii) $a_{ij} = 0$ if $j > N$ and $j > i$. Show that S is a subring of R.

5. Let $R = \mathbf{C}$ and let σ be complex conjugation. Find the center of the skew polynomial ring $\mathbf{C}[x; \sigma]$.

6. Let σ be an automorphism of the ring R.
   (a) Show that σ^{-1} is an automorphism of R^{op}.
   (b) Show that $(R[x; \sigma])^{op}$ is isomorphic to $R^{op}[x; \sigma^{-1}]$.

7. Assume that δ is a derivation of the ring R. Use induction to verify the formula $x^n a = \sum_{i=0}^{n} \binom{n}{i} \delta^{n-i}(a) x^i$ for the ring of differential polynomials defined in Definition 1.5.12.

8. Let S and T be rings, and let V be an abelian group. Show that formal matrix addition and multiplication define a ring structure on the set $\begin{bmatrix} T & 0 \\ V & S \end{bmatrix}$ of formal triangular matrices if and only if V is an S-T-bimodule.

9. Let S and T be rings, and let V be an abelian group Show that $I = \begin{bmatrix} T & 0 \\ V & 0 \end{bmatrix}$ is an ideal of the ring $R = \begin{bmatrix} T & 0 \\ V & S \end{bmatrix}$ of formal triangular matrices, and that $R/I \cong S$.

10. Let V be a vector space over the field F. Let R be the ring $\begin{bmatrix} F & 0 \\ V & F \end{bmatrix}$ of formal matrices, and let $I = \begin{bmatrix} 0 & 0 \\ V & 0 \end{bmatrix}$. Show that I is an ideal of R, and that the ideals of R that are contained in I are in one-to-one correspondence with the F-subspaces of V.

Chapter 2

MODULES

If A is an abelian group, $a \in A$, and $n \in \mathbf{Z}^+$, then na is defined to be the sum of a with itself n times. This 'scalar multiplication' can be extended to negative integers in the obvious way. The properties that hold for this multiplication are the same as those for a vector space, the difference being that the scalars belong to the ring $\mathbf{Z}$, rather than to a field. Relaxing the conditions on the set of scalars leads to the definition of a *module*.

The objects of study in this chapter are modules over arbitrary rings, and they can be thought of as generalizations of vector spaces and abelian groups. We will also see that they can be regarded as 'representations' of a ring, in the same sense that representations of a group are defined by homomorphisms from the group into various groups of permutations or groups of matrices. Since we have already seen that in a sense the 'generic' rings are rings of endomorphisms of abelian groups, a representation of a ring should be a ring homomorphism into the ring of all endomorphisms of some abelian group.

In Section 2.1 we introduce some basic definitions, including those of a module and a module homomorphism, and prove many of the theorems that correspond to the elementary theorems of group theory. The following sections extend familiar properties of vector spaces and abelian groups. Every vector space has a basis; in the general situation, a module with a basis is called a *free* module, and these are studied in Section 2.2, along with the necessary definitions and results about direct sums of modules. Every subspace of a vector space has a complement; a module with this property is called a *completely reducible* module. A one-dimensional vector space has no proper nontrivial subspaces; a module with this property is said to be a *simple* module. These are introduced in Section 2.1, and then direct sums of simple modules are studied in Section 2.3, where they are shown to be completely reducible.

The main result on finite abelian groups is the structure theorem: every finite abelian group is isomorphic to a direct sum of cyclic groups of prime power order. To extend this to modules over any principal ideal domain, we study chain conditions and direct sums of indecomposable modules in Sections 2.4 and 2.5. Tensor products are an important tool in later developments, and these are introduced in Section 2.6. Finally, these tools are used to describe modules over principal ideal domains in Section 2.7. As an important application of the structure theorem for modules over a principal ideal domain, it is possible to apply the results on modules to the theory of a single linear transformation of a vector space, obtaining the canonical forms for the corresponding matrix.

In the 1920s, Emmy Noether used modules as an important tool in bringing to light the connection between the study of representations of finite groups via groups of matrices and the study of rings. She utilized the fact that group representations correspond to modules over the corresponding group ring. In Section 2.3 we will prove Maschke's theorem, which yields information about the structure of certain group rings. In Chapter 3, we will use the Artin–Wedderburn theorem to complete this description. Finally, in Chapter 4 we will use this information to study group representations over the field of complex numbers.

2.1   Basic definitions and examples

Just as permutation groups form the generic class of groups, endomorphism rings form the generic class of rings. In fact, Proposition 1.2.8 shows that any ring is isomorphic to a subring of an endomorphism ring $\text{End}(M)$, for some abelian group M. This is the ring theoretic analog of Cayley's theorem.

One way to study the ring R is to study its 'representations' via more 'concrete' rings. That is, we can consider all ring homomorphisms of the form $\rho : R \to \text{End}(M)$, where M is an abelian group. It turns out to be better to consider these representations of R from a complementary point of view, one which generalizes the familiar notion of a vector space.

We first note some properties of endomorphism rings. Let M be an abelian group, and let $R = \text{End}(M)$. For $r \in R$, we have

$$r(m_1 + m_2) = r(m_1) + r(m_2)$$

for all $m_1, m_2 \in M$, since r is a group homomorphism. In the ring R, the operations are defined by addition and composition of functions. Thus for any $r_1, r_2 \in R$ and any $m \in M$ we have

$$[r_1 + r_2](m) = r_1(m) + r_2(m)$$

and

$$[r_1 r_2](m) = r_1(r_2(m)) \ .$$

Finally, since the identity element of R is the identity function, we have

$$1(m) = m \ ,$$

for any $m \in M$.

Now suppose that R is any ring, and that $\rho : R \to \text{End}(M)$ is the representation of R given by a ring homomorphism ρ from R into the ring of endomorphisms of an abelian group M. We can use the representation ρ to define an 'action' of the ring on the given abelian group. (This is similar to the result from group theory which uses any representation of a group via a group of permutations to define an action of the group on a set.)

For any ring homomorphism $\rho : R \to \text{End}(M)$, denote the image of $r \in R$ by $\rho(r)$ and define a 'scalar multiplication' on M by $r \cdot m = [\rho(r)](m)$, for all $r \in R$ and all $m \in M$. Since $\rho(r) \in \text{End}(M)$, for any $m_1, m_2 \in M$ we have

$$\begin{aligned} r \cdot (m_1 + m_2) &= [\rho(r)](m_1 + m_2) = [\rho(r)](m_1) + [\rho(r)](m_2) \\ &= r \cdot m_1 + r \cdot m_2 \ . \end{aligned}$$

Since ρ is a ring homomorphism, for any $r_1, r_2 \in R$ and any $m \in M$ we have

$$\begin{aligned} (r_1 + r_2) \cdot m &= [\rho(r_1 + r_2)](m) = [\rho(r_1) + \rho(r_2)](m) \\ &= [\rho(r_1)](m) + [\rho(r_2)](m) = r_1 \cdot m + r_2 \cdot m \ , \end{aligned}$$

$$\begin{aligned} (r_1 r_2) \cdot m &= [\rho(r_1 r_2)](m) = [\rho(r_1)\rho(r_2)](m) \\ &= [\rho(r_1)]([\rho(r_2)](m)) = r_1 \cdot (r_2 \cdot m) \ , \end{aligned}$$

and

$$1 \cdot m = [\rho(1)](m) = m \ .$$

This motivates the following definition of an R-module.

Definition 2.1.1 *Let R be a ring, and let M be an abelian group. Then M is called a* left R-module *if there exists a scalar multiplication $\mu : R \times M \to M$ denoted by $\mu(r, m) = rm$, for all $r \in R$ and all $m \in M$, such that for all $r, r_1, r_2 \in R$ and all $m, m_1, m_2 \in M$,*

(i) $r(m_1 + m_2) \;\; = \;\; rm_1 + rm_2$,
(ii) $(r_1 + r_2)m \;\; = \;\; r_1 m + r_2 m$,
(iii) $r_1(r_2 m) \;\; = \;\; (r_1 r_2)m$,
(iv) $1m \;\; = \;\; m$.

The fact that the abelian group M is a left R-module will be denoted by $M = {}_R M$.

The discussion leading up to the definition of a module shows that any abelian group M is a left module over $\text{End}(M)$. There is an analogous definition of a *right* R-module. In this case, with a product $\mu : M \times R \to M$, we use the notation $M = M_R$.

The next examples show that several familiar concepts can be unified by the notion of a module.

Example 2.1.1 (The regular representation of a ring)

> For any ring R, the most basic left R-module is R itself, with the multiplication of R determining the action of R on itself. Note that R can also be viewed as a right R-module. We denote these two modules by ${}_R R$ and R_R, respectively. When thought of as a representation of the ring R, the module ${}_R R$ is often called the *regular representation* of R.

Example 2.1.2 (Vector spaces over a field F are F-modules)

> If V is a vector space over a field F, then it is an abelian group under addition of vectors. The familiar rules for scalar multiplication are precisely those needed to show that V is a module over the ring F.

Example 2.1.3 (Abelian groups are Z-modules)

> If A is an abelian group with its operation denoted additively, then for any element $a \in A$ and any positive integer n, we can define $n \cdot a$ to be the sum of a with itself n times. If n is a negative integer, we can define $n \cdot a = -|n| \cdot a$. With this familiar multiplication, it is easy to check that A is a **Z**-module.

Another way to show that A is a $\mathbf{Z}$-module is to define a ring homomorphism $\rho : \mathbf{Z} \to \mathrm{End}(A)$ by letting $\rho(n) = n \cdot 1$, for all $n \in \mathbf{Z}$. This is the familiar mapping that is used to determine the characteristic of the ring $\mathrm{End}(A)$. The action of $\mathbf{Z}$ on A determined by this mapping is the same one that is used in the previous paragraph.

If M is any left R-module, then left multiplication by $r \in R$ corresponds to a mapping $\lambda_r : M \to M$ defined by $\lambda_r(m) = rm$. Since $r(m_1 + m_2) = rm_1 + rm_2$ for all $m_1, m_2 \in M$, we have $\lambda_r \in \mathrm{End}\, M$, and it can be shown that the function $\rho : R \to \mathrm{End}\, M$ defined by $\rho(r) = \lambda_r$ is a ring homomorphism. The following proposition summarizes our discussion of representations and modules.

Proposition 2.1.2 *Let R be a ring. There is a one-to-one correspondence between left R-modules and representations of R via endomorphism rings of abelian groups.*

It is valuable to be able to use these two viewpoints interchangeably. For instance, the next example shows that a ring homomorphism $\phi : R \to S$ can be used to convert any S-module into an R-module. This becomes crystal clear when approached from the representation theoretic point of view.

Example 2.1.4 (Change of rings)

Let S be a ring, and let $\phi : R \to S$ be a ring homomorphism. Then any left S-module M can be given the structure of a left R-module, by defining $r \cdot m = \phi(r)m$, for all $r \in R$ and all $m \in M$. The easiest way to verify that this is indeed a module structure is to use the defining representation $\rho : S \to \mathrm{End}(M)$, and consider the composition $\rho\phi$ of the two ring homomorphisms.

For a module $_R M$ defined by the representation $\rho : R \to \mathrm{End}(M)$, we have $[\rho(r)](m) = rm$, for all $m \in M$. Thus the kernel of the representation ρ is the set $\{r \in R \mid rm = 0 \text{ for all } m \in M\}$. Since it is the kernel of a ring homomorphism, this set is an ideal of the ring R. The modules $_R M$ for which the representation $\rho : R \to \mathrm{End}(M)$ is one-to-one form an important class. The representation is one-to-one if and only if for each $r \in R$ with $r \neq 0$ there exists $m \in M$ with $rm \neq 0$.

Definition 2.1.3 *Let R be a ring, and let M be a left R-module.*
(a) *The ideal*

$$\mathrm{Ann}(M) = \{r \in R \mid rm = 0 \text{ for all } m \in M\}$$

is called the annihilator *of the module M.*

(b) *For any element* $m \in M$*, the set*

$$\mathrm{Ann}(m) = \{r \in R \mid rm = 0\}$$

is called the annihilator *of m.*

(c) *The module M is called* faithful *if* $\mathrm{Ann}(M) = (0)$.

If $_R M$ is a left R-module, defined by the representation $\rho : R \to \mathrm{End}(M)$, then ρ factors through $R/\mathrm{Ann}(M)$, which shows that M has a natural structure as a left module over $R/\mathrm{Ann}(M)$. In fact, Exercise 5 shows that M is a faithful left module over $R/\mathrm{Ann}(M)$.

If V is a nonzero vector space over a field F, then V is certainly a faithful F-module, since F has no proper nontrivial ideals. If A is an abelian group, when it is viewed as a $\mathbf{Z}$-module the annihilator of an element of order $n \in \mathbf{Z}^+$ is $n\mathbf{Z}$, while the annihilator of an element of infinite order is (0). If $\mathrm{Ann}(A) = n\mathbf{Z}$, for a positive integer n, then n is the exponent of the group A.

In general, for any module M we have

$$\mathrm{Ann}(M) = \bigcap_{m \in M} \mathrm{Ann}(m) \ .$$

Thus in a faithful abelian group A, with $\mathrm{Ann}(A) = (0)$, we have one of two cases: either A has an element of infinite order or else there is no bound on the orders of its elements of finite order.

If M is a left R-module, then there is an obvious definition of a *submodule* of M: any subset of M that is a left R-module under the operations induced from M. The subset $\{0\}$ is called the *trivial submodule*, and is denoted by (0). The module M is a submodule of itself, an *improper* submodule.

It can be shown that if M is a left R-module, then a subset $N \subseteq M$ is a submodule if and only if it is nonempty, closed under sums, and closed under multiplication by elements of R. It is also true that for the ring $S = R/\mathrm{Ann}(M)$, the S-submodules of M are precisely the same as the R-submodules of M.

A submodule of $_R R$ is called a *left ideal* of R. A submodule of R_R is called a *right ideal* of R, and it is clear that a subset of R is an ideal if and only if it is both a left ideal and a right ideal of R.

If M is a left R-module, and $m \in M$, then $\{r \in R \mid rm = 0\}$ is a left ideal of R. This follows from the observations that if $a_1 m = 0$ and $a_2 m = 0$, then $(a_1 + a_2)m = a_1 m + a_2 m = 0$, and if $am = 0$ then $(ra)m = r(am) = 0$, for all $r \in R$.

If N is a submodule of $_R M$, then we can form the factor group M/N. There is a natural multiplication defined on the cosets of N: for any $r \in R$ and any $x \in M$, let $r \cdot (x + N) = rx + N$. If $x + N = y + N$, then $x - y \in N$, and so $rx - ry = r(x-y) \in N$, and this shows that scalar multiplication is well-defined. It follows that M/N is a left R-module, called a *factor module* of M.

Proposition 2.1.4 *Let M be a left R-module, and let N be any submodule of M. Then there is a one-to-one correspondence between the submodules of M/N and the submodules of M that contain N.*

Proof. Considering M as an abelian group, the submodule N is in particular a subgroup of M. There is a one-to-one correspondence between the subgroups of M/N and the subgroups of M that contain N. (See Theorem 2.7.2 of [11] or Proposition 3.8.6 of [4].) We only need to show that under this correspondence the image of a submodule is a submodule.

The correspondence between subgroups is defined in the following way. If L is a subgroup of M/N, then $\{m \in M \mid m + N \in L\}$ is a subgroup of M that contains N. In this case, if L is a submodule of M/N, then for any $r \in R$ and $m \in M$ with $m + N \in L$, we have $rm + N = r(m + N) \in L$, and so rm belongs to the appropriate submodule. On the other hand, if K is a subgroup of M that contains N, then $K/N = \{m + N \mid m \in K\}$ is the corresponding subgroup of M/N. If K is a submodule of M, and $r \in R$, then $r(m + N) = rm + N \in K/N$ since $rm \in M$. $\square$

Our next examples emphasize the multiplication used to define a module structure.

Example 2.1.5 (Matrix rings)

Let F be a field, and let R be the ring $\mathrm{M}_n(F)$ of all $n \times n$ matrices over F. Consider the kth column in the ring of matrices. That is, let
$$A_k = \{[a_{ij}] \in \mathrm{M}_n(F) \mid a_{ij} = 0 \text{ for all } j \neq k\} \ .$$

Then A_k is a left ideal of R, since it is closed under addition and multiplication (on the left) by any element of R. Similarly, any row of R is a right ideal.

If V is any n-dimensional vector space over F, then viewing V as the set of all column vectors with n entries shows that multiplication of the vectors by $n \times n$ matrices is well-defined. The standard laws for matrix multiplication show that V is a left R-module. From the alternative viewpoint of representations, to show that V is a left R-module we only need to observe that R is isomorphic to the ring of all F-linear transformations on V, which is a subring of $\mathrm{End}(V)$.

Example 2.1.6 (The action of a single linear transformation)

Let F be a field, let V be a vector space over F, and let T be a linear transformation on V. For any polynomial $f(x) \in F[x]$, the

mapping $f(T) : V \to V$ is a linear transformation. This allows us
to make the following definition, for all $f(x) \in F[x]$ and all $v \in V$:

$$f(x) \cdot v = [f(T)](v) \ .$$

With this multiplication, V can be viewed as a left $F[x]$-module.
In Section 2.7, the structure theorem for modules over a principal
ideal domain will allow us to obtain significant information about T,
regarding a canonical form for the matrix associated with T. (See
Exercises 2–5 of Section 2.7, or Chapter 11 of [23].)

We next consider several ways of constructing submodules. Let M be a left
R-module, and let $\{M_\alpha\}_{\alpha \in I}$ be a collection of submodules of M. First, the
intersection $\bigcap_{\alpha \in I} M_\alpha$ is a submodule of M, as can be checked easily. Next, the
sum of the submodules $\{M_\alpha\}_{\alpha \in I}$, denoted by $\sum_{\alpha \in I} M_\alpha$, consists of all finite
sums of the form $\sum_{\alpha \in F} m_\alpha$, such that $m_\alpha \in M_\alpha$ for each $\alpha \in F$, where F is
any finite subset of I. It is also easy to check that $\sum_{\alpha \in I} M_\alpha$ is a submodule of
M.

For any element m of the module M, we can construct the submodule

$$Rm = \{x \in M \mid x = rm \text{ for some } r \in R\} \ .$$

This is the smallest submodule of M that contains m, so it is called the *cyclic
submodule* generated by m. This is the most basic way to construct a submodule.

More generally, if X is any subset of M, then the intersection of all submodules of M which contain X is the smallest submodule of M which contains
X We will use the notation $\langle X \rangle$ for this submodule, and call it the submodule
generated by X. We must have $Rx \subseteq \langle X \rangle$ for all $x \in X$, and then it is not
difficult to show that

$$\langle X \rangle = \sum_{x \in X} Rx \ .$$

Definition 2.1.5 *Let M be a left R-module.*

(a) *The module M is called* cyclic *if there exists $m \in M$ such that $M = Rm$.*

(b) *The module M is said to be* finitely generated *if there exist elements
$m_1, m_2, \ldots, m_n \in M$ such that $M = \sum_{j=1}^{n} Rm_j$. In this case, we say that
$\{m_1, m_2, \ldots, m_n\}$ is a set of generators for M.*

We note that if N is a submodule of M such that N and M/N are finitely
generated, then M is finitely generated. In fact, if $x_1, x_2, \ldots, x_n$ generate N
and $y_1 + N$, $y_2 + N$, $\ldots, y_m + N$ generate M/N, then $x_1, \ldots, x_n, y_1, \ldots, y_m$
generate M.

We now consider the appropriate generalization of the notion of a group homomorphism to the case of R-modules. If R is a field, the general definition of an R-homomorphism reduces to the familiar definition of a linear transformation.

Definition 2.1.6 *Let M and N be left R-modules.*

(a) *A function $f : M \to N$ is called an* R-homomorphism *if*

$$f(m_1 + m_2) = f(m_1) + f(m_2) \qquad and \qquad f(rm) = rf(m)$$

for all $r \in R$ and all $m, m_1, m_2 \in M$. The set of all R-homomorphisms from M into N is denoted by

$$\text{Hom}_R(M, N) \, .$$

(b) *For an R-homomorphism $f \in \text{Hom}_R(M, N)$ we define its* kernel *as*

$$\ker(f) = \{m \in M \mid f(m) = 0\} \, .$$

(c) *We say that f is an* isomorphism *if it is both one-to-one and onto.*

(d) *Elements of $\text{Hom}_R(M, M)$ are called* endomorphisms, *and isomorphisms in $\text{Hom}_R(M, M)$ are called* automorphisms. *The set of endomorphisms of $_RM$ will be denoted by*

$$\text{End}_R(M) \, .$$

If M is a left R-module, then we can view it as an abelian group. Rather than continuing to use the previous notation $\text{End}(M)$ for the set of group homomorphisms mapping M into M, we will use the new notation $\text{End}_Z(M)$. More generally, we may use $\text{Hom}(_RM, _R N)$ rather than $\text{Hom}_R(M, N)$, or even $\text{Hom}(M, N)$, when the context is clear.

For any pair of elements in $\text{End}_R(M)$ the sum, difference, and composition belong to $\text{End}_R(M)$. Furthermore, the identity function 1_M belongs to $\text{End}_R(M)$, and so it follows that $\text{End}_R(M)$ is a subring of $\text{End}_Z(M)$.

We next observe that any R-homomorphism is a group homomorphism of the underlying abelian groups, and its kernel is just the kernel of this group homomorphism. It follows from elementary results on group homomorphisms that an R-homomorphism is one-to-one if and only if its kernel is trivial. This is really just the observation that for any R-modules M and N, any R-homomorphism $f : M \to N$, and any elements $m_1, m_2 \in M$ we have $f(m_1) = f(m_2)$ if and only if $f(m_1 - m_2) = 0$.

For a module $_RM$, defined by the representation $\rho : R \to \text{End}_Z(M)$, we can consider the underlying abelian group as a $\mathbf{Z}$-module. Then the set of R-endomorphisms of M is the subring of $\text{End}_Z(M)$ defined as follows:

$$\text{End}_R(M) = \{f \in \text{End}_Z(M) \mid f(rm) = rf(m) \text{ for all } r \in R \text{ and } m \in M\}$$

$$= \{f \in \text{End}_Z(M) \mid f \cdot \rho(r) = \rho(r) \cdot f \text{ for all } r \in R\}.$$

Thus $\text{End}_R(M)$ is the centralizer in $\text{End}_Z(M)$ of the image of the representation $\rho : R \to \text{End}_Z(M)$. (The *centralizer* of a subset X of a ring S is $\{s \in S \mid sx = xs \text{ for all } x \in X\}$.) Since $\text{End}_R(M)$ is a subring of $\text{End}_Z(M)$, the group M also defines a representation of $\text{End}_R(M)$, and so M has the structure of a left module over $\text{End}_R(M)$.

The fundamental homomorphism theorem holds for R-homomorphisms. It provides a basic tool for our study of modules.

Theorem 2.1.7 (Fundamental homomorphism theorem) *Let M and N be left R-modules. If $f : M \to N$ is any R-homomorphism, then*

$$f(M) \cong M/\ker(f).$$

Proof. Given an R-homomorphism $f : M \to N$ with $\ker(f) = K$, define $\overline{f} : M/K \to N$ by $\overline{f}(m + K) = f(m)$, for each coset $m + K$ in M/K. The function $\overline{f}$ is well-defined since if $m_1 + K = m_2 + K$, then $m_1 - m_2 \in \ker(f)$, and so $f(m_1) = f(m_2)$. If $\overline{f}(m_1 + K) = \overline{f}(m_2 + K)$, then $f(m_1) = f(m_2)$, so $f(m_1 - m_2) = 0$ and hence $m_1 + K = m_2 + K$. Thus $\overline{f}$ is one-to-one, and since it is by definition onto, we have defined an isomorphism from M/K onto $f(M)$. For any $r \in R$ and any $m \in M$ we have

$$r\overline{f}(m + K) = rf(m) = f(rm) = \overline{f}(rm + K) = \overline{f}(r(m + K)),$$

so $\overline{f}$ is an R-isomorphism.   □

If A is a left ideal of the ring R, then R/A is a left R-module, and modules of this type provide some of the most basic examples. As an immediate application of the fundamental homomorphism theorem, we can show that any cyclic module is of this form.

Proposition 2.1.8 *Let M be a left R-module.*
(a) *For any element $m \in M$, we have $Rm \cong R/\text{Ann}(m)$.*
(b) $\text{Ann}(M) = \bigcap_{f \in \text{Hom}(R,M)} \ker(f)$.

Proof. (a) For any element $m \in M$, an R-homomorphism $f_m : R \to M$ can be defined by setting $f_m(r) = rm$, for all $r \in R$. Then $\ker(f_m) = \text{Ann}(m)$, and $\text{Im}(f_m) = Rm$, so the fundamental homomorphism theorem shows immediately that the cyclic submodule Rm is isomorphic to $R/\text{Ann}(m)$.

(b) Since R has an identity element, every element $f \in \text{Hom}_R(R, M)$ is of the form f_m, for some $m \in M$, because $f(r) = rf(1)$ for all $f \in \text{Hom}_R(R, M)$. Thus we have

$$\text{Ann}(M) = \bigcap_{m \in M} \text{Ann}(m) = \bigcap_{f \in \text{Hom}(R, M)} \ker(f) .$$

This completes the proof. $\square$

Recall that a group is said to be *simple* if it has no proper nontrivial normal subgroups. The simple abelian groups correspond to the cyclic groups $\mathbf{Z}_p$, where p is a prime number. The same definition for modules over arbitrary rings will prove to be extremely important in our later work.

Definition 2.1.9 *A nonzero module $_R M$ is called* simple *(or* irreducible*) if its only submodules are* (0) *and* M.

The notion of simple module is closely connected to the following definitions for submodules. A submodule $N \subseteq M$ is called a *maximal submodule* if $N \neq M$ and for any submodule K with $N \subseteq K \subseteq M$, either $N = K$ or $K = M$. A submodule $N \subseteq M$ is called a *minimal submodule* if $N \neq (0)$ and for any submodule K with $(0) \subseteq K \subseteq N$, either $(0) = K$ or $K = N$.

Consistent with this terminology, a left ideal A of R is called a *maximal left ideal* if $A \neq R$ and for any left ideal B with $A \subseteq B \subseteq R$, either $A = B$ or $B = R$.

We first note that with this terminology a submodule $N \subseteq M$ is maximal if and only if M/N is a simple module. Furthermore, a submodule N is minimal if and only if it is simple when considered as a module in its own right.

Since a field has no proper nontrivial ideals, it is simple when viewed as a module over itself. As shown in the next example, very important cases of simple modules are obtained by viewing any vector space $_F V$ as a module over its endomorphism ring $\text{End}_F(V)$.

Example 2.1.7 ($_F V$ is a faithful simple module over $\text{End}_F(V)$)

Let V be a vector space over the field F, and consider V as a module over the ring $R = \text{End}_F(V)$ of all linear transformations of V. If v is any nonzero vector in V, then it is part of a basis $\mathcal{B}$ for V (as shown in Theorem A.2.5 of the appendix). Given any other vector $w \in V$, it is possible to define a linear transformation $f : V \to V$ for which $f(v) = w$ and $f(x) = 0$ for all other basis vectors $x \in \mathcal{B}$. This shows that $V = Rv$, and so V has no proper nontrivial R-submodules, and is therefore a simple R-module.

By definition, if $f \in R$ and $f(v) = 0$ for all $v \in V$, then $f = 0$. This shows that V is faithful when viewed as a left module over R.

As a special case of this example, let S be the ring $M_n(F)$, for some positive integer n. As in Example 2.1.5, any column A of S is a left ideal of S. Since the column A can be viewed as an n-dimensional vector space, and $M_n(F)$ is isomorphic to the ring of all linear transformations of A, it follows that A is a faithful simple left S-module, and hence a minimal left ideal of S.

The next result, known as Schur's lemma, provides one reason that simple modules play an important role in the elementary structure theory of rings. Recall that a ring R is a *division ring* if each nonzero element of R has a multiplicative inverse. Since division rings are just the noncommutative fields, we usually call a module over a division ring a vector space, and refer to linear transformations rather than ring homomorphisms.

Lemma 2.1.10 (Schur) *If $_RM$ is simple, then* $\mathrm{End}_R(M)$ *is a division ring.*

Proof. If $f \in \mathrm{End}_R(M)$ and $f \neq 0$, then $\ker(f) \neq M$, which implies that $\ker(f) = 0$ since M is simple. Furthermore, $\mathrm{Im}(f) \neq 0$ and so $\mathrm{Im}(f) = M$ since M is simple. Thus f is one-to-one and onto, and so it is invertible. □

Proposition 2.1.11 *The following conditions hold for a left R-module M.*

(a) *The module M is simple if and only if $Rm = M$, for each nonzero element $m \in M$.*

(b) *If M is simple, then $\mathrm{Ann}(m)$ is a maximal left ideal, for each nonzero element $m \in M$.*

(c) *If M is simple, then it has the structure of a left vector space over the division ring $\mathrm{End}_R(M)$.*

Proof. (a) If M is simple and $0 \neq m \in M$, then Rm is a nonzero submodule, and so we must have $Rm = M$. Conversely, if N is any nonzero submodule of M, then N must contain some nonzero element m, and so $M = Rm \subseteq N$ shows that $N = M$.

(b) If M is simple, and $0 \neq m \in M$, then $R/\mathrm{Ann}(m) \cong Rm = M$, and so $\mathrm{Ann}(m)$ must be a maximal left ideal since the submodules of M correspond to the left ideals of R that contain $\mathrm{Ann}(m)$.

(c) We have already observed that any left R-module M is a left module over the ring $\mathrm{End}_R(M)$. Thus Schur's lemma implies that any simple left R-module has the structure of a left vector space over a division ring. □

The fundamental homomorphism theorem is used to prove parts (a) and (b) of the next theorem, which are (respectively) the first and second isomorphism theorems for modules. Part (a) shows that in the factor module M/K the submodule $(N + K)/K$ generated by N is closely related to N itself.

Part (c) of the next theorem is sometimes called the *modular law*, and can be remembered as stating that $\cap$ distributes over $+$, under appropriate circumstances. That is, $L \cap (K + N) = (L \cap K) + (L \cap N)$, whenever $K \subseteq L$.

Theorem 2.1.12 *Let M be a left R-module, with a submodule N.*

(a) *If K is any submodule of M, then $(N + K)/K \cong N/(N \cap K)$.*

(b) *If $K \subseteq L$ are any submodules of M, then $(M/K)/(L/K) \cong M/L$.*

(c) *If $K \subseteq L$ are any submodules of M, then $L \cap (K + N) = K + (L \cap N)$.*

Proof. (a) The homomorphism $N \to N + K \to (N + K)/K$ defined as the composition of the natural inclusion followed by the natural projection has image $(N+K)/K$ and kernel $N \cap K$. Applying the fundamental homomorphism theorem (Theorem 2.1.7) gives the desired isomorphism.

(b) Applying the fundamental homomorphism theorem to the natural mapping from M/K onto M/L gives the desired isomorphism, since the kernel is $\{x + K \mid x \in L\}$.

(c) It is clear that the right hand side is contained in the left hand side. To prove the reverse inclusion, let $x \in L \cap (K + N)$. Then $x \in L$ and there exist $y \in K$, $z \in N$ with $x = y + z$. Thus $z = x - y \in L$ (since $y \in K$ and $K \subseteq L$), and so $y + z \in K + (L \cap N)$. $\square$

Our next result requires the use of Zorn's lemma, which provides a very useful condition that is equivalent to the axiom of choice. (See Section A.2 of the appendix.) Recall that a partial order on a set S is a relation $\leq$ such that (i) $x \leq x$ for all $x \in S$, (ii) if $x, y, z \in S$ with $x \leq y$ and $y \leq z$, then $x \leq z$, (iii) if $x, y \in S$ with $x \leq y$ and $y \leq x$, then $x = y$. We note that both $\subseteq$ and $\supseteq$ are partial orders on the set of all submodules of a given module. A subset of S is called a chain if for any x, y in the subset either $x \leq y$ or $y \leq x$. Zorn's lemma states that if $(S, \leq)$ is a nonempty partially ordered set such that every chain in S has an upper bound in S, then S has a maximal element.

Lemma 2.1.13 *Let X be any subset of the module $_R M$. Any submodule N with $N \cap X \subseteq (0)$ is contained in a submodule maximal with respect to this property.*

Proof. Consider $\{\,_R K \subseteq M \mid K \supseteq N \text{ and } K \cap X \subseteq (0)\}$. This set contains N, so it is nonempty, and it is partially ordered by inclusion. For any chain

$\{K_\alpha \mid \alpha \in I\}$, the set $\bigcup_{\alpha \in I} K_\alpha$ is a least upper bound. It is a submodule and intersects X trivially because the submodules $\{K_\alpha\}_{\alpha \in I}$ are linearly ordered. By Zorn's lemma, the set under consideration has a maximal element. $\square$

The left ideal A is a maximal left ideal precisely when it is a maximal element in the set of proper left ideals of R, ordered by inclusion. It is an immediate consequence of Lemma 2.1.13 that every left ideal of the ring R is contained in a maximal left ideal, by applying the proposition to the set $X = \{1\}$. Furthermore, any left ideal maximal with respect to not including the multiplicative identity is in fact a maximal left ideal.

Proposition 2.1.14 *For any nonzero element m of the module $_RM$ and any submodule N of M with $m \notin N$, there exists a submodule $\widehat{N}$ maximal with respect to the conditions $\widehat{N} \supseteq N$ and $m \notin \widehat{N}$. Moreover, $M/\widehat{N}$ has a minimal submodule contained in every nonzero submodule.*

Proof. We can apply Lemma 2.1.13 to the set $X = \{m\}$, to obtain the submodule $\widehat{N}$. Any nonzero submodule of $M/\widehat{N}$ corresponds to a submodule K that properly contains $\widehat{N}$, and thus $m \in K$. It follows that $K/\widehat{N}$ contains the cyclic submodule generated by the coset $\overline{m}$ determined by m. Since this cyclic submodule is contained in every nonzero submodule of M/N, it is minimal in M/N. $\square$

Corollary 2.1.15 *Any proper submodule of a finitely generated module is contained in a maximal submodule.*

Proof. Let $_RM$ be generated by $m_1, m_2, \ldots, m_n$, and let N be any proper submodule of M. Since N is a proper submodule, it cannot contain all of the generators $m_1, m_2, \ldots, m_n$. By reordering the generators, if necessary, it is possible to find $m_1, m_2, \ldots, m_k$ such that $\sum_{j=1}^{k} Rm_j + N = M$ but $\sum_{j=2}^{k} Rm_j + N \neq M$. Applying the previous proposition to the element m_1 and the submodule $\sum_{j=2}^{k} Rm_j + N$ yields a maximal submodule. $\square$

EXERCISES: SECTION 2.1

1. Let M be a left R-module, let X be a nonempty subset of M, and let A be a left ideal of R.

   (a) Show that $AX = \{\sum_{i=1}^{k} a_i x_i \mid a_i \in A, x_i \in X, k \in \mathbf{Z}^+\}$ is a submodule of M.

   (b) Show that X is a submodule of M if and only if $RX = X$.

2. Let M be a left R-module. Write out careful proofs of the following statements from the text.

   (a) Let $\{M_\alpha\}_{\alpha \in I}$ be a collection of submodules of M. Then $\bigcap_{\alpha \in I} M_\alpha$ and $\sum_{\alpha \in I} M_\alpha$ are submodules of M.

   (b) Let X be any subset of M. Then $\langle X \rangle = \sum_{x \in X} Rx$.

3. Let M be a left R-module.

   (a) Prove, directly from the definition, that $\mathrm{Ann}(M)$ is a two-sided ideal of R.

   (b) Prove that if $M \cong R/A$, for a left ideal A of R, then $\mathrm{Ann}(M)$ is the largest two-sided ideal of R that is contained in A.

4. Let M, N be left R-modules, and let I be an ideal of R with $I \subseteq \mathrm{Ann}(M)$.

   (a) Show that M has a natural structure as a left (R/I)-module.

   (b) Show that if N is a left (R/I)-module, then $\mathrm{Hom}_{R/I}(M, N) = \mathrm{Hom}_R(M, N)$.

5. Let M be a left R-module. Prove that M is a faithful module over the factor ring $R/\mathrm{Ann}(M)$.

6. Let D be an integral domain, and let M be a module over D. The *torsion submodule* of M is defined to be $\mathrm{tor}(M) = \{m \in M \mid \mathrm{Ann}(m) \neq (0)\}$.

   (a) Prove that $\mathrm{tor}(M)$ is in fact a submodule of M.

   (b) Prove that $\mathrm{tor}(M/\mathrm{tor}(M)) = (0)$.

   (c) Prove that if N is any D-module, and $f : M \to N$ is any D-homomorphism, then $f(\mathrm{tor}(M)) \subseteq \mathrm{tor}(N)$.

7. Show that any ring that has a faithful, simple module is isomorphic to a subring of the ring of all linear transformations of a vector space over a division ring.

8. Prove that $\mathrm{Hom}_Z(\mathbf{Z}_m, \mathbf{Z}_n) \cong \mathbf{Z}_d$, where $d = \gcd(m, n)$.

9. Show that if M is a left R-module, then $\mathrm{Hom}_R(R, M)$ has a natural structure as a left R-module. Show that $\mathrm{Hom}_R(R, M) \cong M$, as left R-modules.

10. The *opposite ring* of R is defined on the set R, using the addition of R but defining $a \cdot b = ba$, for all $a, b \in R$, where ba is defined by the original multiplication in R. (Refer to Exercise 6 of Section 1.2.) Using the notation R^{op} for the opposite ring of R, prove that $\mathrm{End}_R(R) \cong R^{op}$.

    Hint: For any $a \in R$, define $\mu_a : R \to R$ by $\mu_a(x) = xa$, for all $x \in R$. Show that $\mu_a \in \mathrm{End}_R(R)$.

11. Let R be the ring of lower triangular 2×2 matrices over the field F, denoted by $\begin{bmatrix} F & 0 \\ F & F \end{bmatrix}$. Let $A = \begin{bmatrix} F & 0 \\ F & 0 \end{bmatrix}$, $S_1 = \begin{bmatrix} 0 & 0 \\ F & 0 \end{bmatrix}$, and $S_2 = \begin{bmatrix} 0 & 0 \\ 0 & F \end{bmatrix}$.

    (a) Show that S_1, S_2, and A/S_1 are simple left R-modules.

    (b) Show that $S_1 \cong S_2$, but $S_1 \not\cong A/S_1$.

12. Let S be a commutative ring, and let M be an S-module. We define the *trivial extension* of M by S to be the set of all formal matrices

$$R = \left\{ \begin{bmatrix} s & 0 \\ m & s \end{bmatrix} \,\middle|\, s \in S \text{ and } m \in M \right\} .$$

The notation $R = S \ltimes M$ is used, and this ring is sometimes called the *idealization* of M.

(a) Show that R is a commutative ring.

(b) If N is an S-submodule of M, and I is an ideal of S such that $IM \subseteq N$, we define

$$I \ltimes N = \left\{ \begin{bmatrix} a & 0 \\ x & a \end{bmatrix} \middle| a \in I \text{ and } x \in N \right\} .$$

Show that $I \ltimes N$ is an ideal of R.

(c) Show that the correspondence $N \leftrightarrow (0) \ltimes N$ determines a one-to-one correspondence between the set of S-submodules N of M and the set of ideals of R that are contained in $(0) \ltimes M$.

2.2   Direct sums and products

In this section we study direct sums and their connection with free and projective modules. We introduce free modules, together with several other classes of modules, including completely reducible modules. The latter definition is motivated by the familiar result that any subspace of a vector space has a complement, i.e., that any subspace is a direct summand.

The most fundamental example of a module is that of a vector space over a field. The most important property of a vector space is that it has a basis (see Theorem A.2.5). In general, over an arbitrary ring, there is no guarantee that a module has a basis. Those that do have a basis form an important class, which is the subject of the next definition.

Definition 2.2.1 *The left R-module M is called a* free module *if there exists a subset $X \subseteq M$ such that each element $m \in M$ can be expressed uniquely as a finite sum $m = \sum_{i=1}^{n} a_i x_i$, with $a_1, \ldots, a_n \in R$ and $x_1, \ldots, x_n \in X$.*

The module $_R R$ is the prototype of a free module, with generating set $\{1\}$. If $_R M$ is a module, and $X \subseteq M$, we say that the set X is *linearly independent* if $\sum_{i=1}^{n} a_i x_i = 0$ implies $a_i = 0$ for $i = 1, \ldots, n$, for any distinct $x_1, x_2, \ldots, x_n \in X$ and any $a_1, a_2, \ldots, a_n \in R$. Then a linearly independent generating set for M is called a *basis* for M, and so M is a free module if and only if it has a basis.

In the next proposition we show that free R-modules are characterized by the fact that any R-homomorphism out of the module is completely determined by its action on the basis elements.

Proposition 2.2.2 *Let M be a free left R-module, with basis X. For any left R-module N and any function $\eta : X \to N$ there exists a unique R-homomorphism $f : M \to N$ with $f(x) = \eta(x)$, for all $x \in X$.*

Proof. If $m \in M$, then since X is a basis for M there exist $x_1, x_2, \ldots, x_n \in X$ and $a_1, a_2, \ldots, a_n \in R$ with $m = \sum_{i=1}^{n} a_i x_i$.

We first show that there is only one way to extend η to M. If $f : M \to N$ is any R-homomorphism with $f(x) = \eta(x)$, for all $x \in X$, we must have

$$f(m) = f(\sum_{i=1}^{n} a_i x_i) = \sum_{i=1}^{n} a_i f(x_i) = \sum_{i=1}^{n} a_i \eta(x_i) \, .$$

On the other hand, given any function $\eta : X \to N$, defining $f(m) = \sum_{i=1}^{n} a_i \eta(x_i)$, for all $m \in M$, yields an R-homomorphism with the required property. We note that f is well-defined since each element of M can be written uniquely as a linear combination of finitely many elements of the basis X. □

If $_R M$ and $_R N$ are modules, we define their direct sum as follows:

$$M \oplus N = \{(x, y) \mid x \in M \text{ and } y \in N\}$$

with addition $(x_1, y_1) + (x_2, y_2) = (x_1 + x_2, y_1 + y_2)$ and scalar multiplication $r(x, y) = (rx, ry)$, for all $x, x_1, x_2 \in M$, $y, y_1, y_2 \in N$, and $r \in R$. The addition in $M \oplus N$ agrees with the operation defined on the direct sum of M and N viewed as abelian groups, and the scalar multiplication is the natural one. We refer to these operations as componentwise addition and scalar multiplication. In the next definition, we extend this concept to the case of infinitely many modules. The Cartesian product of the modules $\{M_\alpha\}_{\alpha \in I}$ can be described as the set of functions $m : I \to \bigcup_{\alpha \in I} M_\alpha$, where for each $\alpha \in I$ we have $m(\alpha) \in M_\alpha$. The element $m(\alpha)$ is usually written m_α.

Definition 2.2.3 *Let $\{M_\alpha\}_{\alpha \in I}$ be a collection of left R-modules indexed by the set I.*

(a) *The* direct product *of the modules $\{M_\alpha\}_{\alpha \in I}$ is the Cartesian product*

$$\prod_{\alpha \in I} M_\alpha \, ,$$

with componentwise addition and scalar multiplication. If $x, y \in \prod_{\alpha \in I} M_\alpha$, with components $x_\alpha, y_\alpha \in M_\alpha$ for all $\alpha \in I$, then $x + y$ is defined to be the element with components $(x + y)_\alpha = x_\alpha + y_\alpha$, for all $\alpha \in I$. If $r \in R$, then rx is defined to be the element with components $(rx)_\alpha = rx_\alpha$, for all $\alpha \in I$.

(b) The submodule of $\prod_{\alpha \in I} M_\alpha$ consisting of all elements m such that $m_\alpha = 0$ for all but finitely many components m_α is called the direct sum of the modules $\{M_\alpha\}_{\alpha \in I}$, and is denoted by

$$\bigoplus\nolimits_{\alpha \in I} M_\alpha \, .$$

It is left as an exercise to show that $\prod_{\alpha \in I} M_\alpha$ and $\bigoplus_{\alpha \in I} M_\alpha$ are left R-modules. If the index set I is finite, then the direct sum coincides with the direct product.

For each index $\gamma \in I$ we have the projection $p_\gamma : \prod_{\alpha \in I} M_\alpha \to M_\gamma$ defined by $p_\gamma(m) = m_\gamma$, and this is an R-homomorphism. For each $\gamma \in I$, we also have an inclusion mapping $i_\gamma : M_\gamma \to \prod_{\alpha \in I} M_\alpha$ defined for all $x \in M_\gamma$ by $i_\gamma(x) = m$, where $m_\gamma = x$ and $m_\alpha = 0$ for all $\alpha \neq \gamma$. The mapping i_γ is also an R-homomorphism. For each element in the image of i_γ in $\prod_{\alpha \in I} M_\alpha$, all but one component must be zero. Thus the image of i_γ is contained in the direct sum, so for each $\gamma \in I$ we also have the inclusion $i_\gamma : M_\gamma \to \bigoplus_{\alpha \in I} M_\alpha$. The mapping that we mean to use will always be clear from the context, so it does not make sense to use two different names for the two inclusion mappings, although they are different functions.

Proposition 2.2.4 Let $\{M_\alpha\}_{\alpha \in I}$ be a collection of left R-modules indexed by the set I, and let N be a left R-module.

For each $\gamma \in I$ let $p_\gamma : \prod_{\alpha \in I} M_\alpha \to M_\gamma$ be the projection defined by $p_\gamma(m) = m_\gamma$, for all $m \in \prod_{\alpha \in I} M_\alpha$, and let $i_\gamma : M_\gamma \to \bigoplus_{\alpha \in I} M_\alpha$ be the inclusion defined for all $x \in M_\alpha$ by $i_\gamma(x) = m$, where $m_\gamma = x$ and $m_\alpha = 0$ for all $\alpha \neq \gamma$.

(a) For any set $\{f_\alpha\}_{\alpha \in I}$ of R-homomorphisms such that $f_\alpha : N \to M_\alpha$ for each $\alpha \in I$, there exists a unique R-homomorphism $f : N \to \prod_{\alpha \in I} M_\alpha$ such that $p_\alpha f = f_\alpha$ for all $\alpha \in I$.

(b) For any set $\{f_\alpha\}_{\alpha \in I}$ of R-homomorphisms such that $f_\alpha : M_\alpha \to N$ for each $\alpha \in I$, there exists a unique R-homomorphism $f : \bigoplus_{\alpha \in I} M_\alpha \to N$ such that $f i_\alpha = f_\alpha$ for each $\alpha \in I$.

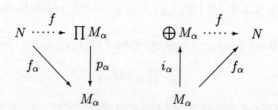

Proof. (a) If $x \in N$, we define $f(x)$ by letting its components be $(f(x))_\alpha = f_\alpha(x)$, for each $\alpha \in I$. This is the only way in which f can be defined to satisfy $p_\alpha f = f_\alpha$ for all $\alpha \in I$.

(b) If $m \in \bigoplus_{\alpha \in I} M_\alpha$, there are only finitely many components m_α that are nonzero, and we can define $f(m) = \sum_{\alpha \in I} f_\alpha(m_\alpha)$. □

In many situations we need to consider a direct sum or direct product in which each summand or factor is isomorphic to a fixed module. If $M_\alpha \cong M$ for all $\alpha \in I$, then we will use the following notation:

$$\bigoplus_{j=1}^{n} M_j = M^n , \qquad \bigoplus_{\alpha \in I} M_\alpha = M^{(I)} , \qquad \prod_{\alpha \in I} M_\alpha = M^I .$$

Part (a) of the next proposition provides a proof of the existence of free modules.

Proposition 2.2.5 *Let M be a left R-module.*

(a) *The module M is free if and only if it is isomorphic to a direct sum $R^{(I)}$, for some index set I.*

(b) *The module M is a homomorphic image of a free module.*

Proof. (a) Let I be any index set. We define the 'standard basis' for $R^{(I)}$ as follows: for each $\alpha \in I$, let e_α be the element of $R^{(I)}$ which has 1 in the αth component and 0 in every other component. For any element $m \in R^{(I)}$, there are only finitely many nonzero components, say $r_\alpha \neq 0$ for each $\alpha \in F$, where F is a finite subset of I. Then $m = \sum_{\alpha \in F} r_\alpha e_\alpha$, and since this representation is unique, we have shown that $\{e_\alpha\}_{\alpha \in I}$ is a basis for $R^{(I)}$. Any module isomorphic to $R^{(I)}$ must therefore also have a basis.

Conversely, let $_R M$ be any free module with basis X. Consider $R^{(I)}$, where the index set I is chosen to be X. For each $x \in X$, we define $f_x : R \to M$ by $f_x(r) = rx$. By Proposition 2.2.4 there is a unique R-homomorphism $f : R^{(I)} \to M$ which agrees with each f_x, and it is clear that $f(e_x) = x$ for each index $x \in I = X$. The uniqueness of representations in M and in $R^{(I)}$ shows that f is an isomorphism.

(b) We can always find a generating set X for M. (If necessary, we can simply take all elements of M.) The proof in part (a) produces an R-homomorphism $f : R^{(X)} \to M$ that maps the free module $R^{(X)}$ onto M. □

We have already noted that the direct sum of finitely many modules is the same as their direct product. The next proposition characterizes finite sums in terms of the relevant inclusion and projection mappings. In the proof we use the Kronecker delta functions, defined for integers $1, \ldots, n$ by letting $\delta_{jk} = 0$ if $j \neq k$ and $\delta_{jk} = 1$ if $j = k$.

Proposition 2.2.6 *Let M and $M_1, \ldots, M_n$ be left R-modules. Then*

$$M \cong M_1 \oplus M_2 \oplus \cdots \oplus M_n$$

if and only if there exist R-homomorphisms $i_j : M_j \to M$ and $p_j : M \to M_j$ for $j = 1, \ldots, n$ such that

$$p_j i_k = \delta_{jk} \cdot 1_{M_k} \quad and \quad i_1 p_1 + \cdots + i_n p_n = 1_M \ .$$

Proof. Assume that $M \cong M_1 \oplus M_2 \oplus \cdots \oplus M_n$. We first consider the case in which M is the direct sum itself, so that $M = \bigoplus_{j=1}^{n} M_j$. By definition of the direct sum we have inclusion mappings $i_k : M_k \to \bigoplus_{j=1}^{n} M_j$, for $1 \le k \le n$. Since there are only finitely many components, M is also the direct product of the modules $\{M_j\}_{j=1}^{n}$, and so by the definition of the direct product we have projection mappings $p_k : \bigoplus_{j=1}^{n} M_j \to M_k$, for $1 \le k \le n$. Then for each $x \in M$ we have

$$
\begin{aligned}
x = (x_1, x_2, \ldots, x_n) &= (p_1(x), p_2(x), \ldots, p_n(x)) \\
&= (p_1(x), 0, \ldots, 0) + \cdots + (0, 0, \ldots, p_n(x)) \\
&= i_1 p_1(x) + \cdots + i_n p_n(x) \ .
\end{aligned}
$$

It is also clear that $p_k i_k = 1$ for $1 \le k \le n$, and that $p_j i_k = 0$ if $j \ne k$. Finally, if instead of equality we are given an isomorphism $f : M \to \bigoplus_{j=1}^{n} M_j$, then we simply replace p_j by $p_j f$ and i_j by $f^{-1} i_j$, and it is easy to check that the necessary equations hold.

Conversely, assume that we are given mappings $\{i_j\}_{j=1}^{n}$ and $\{p_j\}_{j=1}^{n}$ which satisfy the given conditions. Define $f : M \to \bigoplus_{j=1}^{n} M_j$ by

$$f(x) = (p_1(x), p_2(x), \ldots, p_n(x))$$

for all $x \in M$ and define $g : \bigoplus_{j=1}^{n} M_j \to M$ by

$$g(x_1, x_2, \ldots, x_n) = i_1(x_1) + i_2(x_2) + \cdots + i_n(x_n)$$

for all $(x_1, x_2, \ldots, x_n) \in \bigoplus_{j=1}^{n} M_j$. Then $gf(x) = i_1 p_1(x) + \cdots + i_n p_n(x) = x$ for all $x \in M$, and $fg((x_1, \ldots, x_n)) = (x_1, \ldots, x_n)$ for all $(x_1, \ldots, x_n) \in \bigoplus_{j=1}^{n} M_j$ since for each j we have $p_j(g((x_1, \ldots, x_n))) = p_j i_j(x_j) = x_j$. $\square$

It is interesting to consider left ideals that are direct summands of R. An element $e \in R$ is called *idempotent* if $e^2 = e$. A set of idempotent elements is said to be *orthogonal* if $ef = 0$ for all distinct idempotents e, f in the set. The next proposition shows that direct summands of $_R R$ correspond to certain orthogonal sets of idempotent elements. As a special case of the proposition, the left ideal A is a direct summand of $_R R$ if and only if $A = Re$ for an idempotent element $e \in R$, since if $e \in R$ is idempotent then $1 - e$ is an orthogonal idempotent of the required type.

Proposition 2.2.7 *Let $A_1, A_2, \ldots, A_n$ be left ideals of the ring R.*

(a) *The ring R can be written as a direct sum $R = A_1 \oplus A_2 \oplus \cdots \oplus A_n$ if and only if there exists a set $e_1, e_2, \ldots, e_n$ of orthogonal idempotent elements of R such that $A_j = Re_j$ for $1 \leq j \leq n$ and $e_1 + e_2 + \cdots + e_n = 1$.*

(b) *The left ideals A_j in part (a) are two-sided ideals if and only if the corresponding idempotent elements belong to the center of R.*

(c) *If condition (b) holds, then every left R-module M can be written as a direct sum $M = M_1 \oplus M_2 \oplus \cdots \oplus M_n$, where $M_j = e_j M$ is a module over the ring A_j, for $1 \leq j \leq n$.*

Proof. (a) First assume that $R = A_1 \oplus A_2 \oplus \cdots \oplus A_n$. Then the identity element of R can be written uniquely as $1 = e_1 + e_2 + \cdots + e_n$ for elements $e_j \in A_j$. For any $a \in A_j$, multiplying this equation by a yields

$$a = ae_1 + \cdots + ae_j + \cdots + ae_n .$$

Uniqueness of representations in the direct sum implies that $ae_j = a$ and $ae_k = 0$ for $k \neq j$. This implies that $A_j = Re_j$, and shows that $e_1, e_2, \ldots, e_n$ form an orthogonal set of idempotents.

Conversely, assume that $e_1, e_2, \ldots, e_n$ form an orthogonal set of idempotent elements such that $e_1 + e_2 + \cdots + e_n = 1$. For any $r \in R$ we have $r = re_1 + re_2 + \cdots + re_n$, and so it is clear that $Re_1 + Re_2 + \cdots + Re_n = R$. If

$$a_1 e_1 + a_2 e_2 + \cdots + a_n e_n = b_1 e_1 + b_2 e_2 + \cdots + b_n e_n ,$$

then multiplying on the right by e_j shows that $a_j e_j = b_j e_j$, and so these representations are unique and we have a direct sum.

(b) If the idempotents in part (a) are central, it follows immediately that Re_j is a two-sided ideal. Conversely, if Re_j is a two-sided ideal, then for any $r \in R$ we have $r = e_1 r + \cdots + e_n r$, with $e_j r \in Re_j$. This implies that $e_j r = re_j$ for each j, and so the idempotents are central.

(c) If the idempotents e_j are central, then the subsets $e_j M$ are submodules of M, for $1 \leq j \leq n$. Using the fact that the idempotents are central, an argument similar to that of part (a) shows that for any $m \in M$ we have a unique representation $m = e_1 m + \cdots + e_n m$. It can be shown that Re_j is a ring, with identity e_j, and that $e_j M$ is a module over Re_j. $\square$

We now consider conditions under which we can determine that a submodule or a factor module is in fact a direct summand. If M_1 and M_2 are submodules of M, then the two conditions $M_1 + M_2 = M$ and $M_1 \cap M_2 = (0)$ are equivalent to the condition that each element of M can be represented uniquely as a sum of elements of M_1 and M_2, and so they are equivalent to the condition that $M = M_1 \oplus M_2$. In this case we say that M_2 is a *complement* of M_1.

Definition 2.2.8 *Let L, M, N be left R-modules.*

(a) *An onto R-homomorphism $f : M \to N$ is said to be* split *if there exists an R-homomorphism $g : N \to M$ with $fg = 1_N$. In this case g is called a* splitting map *for f.*

(b) *A one-to-one R-homomorphism $g : L \to M$ is said to be* split *if there exists an R-homomorphism $f : M \to L$ such that $fg = 1_L$. In this case f is called a* splitting map *for g.*

Proposition 2.2.9 *Let M, N be left R-modules.*

(a) *Let $f : M \to N$ and $g : N \to M$ be R-homomorphisms such that $fg = 1_N$. Then $M = \ker(f) \oplus \operatorname{Im}(g)$.*

(b) *A one-to-one R-homomorphism $g : N \to M$ splits if and only if $\operatorname{Im}(g)$ is a direct summand of M.*

(c) *An onto R-homomorphism $f : M \to N$ splits if and only if $\ker(f)$ is a direct summand of M.*

Proof. (a) Assume that $f : M \to N$ and $g : N \to M$ are homomorphisms with $fg = 1_N$. If $m \in \ker(f) \cap \operatorname{Im}(g)$, then there exists $x \in N$ with $g(x) = m$, and so $x = fg(x) = f(m) = 0$, which implies $m = g(0) = 0$. Thus $\ker(f) \cap \operatorname{Im}(g) = (0)$.

Given any $m \in M$, consider the element $m - gf(m)$. We have $gf(m) \in \operatorname{Im}(g)$ and $m - gf(m) \in \ker(f)$ since $f(m - gf(m)) = f(m) - fg(f(m)) = 0$. Thus

$$m = (m - gf(m)) + (gf(m)) \in \ker(f) + \operatorname{Im}(g) .$$

It follows that $M = \ker(f) \oplus \operatorname{Im}(g)$.

(b) Assume that $g : N \to M$ is a one-to-one homomorphism. If g has a splitting map $f : M \to N$ such that $fg = 1_N$, then $M = \operatorname{Im}(g) \oplus \ker(f)$ by part (a).

Conversely, assume that g is one-to-one and $\operatorname{Im}(g)$ is a direct summand of M, say $M = \operatorname{Im}(g) \oplus M_2$. Then each $m \in M$ is uniquely represented as $g(x) + y$, for some $x \in N$ and some $y \in M_2$. Since g is one-to-one, the element x is uniquely determined by m. To show this, suppose that $g(x_1) + y_1 = g(x_2) + y_2$. Since $\operatorname{Im}(g) \cap M_2 = (0)$, this implies that $g(x_1) = g(x_2)$, and so $x_1 = x_2$ since g is one-to-one. Thus we have a well-defined function $f : M \to N$ given by $f(m) = x$, for all $m \in M$. Since f is well-defined, it follows easily that f is an R-homomorphism, and it is clear that $fg = 1_N$.

(c) Assume that $f : M \to N$ is an onto homomorphism that has a splitting map $g : N \to M$ with $fg = 1_N$. Applying part (a) gives $M = \ker(f) \oplus \operatorname{Im}(g)$.

Conversely, assume that $M = \ker(f) \oplus M_2$ for some submodule $M_2 \subseteq M$. Define $h : M_2 \to N$ by letting $h(x) = f(x)$, for all $x \in M_2$. Then h is an onto mapping, since if $y \in N$ then there exists $m \in M$ with $f(m) = y$, where $m = m_1 + m_2$ for some $m_1 \in \ker(f)$ and $m_2 \in M_2$, and so $y = f(m) = f(m_2) = h(m_2)$. Furthermore, h is one-to-one, since $\ker(h) = \ker(f) \cap M_2 = (0)$. It is clear that h is an R-homomorphism, and if we let $g = h^{-1}$, then $fg = 1_N$. $\square$

Proposition 2.2.10 *Let* L, M, N *be left* R-*modules. Let* $g : L \to M$ *be a one-to-one* R-*homomorphism, and let* $f : M \to N$ *be an onto* R-*homomorphism such that* $\text{Im}(g) = \ker(f)$. *Then* g *is split if and only if* f *is split, and in this case* $M \cong L \oplus N$.

Proof. By Proposition 2.2.9, g is split if and only if $\text{Im}(g)$ is a direct summand of M, and f is split if and only if $\ker(g)$ is a direct summand of M. The two conditions are equivalent since $\text{Im}(g) = \ker(f)$. Now assume that $p : M \to L$ with $pg = 1_L$. Then $M = \text{Im}(g) \oplus \ker(p)$, and $\text{Im}(g) \cong L$ since g is one-to-one. On the other hand, $\ker(p) \cong M/\text{Im}(g)$, and $M/\text{Im}(g) \cong N$ since $\text{Im}(g) = \ker(f)$, so we have $M \cong L \oplus N$. $\square$

Corollary 2.2.11 *The following conditions are equivalent for* $_RM$:

(1) *every submodule of* M *is a direct summand;*

(2) *every one-to-one* R-*homomorphism into* M *splits;*

(3) *every onto* R-*homomorphism out of* M *splits.*

An extremely important property of a vector space is that every subspace has a complement. That is, every subspace is a direct summand. We now consider modules that have the same property.

Definition 2.2.12 *A module* $_RM$ *is called* completely reducible *if every submodule of* M *is a direct summand of* M.

If F is a free left R-module with basis X, and $p : M \to F$ is an R-homomorphism that maps M onto F, then for each basis element $x \in X$ we can choose $m \in M$ with $p(m) = x$. Since F is free, there exists an R-homomorphism $f : F \to M$ with $f(x) = m$, for all $x \in X$. Because $pf(x) = x$ for all $x \in X$, we must have $pf = 1_F$, and so p splits. This splitting property can be extended to direct summands of free modules, and characterizes an important class of modules.

Proposition 2.2.13 *The following conditions are equivalent for $_RP$:*

(1) *every R-homomorphism onto P splits;*

(2) *P is isomorphic to a direct summand of a free module;*

(3) *for any onto R-homomorphism $p : M \to N$ and any R-homomorphism $f : P \to N$ there exists a lifting $\widehat{f} : P \to M$ such that $p\widehat{f} = f$.*

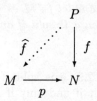

Proof. (1) $\Rightarrow$ (2) We have observed that every module is a homomorphic image of a free module, so let $p : F \to P$ be a mapping from a free module F onto P. Since p splits by assumption, this shows that P is isomorphic to a direct summand of F.

(2) $\Rightarrow$ (3) Let $p : M \to N$ be an onto R-homomorphism, and let $f : P \to N$ be any R-homomorphism. By assumption there exists a free module F which contains P as a direct summand, and so there exist mappings $i : P \to F$ and $q : F \to P$ with $qi = 1_P$. The following diagram summarizes the information at our disposal.

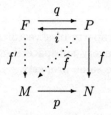

Since F is a free module, it has a basis $\{x_\alpha\}_{\alpha \in I}$. For each $\alpha \in I$, choose $m_\alpha \in M$ such that $p(m_\alpha) = fq(x_\alpha)$. Since F is free, defining $f' : F \to M$ by $f'(x_\alpha) = m_\alpha$, for each $\alpha \in I$, defines an R-homomorphism with $pf' = fq$. If $\widehat{f} = f'i$, then $f = fqi = pf'i = p\widehat{f}$, and so P satisfies the requirements of condition (3).

(3) $\Rightarrow$ (1) Assume that the module P satisfies condition (3). For any onto R-homomorphism $p : M \to P$ we have the following diagram.

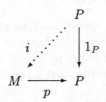

By assumption there exists an R-homomorphism $i : P \to M$ such that $pi = 1_P$, and so p is split. $\square$

Definition 2.2.14 *A module $_RM$ is called* projective *it it is a direct summand of a free R-module.*

EXERCISES: SECTION 2.2

1. Let D be an integral domain. With reference to Exercise 6 of Section 2.1, a module $_DM$ is called a *torsion module* if $\tau(M) = M$. Prove that any direct sum of torsion modules is again a torsion module.

2. Let D be an integral domain. With reference to Exercise 6 of Section 2.1, a module $_DM$ is called a *torsionfree module* if $\tau(M) = (0)$. Prove that any direct product of torsionfree modules is again a torsionfree module.

3. Let R be a commutative ring. Let A_k be the left ideal of matrices $[a_{ij}]$ in $\mathrm{M}_n(R)$ with $a_{ij} = 0$ for $j \neq k$. Show that $\mathrm{M}_n(R) = \bigoplus_{k=1}^n A_k$.

4. Let $f \in \mathrm{End}_R(M)$ with $f^2 = f$. Prove that $M = \ker(f) \oplus \mathrm{Im}(f)$.

5. Let $\{M_\alpha\}_{\alpha \in I}$ be a set of left R-modules, and let N be any left R-module.
   (a) Prove that $\mathrm{Hom}_R(N, \prod_{\alpha \in I} M_\alpha) \cong \prod_{\alpha \in I} \mathrm{Hom}_R(N, M_\alpha)$, as abelian groups.
   (b) Prove that $\mathrm{Hom}_R(\bigoplus_{\alpha \in I} M_\alpha, N) \cong \prod_{\alpha \in I} \mathrm{Hom}_R(M_\alpha, N)$, as abelian groups.

6. Let $\{P_\alpha\}_{\alpha \in I}$ be a collection of left R-modules. Prove that the direct sum $\bigoplus_{\alpha \in I} P_\alpha$ is projective if and only if P_α is projective for each $\alpha \in I$.

7. Give an example of a module over the ring $\mathbf{Z}_4$ that is not projective. Give an example of a module over $\mathbf{Z}_4$ that is not completely reducible.

8. For each prime number p, let $\mathbf{Z}_{p^\infty}$ denote the set of all elements of $\mathbf{Q}/\mathbf{Z}$ whose additive order is a power of p. Let $\mathcal{P}$ denote the set of all prime numbers. Show that $\mathbf{Q}/\mathbf{Z} = \bigoplus_{p \in \mathcal{P}} \mathbf{Z}_{p^\infty}$.

9. Let p be a prime number. Using the definition given in Exercise 8, show that $\mathbf{Z}_{p^\infty}$ contains a unique minimal submodule, which is isomorphic to $\mathbf{Z}_p$.

10. Let $\mathcal{P}$ denote the set of all prime numbers. Show that $\prod_{p \in \mathcal{P}} \mathbf{Z}_p$ is not a completely reducible $\mathbf{Z}$-module.

2.3   Semisimple modules

As we have seen in Section 2.1, simple modules play an important role in studying rings and their representations. As shown in Example 1.1.9, a commutative ring R is a field if and only if it has no proper nontrivial ideals. We can now interpret this proposition as stating that a commutative ring R is a field if and only if $_R R$ is a simple R-module. In this case, when R is a field, any vector space of dimension n is isomorphic to R^n, and so it is a finite direct sum of simple modules. In this section we will consider the noncommutative analog of this result. We are interested in how closely a given module is related to simple modules.

Proposition 2.1.11 (a) shows that a module $_R M$ is simple if and only if $Rm = M$, for all nonzero elements $m \in M$. We should also note that a submodule is minimal if and only if it is simple when considered as a module in its own right.

Definition 2.3.1 *Let M be a left R-module.*

(a) *The sum of all minimal submodules of M is called the* socle *of M, and is denoted by* $\mathrm{Soc}(M)$. *If M has no minimal submodules, then* $\mathrm{Soc}(M) = (0)$.

(b) *The module M is called* semisimple *if it can be expressed as a sum of minimal submodules.*

A homomorphic image of a minimal submodule S_α is either simple or zero, and $f\left(\sum_{\alpha \in I} S_\alpha\right) \subseteq \sum_{\alpha \in I} f(S_\alpha)$. Thus $f(\mathrm{Soc}(M)) \subseteq \mathrm{Soc}(N)$, for any homomorphism $f \in \mathrm{Hom}_R(M, N)$. Furthermore, $\mathrm{Soc}(\mathrm{Soc}(M)) = \mathrm{Soc}(M)$, and so the socle of any module is semisimple. Our first result about the socle is that it is actually a direct sum of minimal modules. In particular, then, a module is semisimple if and only if it is isomorphic to a direct sum of simple modules. Furthermore, a semisimple module $_R M$ behaves like a vector space in that any submodule splits off, or equivalently, that any submodule N has a *complement* N' such that $N + N' = M$ and $N \cap N' = 0$. As might be expected, the proof requires a Zorn's lemma argument.

Theorem 2.3.2 *Any submodule of a semisimple module is semisimple and has a complement that is a direct sum of minimal submodules.*

Proof. Let $_R M$ be a sum of minimal submodules, and let M_0 be any proper submodule of M. If all minimal submodules of M are contained in M_0, then their sum is contained in M_0, forcing $M_0 = M$, a contradiction. Thus there exists a minimal submodule S that is not contained in M_0, and then $S \cap M_0 = (0)$ since S is minimal.

Let I be a set indexing all minimal submodules of M, and consider the subsets of I for which the sum of the submodules indexed is a direct sum that has zero intersection with M_0. The union of any chain J in this collection is again in the collection, since if $m \in \bigcup_{j \in J} M_j$ then $m \in M_j$ for some $j \in J$ and m is a unique sum of finitely many components in minimal submodules. Thus $\bigcup_{j \in J} M_j$ is a direct sum of minimal submodules, and it certainly has zero intersection with M_0. By Zorn's lemma there exists a maximal element in the collection of index sets, and if M_1 is the direct sum of the minimal submodules indexed by this set, then we have $M_0 \cap M_1 = (0)$.

If $M_0 + M_1 \neq M$, then as before there exists a minimal submodule S with $S \cap (M_0 + M_1) = (0)$. It follows immediately that $S \cap M_1 = (0)$, and then we claim that $M_0 \cap (S + M_1) = (0)$, since if $m = s + m_1 \in M_0$ with $s \in S$ and $m_1 \in M_1$, then $s = m - m_1 \in S \cap (M_0 + M_1)$, which implies that $s = 0$ and $m = m_1 = 0$. This contradicts the fact that M_1 is a direct sum of minimal submodules that is maximal with respect to $M_0 \cap M_1 = (0)$. We conclude that $M_0 + M_1 = M$, and we have found a complement for M_0.

Since $M \cong M_0 \oplus M_1$, it follows that M_0 is a homomorphic image of M, and so M_0 is semisimple. This completes the proof. $\square$

Corollary 2.3.3 *The following conditions are equivalent for a module $_R M$:*

(1) M *is semisimple;*

(2) $\mathrm{Soc}(M) = M$*;*

(3) M *is completely reducible;*

(4) M *is isomorphic to a direct sum of simple modules.*

Proof. (1) $\Rightarrow$ (4) We only need to recall that any minimal submodule is simple, and apply Theorem 2.3.2 to the submodule (0) of M.

(4) $\Rightarrow$ (3) Given any submodule N of M, Theorem 2.3.2 shows that we can find a submodule N' such that $N + N' = M$ and $N \cap N' = (0)$. This shows that N is a direct summand of M.

(3) $\Rightarrow$ (2) Assume that M is completely reducible. If there exists an element $m \in M$ that is not in $\mathrm{Soc}(M)$, then by Proposition 2.1.14 there exists a submodule N maximal with respect to containing $\mathrm{Soc}(M)$ but not m, and the factor module M/N has a unique minimal submodule. Since M is completely reducible, there exists a complement N' of N, with $N + N' = M$ and $N \cap N' = (0)$. Then N' contains a minimal submodule since it is isomorphic to M/N. This contradicts the fact that $\mathrm{Soc}(M)$ is the sum of all simple submodules of M, and so $\mathrm{Soc}(M) = M$.

(2) $\Rightarrow$ (1) This follows from the definition of $\mathrm{Soc}(M)$. $\square$

Corollary 2.3.4 *Every vector space over a division ring has a basis.*

Proof. Let D be a division ring, and let V be any left vector space over D. That is, V is a left D-module. If $v \in V$ and $v \neq 0$, the natural mapping from D onto the cyclic submodule Dv must have zero kernel since D has no proper nontrivial left ideals. Thus every cyclic submodule of V is isomorphic to the simple module $_D D$, so each element of V belongs to a minimal submodule, which shows that V is semisimple. Therefore V is isomorphic to a direct sum of simple modules, each isomorphic to D, and this shows that V is a free left D-module. The statement that V is free is equivalent to the statement that V has a basis. $\square$

Our next result is a complete module theoretic characterization of the rings over which every left module is completely reducible. We first need a new definition, which identifies an important class of modules. Corollary 2.2.11 shows that for a module $_R M$ the conditions (1) every one-to-one R-homomorphism into M splits and (2) every onto R-homomorphism out of M splits are each equivalent to the condition that M is completely reducible. Proposition 2.2.13 shows that M is projective if and only if (3) every R-homomorphism onto M splits. There is one condition remaining: (4) every one-to-one R-homomorphism out of M splits. The class of modules with this property turns out to be important in later work, and merits a definition at this stage. For abelian groups, exercises at the end of this section show that this class of modules includes the group **Q** of rational numbers and all of its homomorphic images.

Definition 2.3.5 *Let Q be a left module over the ring R.*

(a) *The module Q is called* injective *if every one-to-one R-homomorphism out of Q splits.*

(b) *The ring R is called* left self-injective *if the module $_R R$ is an injective R-module.*

The next theorem is an example of a situation in which the structure of the ring R can be determined by the structure of its modules (i.e. its representations). Such theorems are one of the goals of module theory, which we can think of as a representation theory for rings.

Theorem 2.3.6 *The following conditions are equivalent for the ring R:*

(1) *R is a direct sum of finitely many minimal left ideals;*

(2) *$_R R$ is a semisimple module;*

(3) *every left R-module is semisimple;*

(4) *every left R-module is projective;*

(5) *every left R-module is injective;*

(6) *every left R-module is completely reducible.*

Proof. (1) $\Rightarrow$ (2) If R is a direct sum of minimal left ideals, then it is certainly semisimple as a left module.

(2) $\Rightarrow$ (3) If $_RR$ is semisimple, then so is every free left R-module, and every homomorphic image of a free left R-module. Since every left R-module is a homomorphic image of a free module, this completes the proof.

(3) $\Rightarrow$ (1) If $_RR$ is semisimple, then Theorem 2.3.2 shows that $_RR$ is a direct sum of minimal left ideals. For any element $r \in R$, we have $r = r \cdot 1$, and since 1 has only finitely many components, each element of R has only finitely many components and thus R is a direct sum of finitely many minimal left ideals.

(3) $\Leftrightarrow$ (6) Corollary 2.3.3 shows that the definitions of semisimple and completely reducible are equivalent.

(4) $\Leftrightarrow$ (6) Assume that every left R-module is completely reducible, and let $_RP$ be any left R-module. To show that P is projective, let M, N be left R-modules, let $p \in \mathrm{Hom}_R(M, N)$ be onto, and let $f \in \mathrm{Hom}_R(P, N)$. Since M is completely reducible by assumption, Corollary 2.2.11 implies that p has a splitting R-homomorphism $i : N \to M$ with $pi = 1_N$. If we let $\widehat{f} = if$, then $p\widehat{f} = p(if) = 1_N f = f$, which completes the proof that P is projective.

Conversely, assume that every left R-module is projective, and let M be any left R-module. If $f : M \to N$ is any onto R-homomorphism, then Proposition 2.2.13 shows that f splits since N is projective, and so Corollary 2.2.11 implies that M is completely reducible.

(5) $\Leftrightarrow$ (6) Assume that every left R-module is completely reducible, and let $_RQ$ be any left R-module. If $f : Q \to M$ is any one-to-one R-homomorphism out of Q, then $f(M)$ is a direct summand of M since M is completely reducible, and so f has a splitting map.

Conversely, if every left R-module is injective, then it follows immediately that every R-submodule is a direct summand, and so every left R-module is completely reducible. $\square$

Corollary 2.3.7 *Let $R = R_1 \oplus \cdots \oplus R_k$ be a direct sum of rings R_j such that for each j, $1 \le j \le k$, there exist a division ring D and a positive integer n with $R_j \cong \mathrm{M}_n(D)$. Then every left R-module is completely reducible.*

Proof. We will verify that R is a direct sum of finitely many minimal left ideals. It suffices to show that the condition holds for each summand R_j, since if A_j is a minimal left ideal of R_j, then $i_j(A_j)$ is a minimal left ideal of R, where $i_j : R_j \to R$ is the natural inclusion mapping. (Note that $R_1 \oplus \cdots \oplus R_k$ can be viewed as a direct sum of left R-modules, as well as a direct sum of rings.)

For any division ring D, in the ring $\mathrm{M}_n(D)$ let A_j be the left ideal generated by the matrix unit e_{jj} which has a 1 in the (j, j)-entry and zeros elsewhere. Then A_j is the set of all $n \times n$ matrices in which all entries except possibly those in the jth column are zero. It follows that $M_n(D) = \sum_{j=1}^{n} A_j$, and so condition (1) of Theorem 2.3.6 will be satisfied if we can show that A_j is a minimal left ideal. Given any nonzero element x of A_j, we can multiply on the left by an appropriate permutation matrix p, so that px has a nonzero entry c in the (j,j)-position. Since D is a division ring, the element c is invertible, and $c^{-1}e_{jj}px = e_{jj}$. This shows that any nonzero element of A_j generates A_j, and thus it follows from Proposition 2.1.11 (a) that A_j is a minimal left ideal. $\square$

The definition of a group ring was given in Example 1.1.8. That definition was given for any finite group, and allowed coefficients to come from any field. With our current knowledge, it is no more difficult to give the definition for any group and any ring of coefficients.

Let R be a ring, and let G be a group. The *group ring* RG is defined to be a free left R-module with the elements of G as a basis. The multiplication on RG is defined by

$$\left(\sum_{g\in G} a_g g\right)\left(\sum_{h\in G} b_h h\right) = \sum_{k\in G} c_k k \, ,$$

where $c_k = \sum_{gh=k} a_g b_h$.

The crucial property of a group ring is that it converts group homomorphisms from G into the group of units of a ring into ring homomorphisms. To be more precise, let S be a ring, let $\phi : G \to S^\times$ be a group homomorphism, and let $\theta : R \to Z(S)$ be any ring homomorphism. (Recall that $S^\times$ denotes the group of invertible elements of S and $Z(S)$ denotes the center of S.) Then there is a unique ring homomorphism $\Phi : RG \to S$ such that $\Phi(g) = \phi(g)$ for all $g \in G$ and $\Phi(r) = \theta(r)$ for all $r \in R$. We have the following associated diagram.

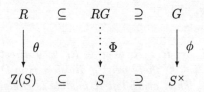

The next theorem gives another important class of rings for which every left module is completely reducible. In fact, certain parts of the representation theory of finite groups are based on Maschke's theorem.

Theorem 2.3.8 (Maschke) *Let G be a finite group and let F be a field such that* char(F) *is not a divisor of* $|G|$. *Then every left FG-module is completely reducible.*

Proof. Let M be a left FG-module. Since F is isomorphic to a subring of FG, its action on M makes M a vector space over F. For any FG-submodule $N \subseteq M$, considering N as a vector space over F yields a splitting map $p : M \to N$ with $pi = 1_N$ for the inclusion $i : N \to M$. We will use the linear transformation p to define an FG-homomorphism that is a splitting map.

Let $|G| = n$, and define $\widehat{p} : M \to N$ by

$$\widehat{p}(x) = \frac{1}{n} \sum_{g \in G} g \cdot p(g^{-1}x)$$

for all $x \in M$. This defines an FG-homomorphism since for all $h \in G$ we have

$$
\begin{aligned}
\widehat{p}(hx) &= \frac{1}{n} \sum_{g \in G} g \cdot p(g^{-1}hx) = \frac{1}{n} \sum_{g \in G} h(h^{-1}g) \cdot p((h^{-1}g)^{-1}x) \\
&= h \cdot \frac{1}{n} \sum_{h^{-1}g \in G} (h^{-1}g) \cdot p((h^{-1}g)^{-1}x) = h \cdot \widehat{p}(x) .
\end{aligned}
$$

Note that we are using the fact that every element of G is uniquely represented in the form $h^{-1}g$, for some $g \in G$. For any $x \in N$ we have $g^{-1}x \in N$, since N is a FG-submodule of M, and so

$$\widehat{p}(x) = \frac{1}{n} \sum_{g \in G} g \cdot p(g^{-1}x) = \frac{1}{n} \sum_{g \in G} g(g^{-1}x) = \frac{1}{n} \sum_{g \in G} x = x .$$

Thus $\widehat{p}i = 1_N$, and so M is a completely reducible FG-module. $\square$

We next give a very useful characterization of injective modules. Condition (2) is usually known as 'Baer's criterion' for injectivity. It can be used (see Exercise 11) to show that over a principal ideal domain the definition of injectivity can be simplified considerably. The statement of condition (4) is dual to that of condition (3) of Proposition 2.2.13, in the sense that mappings are reversed and the notions of one-to-one and onto are interchanged.

Theorem 2.3.9 *For any left R-module Q, the following conditions are equivalent:*

(1) *the module Q is injective;*

(2) *for each left ideal A of R and each R-homomorphism $f : A \to Q$ there exists an element $q \in Q$ such that $f(a) = aq$, for all $a \in A$;*

(3) *for each left ideal A of R and each R-homomorphism $f : A \to Q$ there exists an extension $\widehat{f} : R \to Q$ such that $\widehat{f}(a) = f(a)$ for all $a \in A$;*

(4) *for each one-to-one R-homomorphism* $i : {}_R M \to {}_R N$ *and each R-homomorphism* $f : M \to Q$ *there exists an R-homomorphism* $\widehat{f} : N \to Q$ *such that* $\widehat{f} i = f$.

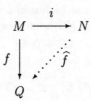

Proof. We first show the equivalence of conditions (2) and (3), and then give a cyclic proof that conditions (1), (2), and (4) are equivalent.

(2) $\Leftrightarrow$ (3) Let A be a left ideal of R, and let $f \in \operatorname{Hom}_R(A, Q)$. If there exists an extension $\widehat{f}$ of f to R, we can let $q = \widehat{f}(1)$, and then for each $a \in A$ we have $f(a) = \widehat{f}(a) = a\widehat{f}(1) = aq$. On the other hand, if there exists $q \in Q$ with $f(a) = aq$ for all $a \in A$, then we can define an extension $\widehat{f}$ of f to R by letting $\widehat{f}(r) = rq$ for all $r \in R$.

(1) $\Rightarrow$ (2) Assume that Q is injective, and let A be a left ideal of R with an R-homomorphism $f : A \to Q$. Let K be the submodule of $R \oplus Q$ defined by $K = \{(a, -f(a)) \mid a \in A\}$, and let P be the factor module $(R \oplus Q)/K$. Define the R-homomorphism $g : Q \to P$ by $g(x) = (0, x) + K$, for all $x \in Q$. (This mapping is just the natural inclusion of Q in $R \oplus Q$ followed by the natural projection onto P.) Then g is one-to-one since if $g(x)$ is zero for some $x \in Q$, then $(0, x) \in K$ and so $(0, x) = (a, -f(a))$ for some $a \in A$. This implies $a = 0$, and therefore $x = -f(a) = 0$.

Because Q is assumed to be injective, the one-to-one function $y : Q \to P$ has a splitting map $h : P \to Q$ with $hg = 1_Q$. Let $q = h((1, 0) + K)$. For any $a \in A$ we have

$$f(a) = [hg](f(a)) = h((0, f(a)) + K) = h((a, 0) + K) - ah((1, 0) + K) = aq$$

since $(a, 0) - (0, f(a)) \in K$. This completes the proof.

(2) $\Rightarrow$ (4) This part of the proof requires an application of Zorn's lemma. Let M, N be left R-modules, let $i : N \to M$ be an R-homomorphism that is one-to-one, and let $f \in \operatorname{Hom}_R(N, Q)$. Since i is one-to-one, it has an inverse defined on the image $\operatorname{Im}(i)$, and so we can view f as an R-homomorphism defined on $\operatorname{Im}(i)$. It suffices to prove that this R-homomorphism can be extended to M, and so without loss of generality we may assume that i is an inclusion mapping and N is a submodule of M.

Consider the set of all ordered pairs $\{(M_\alpha, f_\alpha)\}$ such that M_α is a submodule of M with $N \subseteq M_\alpha \subseteq M$ and $f_\alpha \in \operatorname{Hom}_R(M_\alpha, Q)$ with $f_\alpha(x) = f(x)$ for all

$x \in N$. We define $(M_\alpha, f_\alpha) \le (M_\beta, f_\beta)$ if $M_\alpha \subseteq M_\beta$ and $f_\beta(x) = f_\alpha(x)$ for all $x \in M_\alpha$. Given any chain in this set indexed by $\alpha \in J$, let $M_0 = \bigcup_{\alpha \in J} M_\alpha$, and define $f_0 : M_0 \to Q$ as follows: if $x \in M_0$, then $x \in M_\alpha$ for some α, and so we let $f_0(x) = f_\alpha(x)$. This function is well-defined since if $x \in M_\alpha \cap M_\beta$, then either $(M_\alpha, f_\alpha) \le (M_\beta, f_\beta)$ or $(M_\beta, f_\beta) \le (M_\alpha, f_\alpha)$, and in either case $f_\beta(x) = f_\alpha(x)$. By Zorn's lemma there exists a maximal element in the set $\{(M_\alpha, f_\alpha)\}$, which we denote by (M_1, f_1).

We claim that $M_1 = M$. If there exists $z \in M \setminus M_1$, consider the submodule $M_1 + Rz$, and let $A = \{r \in R \mid rz \in M_1\}$. Define $h : A \to Q$ by $h(r) = f_1(rz)$, for all $r \in A$. This is an R-homomorphism, so there exists $q \in Q$ with $h(r) = rq$, for all $r \in A$. Define $\widehat{f_1} : M_1 + Rz \to Q$ by $\widehat{f_1}(x + rz) = f_1(x) + rq$. This is a well-defined function since if $x_1 + r_1 z = x_2 + r_2 z$, then

$$(r_2 - r_1)q = h(r_2 - r_1) = f_1((r_2 - r_1)z) = f_1(x_1 - x_2) \,,$$

showing that $\widehat{f_1}(x_1 + r_1 z) = \widehat{f_1}(x_2 + r_2 z)$. It is immediate that $\widehat{f_1}$ is an R-homomorphism, and we have

$$(M_1, f_1) < (M_1 + Rz, \widehat{f_1}) \,,$$

a contradiction.

(4) $\Rightarrow$ (1) If $_R N$ is a module for which there is a one-to-one R-homomorphism $i : Q \to N$, then in the statement of condition (4) we can take $M = Q$ and $f = 1_Q$. The extension $\widehat{1_Q}$ of 1_Q to N is a splitting map for i since $\widehat{1_Q} i = 1_Q$. $\square$

Proposition 2.3.10 *Let D be a principal ideal domain, and let Q be the quotient field of R.*

(a) *The module $_D Q$ is injective.*

(b) *If I is any nonzero ideal of D, then the factor ring D/I is a self-injective ring.*

Proof. The proofs of both (a) and (b) will use Baer's criterion for injectivity, as stated in part (2) of Theorem 2.3.9.

(a) Let A be any left ideal of D, and let $f : A \to Q$ be any D-homomorphism. Since D is a principal ideal domain, there exists $d \in D$ with $A = dD$. In Q, for the fraction $q = f(d)/d$ we have $f(d) = dq$, and so for any element $a = rd = dr \in dD$ we have $f(a) = rf(d) = rdq = aq$.

(b) Let I be a nonzero ideal of D, and let R be the ring D/I. Let A be an ideal of R, and let $f : A \to R$ be an R-homomorphism. Since D is a principal ideal domain, we have $I = dD$ and $A = aD/I$ for some $a, d \in D$ with $dD \subseteq aD$, so $d = aq$ for some $q \in D$. If we let $f(a + I) = b + I$ in $R = D/I$, then

$$
\begin{aligned}
(q + I)(b + I) &= (q + I)f(a + I) = f((q + I)(a + I)) \\
&= f(qa + I) = f(d + I) \\
&= f(I) = 0 \, .
\end{aligned}
$$

Thus $qb \in I$, and so $qb = dx$ for some $x \in D$. Substituting $d = aq$ we have $qb = aqx$, and then $b = ax$ since D is an integral domain. For $r \in D$, we have

$$
\begin{aligned}
f(ra + I) &= f((r + I)(a + I)) = (r + I)f(a + I) \\
&= (r + I)(b + I) = rb + I \\
&= r(ax) + I = (ra)x + I \\
&= (ra + I)(x + I) \, .
\end{aligned}
$$

This completes the proof that Baer's criterion is satisfied, and shows that $_R R$ is an injective R-module. $\square$

EXERCISES: SECTION 2.3

1. Let M be a left R-module, and let N be a submodule of M. Show that $\mathrm{Soc}(N) = N \cap \mathrm{Soc}(M)$, and then note that $\mathrm{Soc}(\mathrm{Soc}(M)) = \mathrm{Soc}(M)$.

2. Let A be a finite abelian group. Prove that A is semisimple as a left $\mathbf{Z}$-module if and only if A has no element of order p^2 for any prime p.

3. Prove that any direct sum of semisimple modules is semisimple.

4. A submodule K of the module $_R M$ is said to be *essential* if $K \cap N \neq (0)$ for all nonzero submodules $N \subseteq M$.

   (a) Show that $\mathbf{Z}$ is an essential submodule of $\mathbf{Q}$.

   (b) Referring to Exercise 8 in Section 2.2, show that $\mathbf{Z}_{p^\infty}$ has an essential submodule isomorphic to $\mathbf{Z}_p$.

5. Prove that the socle of a module $_R M$ is the intersection of all essential submodules of M. Conclude that M is semisimple if and only if it has no proper essential submodules.

6. (a) Find the socle of $\mathbf{Q}$, viewed as a $\mathbf{Z}$-module.

   (b) Find the socle of $\mathbf{Q}/\mathbf{Z}$, viewed as a $\mathbf{Z}$-module.

7. A module $_R N$ is said to be an essential extension of $_R M$ if M is an essential submodule of N. Show that M is injective if and only if it has no proper essential extensions.

8. Let $\{Q_\alpha\}_{\alpha \in I}$ be a collection of left R-modules. Prove that the direct product $\prod_{\alpha \in I} Q_\alpha$ is injective if and only if Q_α is injective for each $\alpha \in I$.

9. Let M be a left R-module. We say that M is *divisible* if for each $m \in M$ and each regular element $a \in R$, there exists $x \in M$ such that $m = ax$. Show that any injective module is divisible.

10. Prove the following statements for divisible modules.

(a) Every homomorphic image of a divisible module is divisible.

(b) A direct sum of modules is divisible if and only if each summand is divisible.

(c) A direct product of modules is divisible if and only if every factor is divisible.

11. Let D be a principal ideal domain. Prove that a D-module is injective if and only if it is divisible.

12. Let S be a ring, let $\phi : G \to S^{\times}$ be a group homomorphism, and let $\theta : R \to Z(S)$ be any ring homomorphism. Prove that there is a unique ring homomorphism $\Phi : RG \to S$ such that $\Phi(g) = \phi(g)$ for all $g \in G$ and $\Phi(r) = \theta(r)$ for all $r \in R$.

2.4 Chain conditions

It was Emmy Noether who recognized the importance of the conditions given in the following definition. These conditions are often referred to, respectively, as the ascending chain condition (A.C.C.) and descending chain condition (D.C.C.) for submodules.

Definition 2.4.1 *Let M be a left R-module.*

(a) *The module M is said to be* Noetherian *if every ascending chain*

$$M_1 \subseteq M_2 \subseteq M_3 \subseteq \cdots$$

of submodules of M must terminate after a finite number of steps.

(b) *The module M is said to be* Artinian *if every descending chain*

$$M_1 \supseteq M_2 \supseteq M_3 \supseteq \cdots$$

of submodules of M must terminate after a finite number of steps.

There is a corresponding definition for rings.

Definition 2.4.2 *Let R be any ring.*

(a) *The ring R is said to be* left Noetherian *if the module $_R R$ is Noetherian.*

(b) *The ring R is said to be* left Artinian *if the module $_R R$ is Artinian.*

(c) *If R satisfies the conditions for both right and left ideals, then it is simply said to be* Noetherian *or* Artinian.

Condition (3) of the next proposition is often referred to as the maximum condition for submodules.

Proposition 2.4.3 *The following conditions are equivalent for a module $_R M$:*

(1) *M is Noetherian;*

(2) *every submodule of M is finitely generated;*

(3) *every nonempty set of submodules of M has a maximal element.*

Proof. (1) $\Rightarrow$ (3) If M is Noetherian, then the union of any ascending chain of submodules is actually equal to one of the submodules and therefore provides a least upper bound for the chain. The maximum condition then follows from Zorn's lemma.

(3) $\Rightarrow$ (2) For any submodule N of M, we can apply condition (3) to the set of finitely generated submodules of N, to obtain a maximal element N_0. If $x \in N \setminus N_0$, then $N_0 + Rx$ is a larger finitely generated submodule, so we must have $N = N_0$. Thus N is finitely generated.

(2) $\Rightarrow$ (1) By assumption the union of any ascending chain of submodules must be finitely generated, and since each generator belongs to one of the submodules in the chain, the union is equal to one of these submodules, and hence the chain terminates after finitely many steps. $\square$

There is a similar result for Artinian modules. Our statement does not precisely parallel that of Proposition 2.4.3, since formulating a condition parallel to (2) presents some difficulties. The proof of the proposition is left as an exercise for the reader.

Proposition 2.4.4 *The following conditions are equivalent for a module $_R M$:*

(1) *M is Artinian;*

(2) *every nonempty set of submodules of M has a minimal member.*

Proposition 2.4.5 *The following conditions hold for a module $_R M$ and any submodule N.*

(a) *M is Noetherian if and only if N and M/N are Noetherian.*

(b) *M is Artinian if and only if N and M/N are Artinian.*

Proof. (a) An ascending chain of submodules of N consists of a chain of submodules of M; an ascending chain of submodules of M/N corresponds to a chain of submodules of M that contains N. It is clear that if M is Noetherian, then so are N and M/N.

Conversely, assume that N and M/N are Noetherian, and let M_0 be a submodule of M. Then $M_0 \cap N$ and $M_0/(M_0 \cap N) \cong (M_0 + N)/N$ are both finitely generated, so M_0 is finitely generated.

(b) If M is Artinian, then an argument similar to that in part (a) shows that both N and M/N are Artinian.

Conversely, assume that N and M/N are Artinian. For any descending chain

$$M_1 \supseteq M_2 \supseteq M_3 \supseteq \cdots$$

of submodules of M, consider the descending chain

$$M_1 \cap N \supseteq M_2 \cap N \supseteq M_3 \cap N \supseteq \cdots$$

of submodules of N and the descending chain

$$(M_1 + N)/N \supseteq (M_2 + N)/N \supseteq (M_3 + N)/N \supseteq \cdots$$

of submodules of M/N. By assumption both of these chains become stationary, say $M_n \cap N = M_{n+1} \cap N = \cdots$ and $(M_n + N)/N = (M_{n+1} + N)/N \cdots$. Using Theorem 2.1.12 (c) (the modular law) we have

$$
\begin{aligned}
M_n &= M_n \cap (M_n + N) = M_n \cap (M_{n+1} + N) \\
&= M_{n+1} + (M_n \cap N) = M_{n+1} + (M_{n+1} \cap N) = M_{n+1}.
\end{aligned}
$$

Then $M_{n+1} = M_{n+2} = \cdots$, completing the proof. $\square$

For example, we can apply Proposition 2.4.5 to a direct sum $M_1 \oplus M_2$. If $M_1 \oplus M_2$ is Noetherian, then so is any submodule, so M_1 and M_2 are Noetherian. On the other hand if M_1 and M_2 are Noetherian, then so is $M_1 \oplus M_2$, since $(M_1 \oplus M_2)/M_1$ is isomorphic to M_2. This argument can be extended inductively to any finite direct sum. A similar proof is valid for Artinian modules, and so we obtain the following result.

Corollary 2.4.6 *Let $M = M_1 \oplus \cdots \oplus M_k$ be a direct sum of left R-modules.*

(a) *Then M is Noetherian if and only if M_i is Noetherian for $1 \leq i \leq k$.*

(b) *The module M is Artinian if and only if M_i is Artinian for $1 \leq i \leq k$.*

The next proposition characterizes Noetherian and Artinian rings in terms of their finitely generated modules.

Proposition 2.4.7 *Let R be a ring.*

(a) *The ring R is left Noetherian if and only if every finitely generated left R-module is Noetherian.*

(b) *The ring R is left Artinian if and only if every finitely generated left R-module is Artinian.*

Proof. If the module $_R R$ is Noetherian (or Artinian), then so is every finitely generated free left R-module. Both conditions are preserved in homomorphic images, so the result follows from the fact that every finitely generated module is a homomorphic image of a finitely generated free module.

The converse holds since $_R R$ is finitely generated. $\square$

We now investigate chain conditions for modules over a principal ideal domain. Our goal (achieved in Section 2.7) is to extend the structure theorem for abelian groups to a structure theorem for certain modules over principal ideal domains.

Definition 2.4.8 *Let D be a principal ideal domain, and let M be a D-module. We say that M is a* torsion module *if $\mathrm{Ann}(m) \neq (0)$ for all nonzero elements $m \in M$.*

Proposition 2.4.9 *Let D be a principal ideal domain. Any finitely generated torsion D-module is both Noetherian and Artinian.*

Proof. Since D is a principal ideal domain, every ideal is generated by a single element, so D is certainly Noetherian, and every finitely generated D-module is Noetherian. Let M be a finitely generated torsion module over D, with generators $m_1, m_2, \ldots, m_k$. We must show that M is Artinian.

For each generator m_i we have $\mathrm{Ann}(m_i) \neq (0)$ and $Dm_i \cong D/\mathrm{Ann}(m_i)$. It is sufficient to show that $D/\mathrm{Ann}(m_i)$ is Artinian for $1 \leq i \leq k$, since M is a homomorphic image of the D-module $\bigoplus_{i=1}^{k} (D/\mathrm{Ann}(m_i))$.

If A is any nonzero ideal of D, then $A = aD$ for some nonzero element $a \in D$. Since D is a unique factorization domain, there exist irreducible elements $p_1, p_2, \ldots, p_k$ of D such that $a = p_1^{\alpha_1} p_2^{\alpha_2} \cdots p_k^{\alpha_k}$, for some exponents $\alpha_1, \alpha_2, \ldots, \alpha_k$. The ideals $B = bD$ of D with $A \subseteq B \subseteq D$ correspond to the divisors of a, with the provision that associate elements generate the same ideal. These divisors correspond to the possible combinations of the irreducible elements in the factorization of a, and so (up to associates) there are only finitely many divisors of a. We conclude that the lattice of ideals of D/A is finite, and so every descending chain of submodules of D/A must terminate. $\square$

If R is a commutative Noetherian ring, then the polynomial ring $R[x]$ is also a Noetherian ring. This is usually referred to as the Hilbert basis theorem. We give a proof valid in the noncommutative case.

Theorem 2.4.10 (Hilbert basis theorem) *If R is a left Noetherian ring, then so is the polynomial ring $R[x]$.*

Proof. We will show that if $R[x]$ fails to be Noetherian, then so does R. Let I be an ideal of $R[x]$ that is not finitely generated, and let f_1 be a polynomial of minimal degree in I. We define polynomials $\{f_1, f_2, \ldots\}$ inductively, as follows. If the polynomials f_i have already been chosen, for $1 \leq i < k - 1$, let f_k be a polynomial of minimal degree in I such that f_k does not belong to the ideal $(f_1, \ldots, f_{k-1})$. For $1 \leq i < k$, let $n(i)$ be the degree of f_i, and let $a_i \in R$ be the leading coefficient of f_i. By the choice of f_i we have $n(1) \leq n(2) \leq \cdots$. We will show that the ideals $(a_1) \subset (a_1, a_2) \subset \cdots$ form a strictly ascending chain of ideals that does not become stationary. Suppose that $(a_1, \ldots, a_{k-1}) = (a_1, \ldots, a_k)$. Then $a_k \in (a_1, \ldots, a_{k-1})$, and so there exist elements $r_i \in R$ with $a_k = \sum_{i=1}^{k-1} r_i a_i$. Therefore the polynomial g defined by $g = f_k - \sum_{i=1}^{k-1} r_i x^{n(k)-n(i)} f_i$ belongs to I, but not to the ideal $(f_1, \ldots, f_{k-1})$, and has lower degree than f_k. This contradicts the choice of f_k. $\square$

We can now give some fairly wide classes of examples of Noetherian and Artinian rings. If D is a principal ideal domain, then D is Noetherian since each ideal is generated by a single element. It follows that the polynomial ring $D[x_1, x_2, \ldots, x_n]$ is also Noetherian. If F is a field, then $F[x]/I$ is Artinian, for any nonzero ideal I of $F[x]$, since $F[x]$ is a principal ideal domain. This allows the construction of many interesting examples. Note that $D[x]/I$ need not be Artinian when D is assumed to be a principal ideal domain rather than a field, since $\mathbf{Z}[x]/(x)$ is isomorphic to $\mathbf{Z}$, which is not Artinian.

EXERCISES: SECTION 2.4

1. Show that a module $_RM$ is Artinian if and only if every nonempty set of submodules of M has a minimal member.

2. Let $_RM$ be a completely reducible module. Prove that M is Artinian if and only if M is Noetherian.

3. Prove that if R is a left Noetherian ring, then so is the matrix ring $\mathrm{M}_n(R)$, for any positive integer n.

4. Prove that if R is a left Artinian ring, then so is the matrix ring $\mathrm{M}_n(R)$, for any positive integer n.

5. Let R be the ring $\begin{bmatrix} \mathbf{Q} & 0 \\ \mathbf{R} & \mathbf{R} \end{bmatrix}$ of all lower triangular matrices $\begin{bmatrix} a & 0 \\ b & c \end{bmatrix}$ such that $a \in \mathbf{Q}$ and $b, c \in \mathbf{R}$. Prove that R is left Artinian but not right Artinian.

6. Let R be the ring $\begin{bmatrix} \mathbf{Z} & 0 \\ \mathbf{Q} & \mathbf{Q} \end{bmatrix}$ of all lower triangular matrices $\begin{bmatrix} a & 0 \\ b & c \end{bmatrix}$ such that $a \in \mathbf{Z}$ and $b, c \in \mathbf{Q}$. Prove that R is left Noetherian but not right Noetherian.

7. Prove that if R is a left Noetherian ring, then $ab = 1$ implies $ba = 1$, for all $a, b \in R$.

    Hint: Use the fact that the ascending chain $\operatorname{Ann}(b) \subseteq \operatorname{Ann}(b^2) \subseteq \cdots$ must terminate.

8. Prove that if R is a left Artinian ring, then $ab = 1$ implies $ba = 1$, for all $a, b \in R$.

    Hint: Use the fact that the descending chain $Ra \supseteq Ra^2 \supseteq \cdots$ must terminate.

9. Let R be a commutative ring. Show that R is Noetherian if and only if every prime ideal of R is finitely generated.

    Hint: Use Zorn's lemma to show the existence of an ideal I maximal in the set of all ideals of R that are not finitely generated. Then prove that I is a prime ideal.

10. Use Exercise 9 to show that if R is a commutative Noetherian ring, then so is the ring $R[[x]]$ of all formal power series over R.

2.5 Modules with finite length

In Section 2.3 we considered direct sums of simple modules. We now consider modules that are also built from simple modules, but may have a more complex structure. As the least complicated example, consider the $\mathbf{Z}$-module $\mathbf{Z}_4$. This is not semisimple, since the submodule $2\mathbf{Z}_4$ has no complement. On the other hand, the lattice of submodules includes only $(0) \subseteq 2\mathbf{Z}_4 \subseteq \mathbf{Z}_4$, where $2\mathbf{Z}_4$ and the factor module $\mathbf{Z}_4/2\mathbf{Z}_4$ are both simple, isomorphic to $\mathbf{Z}_2$.

Finite abelian groups are the prototype for the modules that we will study in this section. The fundamental theorem (see Theorem 2.10.3 of [11] or Proposition 7.5.4 of [4]) states that any finite abelian group is isomorphic to a direct sum of cyclic groups of prime power order. As is the case with $\mathbf{Z}_4$, a cyclic group of prime power order cannot be expressed as a sum of proper subgroups, so the fundamental theorem breaks apart a finite group into components that are as small as possible. The subgroups of a cyclic group of prime power order form an ascending chain, for which each factor group is simple. This leads to the notion of a composition series, which the reader may have already met while studying finite groups and their connection with the solution by radicals of polynomial equations.

Definition 2.5.1 *A composition series of length n for a nonzero module M is a chain of $n + 1$ submodules $M = M_0 \supset M_1 \supset \cdots \supset M_n = (0)$ such that M_{i-1}/M_i is a simple module for $i = 1, 2, \ldots, n$.*
These simple modules are called the composition factors *of the series.*

Any module that is both Noetherian and Artinian has a composition series. To see this, if M is Noetherian then it has a maximal submodule M_1, and M_1 is again Noetherian, so it has a maximal submodule M_2, etc. This descending chain must stop after finitely many steps, since M is Artinian, and then each factor M_{i-1}/M_i is simple since M_i is a maximal submodule of M_{i-1}. If G is a finite group, and F is any field, then this condition holds for any finitely generated FG-module M, since M must have finite dimension as a vector space over F.
   If M has a composition series as defined as above and

$$M = N_0 \supseteq N_1 \supseteq \cdots \supseteq N_p = (0)$$

is a second composition series, then the two series are said to be *equivalent* if $n = p$ and there is a permutation σ such that for $i = 0, 1, \ldots, n - 1$

$$M_i/M_{i+1} \cong N_{\sigma(i)}/N_{\sigma(i)+1} .$$

Note that equivalence of the composition series merely means that the two series have the same number of isomorphic factors.
   The Schreier refinement theorem is usually stated as follows: (a) a module $_RM$ is both Noetherian and Artinian if and only if there is an upper bound, say n, on the lengths of chains of submodules of M; (b) in this case every chain of submodules of M can be refined to one of length n in which the composition factors are simple and are uniquely determined up to isomorphism. The proof that we give of the Jordan–Hölder theorem includes a proof of the refinement theorem.

Theorem 2.5.2 (Jordan–Hölder) *If a module M has a composition series, then any other composition series for M is equivalent to it.*

Proof. If M has a composition series, then let $\lambda(M)$ be the minimum length of such a series for M. Note that M is both Noetherian and Artinian since it has a composition series. The proof uses induction on $\lambda(M)$. If $\lambda(M) = 1$, then there is nothing to prove. Assume that $\lambda(M) = n$ and that all composition series are equivalent for any module N with $\lambda(N) < \lambda(M)$. Let

$$M = M_0 \supseteq M_1 \supseteq \cdots \supseteq M_n = (0)$$

be a composition series of minimal length for M and let

$$M = N_0 \supseteq N_1 \supseteq \cdots \supseteq N_p = (0)$$

be a second composition series for M. If $M_1 = N_1$, then the two series are equivalent by the induction hypothesis, so we may assume that $M_1 \neq N_1$. Then since M_1 is a maximal submodule of M, we have $M_1 + N_1 = M$, so we have

$$M/M_1 = (M_1 + N_1)/M_1 \cong N_1/(M_1 \cap N_1)$$

and

$$M/N_1 = (M_1 + N_1)/N_1 \cong M_1/(M_1 \cap N_1) \, .$$

This shows that $M_1 \cap N_1$ is maximal in both M_1 and N_1. Since $M_1 \cap N_1$ is Noetherian and Artinian (as a submodule of M), it has a composition series

$$M_1 \cap N_1 = L_0 \supseteq L_1 \supseteq \cdots \supseteq L_k = (0)$$

which then extends to composition series for both M_1 and N_1. The composition series

$$M_1 \supseteq M_1 \cap N_1 \supseteq L_1 \supseteq \cdots \supseteq L_k = (0)$$

and

$$M_1 \supseteq M_2 \supseteq M_3 \supseteq \cdots \supseteq M_n = (0)$$

must be equivalent by the induction hypothesis, since $\lambda(M_1) < n$. This shows that $k = n - 2$, and so it follows that N_1 has a composition series of length $n - 1$, beginning with $M_1 \cap N_1$. Therefore $\lambda(N_1) < n$, and so by the induction hypothesis the composition series

$$N_1 \supseteq N_2 \supseteq \cdots \supseteq N_p = (0)$$

and

$$N_1 \supseteq M_1 \cap N_1 \supseteq L_1 \supseteq \cdots \supseteq L_k = (0)$$

are equivalent. Finally, as we have already shown, $M/M_1 \cong N_1/L_0$ and $M/N_1 \cong M_1/L_0$, and so the two original composition series for M must be equivalent since they are each equivalent to

$$M = N_0 \supseteq N_1 \supseteq M_1 \cap N_1 \supseteq L_1 \supseteq \cdots \supseteq L_k = (0) \, .$$

This completes the proof. $\square$

As an immediate consequence of the Jordan–Hölder theorem, if a module ${}_R M$ has a composition series, then all composition series for M must have the same length. This justifies the following definition.

Definition 2.5.3 *Let M be a left R-module. We say that M has length n if M has a composition series $M = M_0 \supset M_1 \supset \cdots \supset M_n = (0)$. The length of M will be denoted by $\lambda(M)$. In this case we may also say simply that M has finite length.*

Since any ascending chain of submodules can be refined to a composition series, $\lambda(M)$ gives a uniform bound on the number of terms in any properly ascending chain of submodules. We also note that if M_1 and M_2 have finite length, then $\lambda(M_1 \oplus M_2) = \lambda(M_1) + \lambda(M_2)$.

Proposition 2.5.4 *A module has finite length if and only if it is both Artinian and Noetherian.*

Proof. If ${}_R M$ has finite length, then by the Jordan–Hölder theorem any chain of submodules has length at most $\lambda(M)$.

Conversely, if M is both Artinian and Noetherian, then it contains a minimal submodule M_1, a submodule M_2 minimal in the set of submodules properly containing M_1, and so on. The chain $M_1 \subset M_2 \subset \cdots$ must terminate, since M is Noetherian, and we have constructed a composition series for M. $\square$

Definition 2.5.5 *A nonzero module ${}_R M$ is said to be* indecomposable *if its only direct summands are* (0) *and* M.

As our first example, we note that $\mathbf{Z}$ is indecomposable as a module over itself, since the intersection of any two nonzero ideals is again nonzero. To give additional examples of indecomposable $\mathbf{Z}$-modules, we again recall the fundamental theorem of finite abelian groups. Since any finite abelian group is isomorphic to a direct sum of cyclic groups of prime power order, we see that a finite $\mathbf{Z}$-module is indecomposable if and only if it is isomorphic to $\mathbf{Z}_n$, where $n = p^\alpha$ for some prime p.

After showing that any module of finite length can be decomposed into a finite number of indecomposable summands, we will obtain some information about endomorphisms of indecomposable modules of finite length. This allows us to prove a uniqueness theorem concerning the decomposition.

Proposition 2.5.6 *If ${}_R M$ has finite length, then there exist finitely many indecomposable submodules $M_1, M_2, \ldots, M_n$ such that $M = M_1 \oplus M_2 \oplus \cdots \oplus M_n$.*

Proof. The proof is by induction on the length of M. If M has length 1, then it is a simple module, and hence indecomposable. Now assume that M has length n, and that the result holds for all modules of smaller length. If M itself is indecomposable, then we are done. If not, we can write M as a direct sum of two submodules, say $M = N_1 \oplus N_2$. Since the lengths of N_1 and N_2 are less than that of M, each can be written as a direct sum of indecomposable submodules, and so the same is true for M, completing the proof. $\square$

If M has finite length and we write it as a direct sum of indecomposable submodules, say $M = M_1 \oplus M_2 \oplus \cdots \oplus M_n$, then we can find a composition series for each of the submodules M_j. We can use these series to construct a composition series for M that includes $(0) \subseteq M_1 \subseteq M_1 \oplus M_2 \subseteq \cdots$. Thus the number of indecomposable summands must be less than the length of M.

Lemma 2.5.7 (Fitting) *Let M be a module with length n, and let f be an endomorphism of M. Then*

$$M = \text{Im}(f^n) \oplus \ker(f^n) \,.$$

Proof. Since M has length n, the descending chain

$$\text{Im}(f) \supseteq \text{Im}(f^2) \supseteq \cdots$$

must terminate after at most n steps, and so we have $\text{Im}(f^n) = \text{Im}(f^{2n})$. Given $x \in M$ there exists $z \in M$ such that $f^n(x) = f^{2n}(z)$. It is then clear that

$$x = f^n(z) + (x - f^n(z)) \in \text{Im}(f^n) + \ker(f^n) \,.$$

It follows that $M = \text{Im}(f^n) + \ker(f^n)$.

We next consider the ascending chain

$$\ker(f) \subseteq \ker(f^2) \subseteq \cdots \,,$$

which must terminate after at most n steps, and so we have $\ker(f^n) = \ker(f^{2n})$. If $x \in \text{Im}(f^n) \cap \ker(f^n)$, then $x = f^n(z)$ for some $z \in M$, and $f^{2n}(z) = f^n(x) = 0$. Thus we have $z \in \ker(f^{2n}) = \ker(f^n)$, and it follows that $x = f^n(z) = 0$. This shows that $\text{Im}(f^n) \cap \ker(f^n) = (0)$, and therefore, from the first part, $M = \text{Im}(f^n) \oplus \ker(f^n)$. $\square$

Proposition 2.5.8 *Let M be an indecomposable module of finite length. Then for any endomorphism f of M the following conditions are equivalent:*

(1) *f is one-to-one;*

(2) *f is onto;*

(3) *f is an automorphism;*

(4) *f is not nilpotent.*

Proof. Assume that $\lambda(M) = n$, so that $M = \text{Im}(f^n) \oplus \ker(f^n)$, as guaranteed by Lemma 2.5.7. Note that $\text{Im}(f^n) \subseteq \text{Im}(f)$ and $\ker(f) \subseteq \ker(f^n)$.

If f is an automorphism, then each of the other conditions clearly holds.

If f is one-to-one, then $\ker(f^n) = (0)$, which means that $\text{Im}(f^n) = M$, and this implies that f is onto and hence an automorphism.

If f is onto, then $\text{Im}(f^n) = M$, so $\ker(f^n) = (0)$, which forces $\ker(f) = (0)$, and so f is one-to-one and hence an automorphism.

If f is not nilpotent, then $\text{Im}(f^n) \neq (0)$, and so from Fitting's lemma and the assumption that M is indecomposable, it follows that $\ker(f^n) = (0)$, and so f is one-to-one. By the previous argument it must then be an automorphism. $\square$

Before proving the next proposition, we recall an observation made in Chapter 1 (following Definition 1.1.5). If $a \in R$ is nilpotent, say $a^n = 0$, then $(1 - a)(1 + a + a^2 + \cdots + a^{n-1}) = 1 - a^n = 1$, and thus $1 - a$ is a unit.

Proposition 2.5.9 *Let M be an indecomposable module of finite length, and let f_1, f_2 be endomorphisms of M. If $f_1 + f_2$ is an automorphism, then either f_1 or f_2 is an automorphism.*

Proof. If $f_1 + f_2$ is an automorphism, let g be its inverse, so that $1 = f_1 g + f_2 g$. Then $f_1 g$ is either nilpotent or an automorphism. In the first case, $f_2 g = 1 - f_1 g$ is then an automorphism, which shows that f_2 is an automorphism. In the second case, if $f_1 g$ is an automorphism, then so is f_1. $\square$

Lemma 2.5.10 *Let X_1, X_2, Y_1, Y_2 be left R-modules, and let $f : X_1 \oplus X_2 \to Y_1 \oplus Y_2$ be an isomorphism. Let $i_1 : X_1 \to X_1 \oplus X_2$ and $i_2 : X_2 \to X_1 \oplus X_2$ be the natural inclusion maps, and let $p_1 : Y_1 \oplus Y_2 \to Y_1$ and $p_2 : Y_1 \oplus Y_2 \to Y_2$ be the natural projections. If $p_1 f i_1 : X_1 \to Y_1$ is an isomorphism, then X_2 is isomorphic to Y_2.*

Proof. We note that any R-homomorphism $g : X_1 \oplus X_2 \to Y_1 \oplus Y_2$ can be written in matrix form in the following way, where $g_{jk} = p_j g\, i_k$.

$$g\left(\begin{bmatrix} x_1 \\ x_2 \end{bmatrix} \right) = \begin{bmatrix} g_{11} & g_{12} \\ g_{21} & g_{22} \end{bmatrix} \begin{bmatrix} x_1 \\ x_2 \end{bmatrix}$$

Thus we define $f_{jk} = p_j f\, i_k$, for $j = 1, 2$ and $k = 1, 2$.

By assumption, in the matrix decomposition of f, the induced mapping $f_{11} : X_1 \to Y_1$ is an isomorphism. The matrix $\begin{bmatrix} 1 & 0 \\ -f_{21}f_{11}^{-1} & 1 \end{bmatrix}$ has an inverse $\begin{bmatrix} 1 & 0 \\ f_{21}f_{11}^{-1} & 1 \end{bmatrix}$. Thus if $f_{21} \neq 0$ we can multiply f on the left as follows, and the resulting homomorphism is still an isomorphism from $X_1 \oplus X_2$ into $Y_1 \oplus Y_2$.

$$\begin{bmatrix} 1 & 0 \\ -f_{21}f_{11}^{-1} & 1 \end{bmatrix} \begin{bmatrix} f_{11} & f_{12} \\ f_{21} & f_{22} \end{bmatrix} = \begin{bmatrix} f_{11} & f_{12} \\ 0 & -f_{21}f_{11}^{-1}f_{12} + f_{22} \end{bmatrix}$$

Therefore without loss of generality we can assume that $f_{21} = 0$.

By assumption f has an inverse g, which we can express in matrix form by using inclusions $Y_j \to Y_1 \oplus Y_2$ and projections $X_1 \oplus X_2 \to X_k$, and so we have the following equations:

$$\begin{bmatrix} g_{11} & g_{12} \\ g_{21} & g_{22} \end{bmatrix} \begin{bmatrix} f_{11} & f_{12} \\ 0 & f_{22} \end{bmatrix} = \begin{bmatrix} 1 & 0 \\ 0 & 1 \end{bmatrix},$$

$$\begin{bmatrix} f_{11} & f_{12} \\ 0 & f_{22} \end{bmatrix} \begin{bmatrix} g_{11} & g_{12} \\ g_{21} & g_{22} \end{bmatrix} = \begin{bmatrix} 1 & 0 \\ 0 & 1 \end{bmatrix}.$$

From these equations we first obtain $g_{21}f_{11} = 0$, which implies that $g_{21} = 0$ since f_{11} is invertible. It then follows that $g_{22}f_{22} = 1$ and $f_{22}g_{22} = 1$, proving the desired conclusion that $X_2 \cong Y_2$. $\quad\square$

Theorem 2.5.11 (Krull–Schmidt) *Let $\{X_j\}_{j=1}^m$ and $\{Y_k\}_{k=1}^n$ be sets of indecomposable left R-modules of finite length. If*

$$X_1 \oplus \cdots \oplus X_m \cong Y_1 \oplus \cdots \oplus Y_n,$$

then $m = n$ and there exists a permutation $\pi \in \mathcal{S}_n$ with $X_j \cong Y_{\pi(j)}$, for $1 \le j \le m$.

Proof. The proof is by induction on m. If $m = 1$, then since X_1 is indecomposable it cannot be isomorphic to a direct sum of two or more nonzero modules, so $n = 1$ and $X_1 \cong Y_1$.

In the general case, assume that $f : X_1 \oplus \cdots \oplus X_m \to Y_1 \oplus \cdots \oplus Y_n$ is an isomorphism. We first claim that X_1 is isomorphic to some Y_k. To show this, let $i : X_1 \to X_1 \oplus \cdots \oplus X_m$ be the inclusion mapping, and let $p : X_1 \oplus \cdots \oplus X_m \to X_1$ be the canonical projection, which is the splitting map for i. For $1 \le k \le n$ let $i_k : Y_k \to Y_1 \oplus \cdots \oplus Y_n$ be the inclusion mapping, with splitting map p_k. In $\mathrm{End}_R(X_1)$ we have

$$1 = pi = pf^{-1}fi = pf^{-1}(\textstyle\sum_{k=1}^n i_k p_k)fi = \textstyle\sum_{k=1}^n pf^{-1}i_k p_k fi.$$

By Proposition 2.5.9, one of the terms in this sum must be invertible, say $g = pf^{-1}i_t p_t fi$. Then $p_t fi : X_1 \to Y_t$ has a splitting map $g^{-1}pf^{-1}i_t$, and so X_1 is isomorphic to a direct summand of Y_t. Since Y_t is indecomposable, this implies that $X_1 \cong Y_t$. By permuting the indices, we can assume that X_1 is isomorphic to Y_1.

Since the conditions of Lemma 2.5.10 are satisfied, it follows that we have $X_2 \oplus \cdots \oplus X_m \cong Y_2 \oplus \cdots \oplus Y_n$. Applying the induction hypothesis completes the proof. $\quad\square$

EXERCISES: SECTION 2.5

1. Find a composition series for the **Z**-module $\mathbf{Z}_{360}$.

2. Let F be a field, and let R be the set of all lower triangular 2×2 matrices over F, denoted by $\begin{bmatrix} F & 0 \\ F & F \end{bmatrix}$. Show that the left module $_RR$ and the right module R_R have composition series, but that neither module is completely reducible.

3. For the ring R in Exercise 2, find the socle of $_RR$ and the socle of R_R, and show that they are not equal.

4. Let $_RM$ be a Noetherian module, and let $f \in \mathrm{End}_R(M)$. Prove that if f is onto, then it must be one-to-one.

5. Let R be a left Noetherian ring. Use Exercise 4 to give another proof of Exercise 7 in Section 2.4.

   Hint: For $b \in R$, define $\rho_b : R \to R$ by $\rho_b(x) = xb$, for all $x \in R$. Show that if $ab = 1$, then $\mathrm{Ann}(b) = (0)$, and hence $ba - 1 = 0$.

6. Let $_RM$ be an Artinian module, and let $f \in \mathrm{End}_R(M)$. Prove that if f is one-to-one, then it must be onto.

7. Let R be a left Artinian ring. Use Exercise 6 to give another proof of Exercise 8 in Section 2.4.

   Hint: For $a \in R$, define $\rho_a : R \to R$ by $\rho_a(x) = xa$, for all $x \in R$. Show that if $ab = 1$, then there exists $x \in R$ with $xa = 1$, and hence $ba = 1$.

8. Prove that in a left Artinian ring every regular element is invertible.

2.6 Tensor products

In this section we will study the tensor product of two modules. This construction is extremely useful, although rather abstract. For example, we have already observed that if $\phi : R \to S$ is a ring homomorphism, then any left S-module can be made into a left R-module. We will show in Corollary 2.6.10 that using the tensor product it is possible to take any left R-module and construct a corresponding left S-module.

We need to begin with the definition of a bilinear function. Given modules M_R and $_RN$ over a ring R and an abelian group A, a function $\beta : M \times N \to A$ is said to be R-*bilinear* if

   (i) $\beta(x_1 + x_2, y) = \beta(x_1, y) + \beta(x_2, y)$,
   (ii) $\beta(x, y_1 + y_2) = \beta(x, y_1) + \beta(x, y_2)$,
   (iii) $\beta(xr, y) = \beta(x, ry)$

for all $x, x_1, x_2 \in M$, $y, y_1, y_2 \in N$ and $r \in R$.

Example 2.6.1 (A vector product)

This example gives the bilinear function involved in the definition
of the tensor product of two matrices. Let F be a field, let V be an
m-dimensional vector space over F, and let W be an n-dimensional
vector space over F. We will consider the elements of V as columns,
and the elements of W as rows. This yields a product for elements
$x \in V$ and $y \in W$.

$$
xy = \begin{bmatrix} x_1 \\ x_2 \\ \vdots \\ x_m \end{bmatrix} \begin{bmatrix} y_1 & y_2 & \cdots & y_n \end{bmatrix} = \begin{bmatrix} x_1y_1 & x_1y_2 & \cdots & x_1y_n \\ x_2y_1 & x_2y_2 & \cdots & x_2y_n \\ \vdots & \vdots & & \vdots \\ x_my_1 & x_my_2 & \cdots & x_my_n \end{bmatrix}
$$

Let $T(V,W)$ be the additive group of all $m \times n$ matrices over F, and
define $\tau : V \times W \to T(V,W)$ by $\tau(x,y) = xy$, for all $x \in V$ and $y \in$
W. The distributive and associative laws for matrix multiplication
imply that τ is F-bilinear.

The tensor product of M_R and $_RN$, which we now define, provides a construction that converts R-bilinear functions into homomorphisms of abelian
groups.

Definition 2.6.1 *A* tensor product *of the modules M_R and $_RN$ is an abelian
group $M \otimes_R N$, together with an R-bilinear map $\tau : M \times N \to M \otimes_R N$ such
that for any abelian group A and any R-bilinear map $\beta : M \times N \to A$ there
exists a unique $\mathbf{Z}$-homomorphism $f : M \otimes_R N \to A$ such that $f\tau = \beta$.
For $x \subset M$, $y \in N$ the image $\tau(x,y)$ is denoted by $x \otimes y$.*

The following diagram is associated with the definition of $M \otimes_R N$. In the
diagram, β is any R-bilinear map, and f is an R-homomorphism with $f\tau = \beta$.

$$
\begin{array}{ccc}
M \times N & \xrightarrow{\ \tau\ } & M \otimes_R N \\
& \beta \searrow & \big\downarrow f \\
& & A
\end{array}
$$

Our first result shows that tensor products are unique up to isomorphism.
This proof is easier than the proof that the tensor product exists, so we give it
first.

Proposition 2.6.2 *Let M_R and $_RN$ be modules over the ring R. The tensor product $M \otimes_R N$ is unique up to isomorphism, if it exists.*

Proof. Assume that the tensor product $M \otimes_R N$ exists, with an R-bilinear function $\tau : M \times N \to M \otimes_R N$. Also assume that T is an abelian group with an R-bilinear function $\sigma : M \times N \to T$ which satisfies the conditions of the definition of the tensor product of M and N. Applying the definition to $M \otimes_R N$ and T in turn, first with $\beta = \sigma$ and then $\beta = \tau$, gives the existence of group homomorphisms $f : M \otimes_R N \to T$ and $g : T \to M \otimes_R N$ such that $f\tau = \sigma$ and $g\sigma = \tau$. We have the following diagrams.

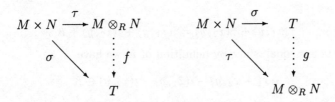

We now have $(fg)\sigma = f(g\sigma) = f\tau = \sigma$ and $(gf)\tau = g(f\tau) = g\sigma = \tau$, and so we apply the definition again, as indicated in the following diagrams.

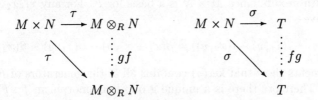

Since the respective identity functions also satisfy the requirements of the diagram in the definition of the tensor product, the uniqueness condition in the definition implies that $gf = 1_{M \otimes N}$ and $fg = 1_T$. This proves that T is isomorphic to $M \otimes_R N$. $\square$

Proposition 2.6.3 *The tensor product $M \otimes_R N$ exists, for any modules M_R and $_RN$.*

Proof. The tensor product of M and N is constructed as follows. Let F be the free **Z**-module with generators $\{(x,y) \mid x \in M \text{ and } y \in N\}$, and let $i : M \times N \to F$ be the inclusion mapping. Let K be the submodule of F generated by all elements of the form

$$(x_1 + x_2, y) - (x_1, y) - (x_2, y) ,$$

$$(x, y_1 + y_2) - (x, y_1) - (x, y_2) ,$$

or

$$(xr, y) - (x, ry) ,$$

where $x, x_1, x_2 \in M$, $y, y_1, y_2 \in N$, and $r \in R$. Let $\pi : F \to F/K$ be the natural projection. If we let $\tau : M \times N \to F/K$ be the composition πi, then F/K satisfies the definition of $M \otimes_R N$.

To show that F/K is the required tensor product, we first need to show that τ is an R-bilinear function. For $x_1, x_2 \in M$ and $y \in N$, we have

$$\tau(x_1 + x_2, y) = i(x_1 + x_2, y) + K$$

and

$$\tau(x_1, y) + \tau(x_2, y) = i(x_1, y) + i(x_2, y) + K .$$

These cosets are equal, since by definition of K we have

$$i(x_1 + x_2, y) - i(x_1, y) - i(x_2, y) \in K .$$

The remaining conditions are verified in the same way, so we conclude that τ is an R-bilinear function.

Let A be any abelian group, and let $\beta : M \times N \to A$ be an R-bilinear function. Since F is a free $\mathbf{Z}$-module, there is a unique group homomorphism $g : F \to A$ with $gi = \beta$, since $M \times N$ is a basis for F. For any $x_1, x_2 \in M$ and $y \in N$, we have

$$g(i(x_1 + x_2, y) - i(x_1, y) - i(x_2, y)) = \beta(x_1 + x_2, y) - \beta(x_1, y) - \beta(x_2, y) = 0 .$$

Similar arguments show that $\ker(g)$ contains all of the generators of K, and so $K \subseteq \ker(g)$. Therefore there is a unique group homomorphism $f : F/K \to A$ with $f\pi = g$. It follows that $f\tau = f(\pi i) = (f\pi)i = gi = \beta$. $\square$

Let M_R and $_R N$ be modules. From the construction of $M \otimes_R N$ as a factor group F/K of the free $\mathbf{Z}$-module generated by $M \times N$, we see that each element of $M \otimes_R N$ can be expressed as a finite sum of the form

$$\sum_{j=1}^{k} n_j i(x_j, y_j) + K$$

for elements $x_j \in M$, $y_j \in N$, and $n_j \in \mathbf{Z}$. We can rewrite $n_j i(x_j, y_j)$ as $i(n_j x_j, y_j)$, and then $n_j x_j \in M$, so in fact we can omit the integers n_j. With the notation $x_j \otimes y_j = \tau(x_j, y_j) = i(x_j, y_j) + K$, this generic expression for elements of $M \otimes_R N$ can be simplified to

$$\sum_{j=1}^{k} x_j \otimes y_j .$$

It is important to note that such an expression is not unique, since it is determined by a choice of representatives for the coset of K to which it belongs.

For $x, x_1, x_2 \in M$, $y, y_1, y_2 \in N$, and $r \in R$, we can write down the following identities.

$$(x_1+x_2)\otimes y = x_1\otimes y+x_2\otimes y \qquad x\otimes(y_1+y_2) = x\otimes y_1+x\otimes y_2 \qquad xr\otimes y = x\otimes ry$$

Since $x \otimes 0 = x \otimes (0+0) = x \otimes 0 + x \otimes 0$, it follows that $x \otimes 0 = 0$ for all $x \in M$.

Example 2.6.2 ($\mathbf{Q} \otimes_Z A = (0)$ if A is finite)

Let A be any finite abelian group. We will show that $\mathbf{Q} \otimes_Z A = (0)$. If $q \in \mathbf{Q}$ and $a \in A$, then a has finite order, and so there exists a positive integer n with $na = 0$. Since we have taken the tensor product over $\mathbf{Z}$, we can freely move integers across the tensor product symbol $\otimes$. Thus

$$\begin{aligned} q \otimes a &= (qn^{-1})n \otimes a = qn^{-1} \otimes na \\ &= qn^{-1} \otimes 0 = 0 \, , \end{aligned}$$

showing that every element of $\mathbf{Q} \otimes_Z A$ must be zero.

Proposition 2.6.4 *Let M, M' be right R-modules, let N, N' be left R-modules, and let $f \in \mathrm{Hom}(M_R, M'_R)$ and $g \in \mathrm{Hom}(\,_R N, \,_R N')$.*

(a) There is a unique $\mathbf{Z}$-homomorphism $f \otimes g : M \otimes_R N \to M' \otimes_R N'$ with $(f \otimes g)(x \otimes y) = f(x) \otimes g(y)$ for all $x \in M$, $y \in N$.

(b) If f and g are onto, then $f \otimes g$ is onto, and $\ker(f \otimes g)$ is generated by all elements of the form $x \otimes y$ such that either $x \in \ker(f)$ or $y \in \ker(g)$.

Proof. (a) Let $\tau : M \times N \to M \otimes_R N$ and $\tau' : M' \times N' \to M' \otimes_R N'$ be the R-bilinear functions that define the respective tensor products. Define $\beta : M \times N \to M' \otimes_R N'$ by $\beta(x,y) = \tau'(f(x), g(y))$, for all $x \in M$, $y \in N$. Then

$$\begin{aligned} \beta(x_1 + x_2, y) &= \tau'(f(x_1 + x_2), g(y)) = \tau'(f(x_1) + f(x_2), g(y)) \\ &= \tau'(f(x_1), g(y)) + \tau'(f(x_2), g(y)) \\ &= \beta(x_1, y) + \beta(x_2, y) \end{aligned}$$

and a similar argument shows that $\beta(x, y_1 + y_2) = \beta(x, y_1) + \beta(x, y_2)$, for all $x, x_1, x_2 \in M$ and $y, y_1, y_2 \in N$. Furthermore, for any $r \in R$ we have

$$\begin{aligned} \beta(xr, y) &= \tau'(f(xr), g(y)) = \tau'(f(x)r, g(y)) \\ &= \tau'(f(x), rg(y)) = \tau'(f(x), g(ry)) \\ &= \beta(x, ry) \, . \end{aligned}$$

By the definition of $M \otimes_R N$, there exists a unique group homomorphism $h :$ $M \otimes_R N \to M' \otimes N'$ with $h\tau = \beta$, and so

$$h(x \otimes y) = h\tau(x,y) = \beta(x,y) = \tau'(f(x), g(y)) = f(x) \otimes g(y)$$

for all $x \in M$ and $y \in N$. We use the notation $h = f \otimes g$.

(b) It is clear that since f and g are onto the generators of $M' \otimes_R N'$ belong to $\text{Im}(f \otimes g)$, and so $f \otimes g$ is onto.

Let K be the subgroup of $M \otimes_R N$ generated by all elements of the form $x \otimes y$ such that either $x \in \ker(f)$ or $y \in \ker(g)$, and let $p : M \otimes_R N \to (M \otimes_R N)/K$ be the natural projection. Define a bilinear mapping $\gamma : M' \times N' \to (M \otimes_R N)/K$ as follows: given $(x', y') \in M' \times N'$, there exist $x \in M$ and $y \in N$ with $f(x) = x'$ and $g(y) = y'$; define $\gamma(x', y') = (x \otimes y) + K$. To show that γ is a well-defined function, suppose that we also have $x_1 \in M$ and $y_1 \in N$ with $f(x_1) = x'$ and $g(y_1) = y'$. Then

$$
\begin{aligned}
x \otimes y - x_1 \otimes y_1 &= x \otimes y - x_1 \otimes y + x_1 \otimes y - x_1 \otimes y_1 \\
&= (x - x_1) \otimes y + x_1 \otimes (y - y_1) \,,
\end{aligned}
$$

and since $x - x_1 \in \ker(f)$ and $y - y_1 \in \ker(g)$, we have $x \otimes y - x_1 \otimes y_1 \in K$. The mapping γ is R-bilinear because of the properties of the tensor product.

From the definition of $M' \otimes_R N'$ there exists a group homomorphism $t :$ $M' \otimes_R N' \to (M \otimes_R N)/K$ with $t(x' \otimes y') = \gamma((x', y'))$ for all $x' \in M'$ and $y' \in N'$. For any $x \otimes y \in M \otimes_R N$, we have

$$t(f \otimes g)(x \otimes y) = t(f(x) \otimes g(y)) = x \otimes y + K \,,$$

and so $t(f \otimes g) = p$. It is clear that $K \subseteq \ker(f \otimes g)$, and so there exists a group homomorphism $k : (M \otimes_R N)/K \to M' \otimes_R N'$ with $kp = f \otimes g$. Now $tkp = t(f \otimes g) = 1p$, which implies that $kt = 1$, since p is onto. Thus k is one-to-one, and it is also onto, since $f \otimes g$ is onto. We conclude that k is an isomorphism, which implies that $K = \ker(f \otimes g)$. $\square$

The following diagram shows the relationship between the various homomorphisms defined in the proof of part (b) of the preceding proposition.

Proposition 2.6.5 *Let M_R be a right R-module, and let $\{N_\alpha\}_{\alpha \in I}$ be a collection of left R-modules. Then*

$$M \otimes_R \left(\bigoplus_{\alpha \in I} N_\alpha \right) \cong \bigoplus_{\alpha \in I} (M \otimes_R N_\alpha) \,.$$

Proof. For each $\alpha \in I$, let $i_\alpha : N_\alpha \to \bigoplus_{\alpha \in I} N_\alpha$ and $p_\alpha : \bigoplus_{\alpha \in I} N_\alpha \to N_\alpha$ be the canonical injection and projection mappings. Let

$$\tau : M \times \left(\bigoplus_{\alpha \in I} N_\alpha \right) \to M \otimes_R \left(\bigoplus_{\alpha \in I} N_\alpha \right)$$

be the defining R-bilinear function. Define

$$\beta : M \times \left(\bigoplus_{\alpha \in I} N_\alpha \right) \to \bigoplus_{\alpha \in I} (M \otimes_R N_\alpha)$$

by $\beta(x,y) = \sum_{\alpha \in I} x \otimes p_\alpha(y)$, for $x \in M$ and $y \in \bigoplus_{\alpha \in I} N_\alpha$. This sum makes sense because $p_\alpha(y) = 0$ for all but finitely many $\alpha \in I$. It is not difficult to check that β is an R-bilinear mapping, and so there exists a unique group homomorphism

$$f : M \otimes_R \left(\bigoplus_{\alpha \in I} N_\alpha \right) \to \bigoplus_{\alpha \in I} (M \otimes_R N_\alpha)$$

with $f\tau = \beta$. Given any $x \in M$ and any $y_\alpha \in N_\alpha$, we have $\beta(x, i_\alpha(y_\alpha)) = x \otimes y_\alpha$, and so f must be onto since $\text{Im}(f)$ contains a subset which generates $\bigoplus_{\alpha \in I} (M \otimes_R N_\alpha)$.

To show that f is one-to-one, we construct a function in the opposite direction. For each $\alpha \in I$, define

$$\gamma_\alpha : M \times N_\alpha \to M \otimes_R \left(\bigoplus_{\alpha \in I} N_\alpha \right)$$

by $\gamma_\alpha(x, y_\alpha) = x \otimes i_\alpha(y_\alpha)$. It can be checked that γ_α is R-bilinear, and so there exists a group homomorphism

$$g_\alpha : M \otimes N_\alpha \to M \otimes_R \left(\bigoplus_{\alpha \in I} N_\alpha \right)$$

such that $g_\alpha(x \otimes y_\alpha) = x \otimes i_\alpha(y_\alpha)$ for all $x \in M$, $y_\alpha \in N_\alpha$. By definition of the direct sum, the group homomorphisms $\{g_\alpha\}_{\alpha \in I}$ determine a unique group homomorphism

$$g : \bigoplus_{\alpha \in I} (M \otimes_R N_\alpha) \to M \otimes_R \left(\bigoplus_{\alpha \in I} N_\alpha \right) \,.$$

For any $x \in N$ and $y \in \bigoplus_{\alpha \in I} N_\alpha$, we have

$$
\begin{aligned}
gf(x \otimes y) &= g \left(\sum_{\alpha \in I} x \otimes p_\alpha(y) \right) = \sum_{\alpha \in I} g(x \otimes p_\alpha(y)) \\
&= \sum_{\alpha \in I} x \otimes i_\alpha p_\alpha(y) = x \otimes \left(\sum_{\alpha \in I} i_\alpha p_\alpha(y) \right) \\
&= x \otimes y \,.
\end{aligned}
$$

It follows that f is one-to-one, since the composition gf is the identity function on the generators of $M \otimes_R \left(\bigoplus_{\alpha \in I} N_\alpha \right)$. This completes the proof that f is an isomorphism of abelian groups. $\square$

A similar result holds in the first variable, so if $\{M_\alpha\}_{\alpha \in I}$ is a collection of right R-modules and N is a left R-module, then

$$\left(\bigoplus_{\alpha \in I} M_\alpha \right) \otimes_R N \cong \bigoplus_{\alpha \in I} (M_\alpha \otimes_R N) .$$

Up to this point we have only considered $M \otimes_R N$ to be an abelian group. To obtain a module structure on $M \otimes_R N$, we need additional structure on the modules M and N. This is the case in Example 2.6.1, where scalar multiplication makes sense on both sides of the vector spaces V and W. Since we know that $V \otimes_F W$ is a vector space over F, the above result on preservation of direct sums verifies that $V \otimes_F W$ should have dimension mn.

Definition 2.6.6 *Let R and S be rings, and let U be a left S-module. If U is also a right R-module such that $(sx)r = s(xr)$ for all $s \in S$, $r \in R$, and $x \in U$, then U is called an S-R-bimodule.*

If U is an S-R-bimodule we use the notation $_S U_R$.

Example 2.6.3 (Ring homomorphisms produce bimodules)

Let R and S be rings, with a ring homomorphism $\phi : R \to S$. We claim that S is an R-S-bimodule. We can give S a left R-module structure by defining $r \cdot s = \phi(r)s$, for all $s \in S$ and $r \in R$. (See Example 2.1.4.) Since S already has a natural structure as a right S-module, we only need to check that

$$r \cdot (s_1 s_2) = \phi(r)(s_1 s_2) = (\phi(r)s_1)s_2 = (rs_1)s_2$$

for all $s_1, s_2 \in S$ and all $r \in R$.

Proposition 2.6.7 *Let R, S, and T be rings, and let $_S U_R$, $_R M_T$, and $_S N_T$ be bimodules.*

(a) *The tensor product $U \otimes_R M$ is a bimodule, over S on the left and T on the right.*

(b) *The set $\mathrm{Hom}_S(U, N)$ is a bimodule, over R on the left and T on the right.*

Proof. (a) For $s \in S$ and $\sum_{i=1}^{n} u_i \otimes x_i \in U \otimes_R M$, we want to have

$$s \cdot \left(\sum_{i=1}^{n} u_i \otimes x_i \right) = \sum_{i=1}^{n} su_i \otimes x_i \, .$$

To verify that this action defines a left S-module, we begin by defining β_s : $U \times M \to U \otimes_R M$ by $\beta_s(u, x) = su \otimes x$, for all $s \in S$, $u \in U$, and $x \in M$. Then for all $u_1, u_2 \in U$ and $x \in M$, we have

$$\begin{aligned}
\beta_s(u_1 + u_2, x) &= s(u_1 + u_2) \otimes x = su_1 \otimes x + su_2 \otimes x \\
&= \beta_s(u_1, x) + \beta_s(u_2, x) \, ,
\end{aligned}$$

and a similar argument shows linearity in the second variable. For all $u \in U$, $x \in M$, and $r \in R$, we have

$$\begin{aligned}
\beta_s(ur, x) &= s(ur) \otimes x = (su)r \otimes x \\
&= (su) \otimes rx = \beta_s(u, rx) \, .
\end{aligned}$$

Since β_s is R-bilinear, by the definition of the tensor product there exists a unique group homomorphism $\lambda_s : U \otimes_R M \to U \otimes_R M$, and this defines a representation of S. To show this, let $s, t \in S$. Then $\lambda_{s+t} = \lambda_s + \lambda_t$ and $\lambda_{st} = \lambda_s \circ \lambda_t$ because these functions agree on the generators of $U \otimes_R M$. Thus the function $\phi : S \to \operatorname{End}_Z(U \otimes_R M)$ defined by $\phi(s) = \lambda_s$ for all $s \in S$ is a ring homomorphism, and since it defines the action $s \cdot u \otimes x = su \otimes x$, for all $s \in S$, $u \in U$, and $x \in M$, we have the desired left S-module structure on $U \otimes_R M$.

For $t \in T$ and $\sum_{i=1}^{n} u_i \otimes x_i \in U \otimes_R M$, we define

$$\left(\sum_{i=1}^{n} u_i \otimes x_i \right) \cdot t = \sum_{i=1}^{n} u_i \otimes x_i t \, .$$

The proof that this multiplication gives $U \otimes_R M$ a right T-module structure is similar to the first part.

(b) For $f \in \operatorname{Hom}_S(U, N)$ and $r \in R$, define $rf : U \to N$ by $[rf](u) = f(ur)$, for all $u \in U$. Then rf is an S-homomorphism, since

$$\begin{aligned}
[rf](x_1 + x_2) &= f((x_1 + x_2)r) = f(x_1 r + x_2 r) \\
&= f(x_1 r) + f(x_2 r) = [rf](x_1) + [rf](x_2)
\end{aligned}$$

for all $x_1, x_2 \in U$ and

$$\begin{aligned}
s \cdot [rf](x) &= s \cdot f(xr) = f(s(xr)) \\
&= f((sx)r) = [rf](sx)
\end{aligned}$$

for all $s \in S$ and $x \in U$. To show that this multiplication defines a left R-module structure, we have the following computations, for all $r, r_1, r_2 \in R$, $f, f_1, f_2 \in \operatorname{Hom}_S(U, N)$, and all $x \in U$.

$$\begin{aligned}
[(r_1 + r_2)f](x) &= f(x(r_1 + r_2)) = f(xr_1) + f(xr_2) \\
&= [r_1 f](x) + [r_2 f](x) = [r_1 f + r_2 f](x)
\end{aligned}$$

$$[r(f_1 + f_2)](x) \;=\; [f_1 + f_2](xr) \;=\; f_1(xr) + f_2(xr)$$
$$\;=\; [rf_1](x) + [rf_2](x) \;=\; [rf_1 + rf_2](x)$$

$$[(r_1 r_2)f](x) \;=\; f(x(r_1 r_2)) \;=\; f((xr_1)r_2)$$
$$\;=\; [r_2 f](xr_1) \;=\; [r_1(r_2 f)](x)$$

For $f \in \mathrm{Hom}_S(U, N)$ and $t \in T$, define $ft : U \to M$ by $[ft](u) = f(u)t$, for all $u \in U$. In general, the proof is similar to the one just given. We check only one condition. If $t_1, t_2 \in T$, then

$$[f(t_1 t_2)](x) \;=\; (f(x))(t_1 t_2) \;=\; (f(x)t_1)t_2$$
$$\;=\; ([ft_1](x))t_2 \;=\; [(ft_1)t_2](x)$$

for all $u \in U$. $\square$

Proposition 2.6.8 *Let M be a left R-module, and let N be a left S-module.*
(a) $R \otimes_R M \cong M$, *as left R-modules.*
(b) $\mathrm{Hom}_S(S, N) \cong N$, *as left S-modules.*

Proof. (a) Define $\beta : R \times M \to M$ by $\beta(a, x) = ax$, for all $a \in R$ and $x \in M$. Since M is a left R-module, we have $(a_1 + a_2)x = a_1 x + a_2 x$, $a(x_1 + x_2) = ax_1 + ax_2$, and $(ar)x = a(rx)$, for all $r, a, a_1, a_2 \in R$ and $x, x_1, x_2 \in M$. These conditions are precisely what we need to prove that β is R-bilinear. Thus there exists a group homomorphism $f : R \otimes_R M \to M$ with $f(a \otimes x) = ax$, for all $a \in R$ and $x \in M$. In fact, f is an R-homomorphism, since

$$rf\left(\sum_{i=1}^n a_i \otimes x_i\right) \;=\; \sum_{i=1}^n rf(a_i \otimes x_i) \;=\; \sum_{i=1}^n r(a_i x_i)$$
$$\;=\; \sum_{i=1}^n (ra_i)x_i \;=\; \sum_{i=1}^n f(ra_i \otimes x_i)$$
$$\;=\; f\left(\sum_{i=1}^n ra_i \otimes x_i\right) \;=\; f\left(r\left(\sum_{i=1}^n a_i \otimes x_i\right)\right)$$

for all $\sum_{i=1}^n a_i \otimes x_i \in R \otimes_R M$.

It is clear that f is onto, since for any $x \in M$ we have $x = f(1 \otimes x)$. To show that f is one-to-one, we define $g : M \to R \otimes_R M$ by $g(x) = 1 \otimes x$, for all $x \in M$. Then for any $\sum_{i=1}^n a_i \otimes x_i \in R \otimes_R M$ we have

$$gf\left(\sum_{i=1}^n a_i \otimes x_i\right) \;=\; g\left(\sum_{i=1}^n a_i x_i\right) \;=\; 1 \otimes \left(\sum_{i=1}^n a_i x_i\right)$$
$$\;=\; \sum_{i=1}^n 1 \otimes a_i x_i \;=\; \sum_{i=1}^n a_i \otimes x_i \,,$$

which shows that gf is the identity, and thus f must be one-to-one.

(b) Define $f : \mathrm{Hom}_S(S, N) \to N$ by $f(\alpha) = \alpha(1)$, for all $\alpha \in \mathrm{Hom}_S(S, N)$. Then we have

$$f(\alpha_1 + \alpha_2) \;=\; [\alpha_1 + \alpha_2](1) \;=\; \alpha_1(1) + \alpha_2(1)$$
$$\;=\; f(\alpha_1) + f(\alpha_2)$$

and

$$
\begin{aligned}
f(s\alpha) &= [s\alpha](1) = \alpha(s) \\
&= s\alpha(1) = sf(\alpha)
\end{aligned}
$$

for any $s \in S$ and any $\alpha, \alpha_1, \alpha_2 \in \mathrm{Hom}_S(S, N)$. Thus f is an S-homomorphism. If $f(\alpha) = 0$ for $\alpha \in \mathrm{Hom}_S(S, N)$, then $\alpha(s) = s\alpha(1) = 0$ for all $s \in S$, so $\alpha = 0$ and f must be one-to-one. For any $y \in N$ we can define $\rho_y \in \mathrm{Hom}_S(S, N)$ with $\rho_y(s) = sy$, for all $s \in S$, and then $y = f(\rho_y)$, showing that f is onto. $\quad\square$

Proposition 2.6.9 *Let $_SU_R$ be a bimodule. For any modules $_RM$ and $_SN$, there is an isomorphism*

$$
\eta : \mathrm{Hom}_S(U \otimes_R M, N) \to \mathrm{Hom}_R(M, \mathrm{Hom}_S(U, N)) \, .
$$

Proof. For any S-homomorphism $f : U \otimes_R M \to N$ and any $x \in M$, we define $f_x : U \to N$ by $f_x(u) = f(u \otimes x)$ for all $u \in U$. We can then define $\eta(f) : M \to \mathrm{Hom}_S(U, N)$ by setting $[\eta(f)](x) = f_x$, for all $x \in M$,

To check that $f_x \in \mathrm{Hom}_S(U, N)$, we have the following equalities, which hold for all $x \in M$, $s \in S$, and $u, u_1, u_2 \in U$:

$$
\begin{aligned}
f_x(u_1 + u_2) &= f((u_1 + u_2) \otimes x) = f(u_1 \otimes x + u_2 \otimes x) \\
&= f(u_1 \otimes x) + f(u_2 \otimes x) = f_x(u_1) + f_x(u_2)
\end{aligned}
$$

and

$$
\begin{aligned}
f_x(su) &= f(su \otimes x) = sf(u \otimes x) \\
&= sf_x(u) \, .
\end{aligned}
$$

We next show that $\eta(f)$ is an R-homomorphism, for all homomorphisms f in $\mathrm{Hom}_S(U \otimes_R M, N)$. For $x, y \in M$, we have

$$
[\eta(f)](x + y) = f_{x+y} = f_x + f_y = [\eta(f)](x) + [\eta(f)](y)
$$

since

$$
\begin{aligned}
f_{x+y}(u) &= f(u \otimes (x + y)) = f(u \otimes x + u \otimes y) \\
&= f(u \otimes x) + f(u \otimes y) = f_x(u) + f_y(u)
\end{aligned}
$$

for all $u \in U$. For $r \in R$ and $x \in M$, we have

$$
[\eta(f)](rx) = f_{rx} = rf_x = r[\eta(f)](x)
$$

since

$$
f_{rx}(u) = f(u \otimes rx) = f(ur \otimes x) = f_x(ur) = rf_x(u)
$$

for all $u \in U$.

We must show that η is a group homomorphism. For any homomorphisms f, g in $\mathrm{Hom}_S(U \otimes_R M, N)$, we have $[\eta(f + g)](x) = (f + g)_x$ and

$$[\eta(f) + \eta(g)](x) = [\eta(f)](x) + [\eta(g)](x) = f_x + g_x .$$

These functions are equal since

$$(f + g)_x(u) = (f + g)(u \otimes x) = f(u \otimes x) + g(u \otimes x) = f_x(u) + g_x(u)$$

for all $u \in U$.

Suppose that $\eta(f) = 0$ for some $f \in \mathrm{Hom}_S(U \otimes_R M, N)$. Then f_x is the zero function, for all $x \in M$, and this implies that $f(u \otimes x) = 0$, for all $u \in U$ and all $x \in M$. Thus $f = 0$, showing that η is one-to-one.

To show that η is onto, we define

$$\theta : \mathrm{Hom}_R(M, \mathrm{Hom}_S(U, N)) \to \mathrm{Hom}_S(U \otimes_R M, N)$$

as follows. Given $f \in \mathrm{Hom}_R(M, \mathrm{Hom}_S(U, N))$, define $\beta_f : U \times M \to N$ by $\beta_f(u, x) = [f(x)](u)$, for all $(u, x) \in U \times M$. To show that β_f is S-bilinear, let $u, u_1, u_2 \in U$, $x, x_1, x_2 \in M$, and $s \in S$. We have the following computations.

$$\begin{aligned}
\beta_f(u_1 + u_2, x) &= [f(x)](u_1 + u_2) = [f(x)](u_1) + [f(x)](u_2) \\
&= \beta_f(u_1, x) + \beta_f(u_2, x)
\end{aligned}$$

$$\begin{aligned}
\beta_f(u, x_1 + x_2) &= [f(x_1 + x_2)](u) = [f(x_1) + f(x_2)](u) \\
&= [f(x_1)](u) + [f(x_2)](u) = \beta_f(u, x_1) + \beta_f(u, x_2)
\end{aligned}$$

$$\begin{aligned}
\beta_f(us, x) &= [f(x)](us) = [sf(x)](u) \\
&= [f(sx)](u) = \beta_f(u, sx)
\end{aligned}$$

Therefore we get a unique group homomorphism $\theta(f) : U \otimes_R M \to N$, characterized by the fact that $[\theta(f)](u \otimes x) = [f(x)](u)$ for all $u \in U$ and $x \in M$.

Given $f \in \mathrm{Hom}_R(M, \mathrm{Hom}_S(U, N))$, we compute $\eta(\theta(f))$. By definition, for all $x \in M$ we have $[\eta(\theta(f))](x) = \theta(f)_x$, where

$$[\theta(f)_x](u) = \theta(f)(u \otimes x) = [f(x)](u)$$

for all $u \in U$. Thus $\theta(f)_x = f(x)$ for all $x \in M$, and so $\eta(\theta(f)) = f$. Then since $\eta\theta$ is the identity, it follows that η is onto. $\square$

Corollary 2.6.10 *Let R and S be rings, and let $\phi : R \to S$ be a ring homomorphism. Let M be any left R-module. Then $S \otimes_R M$ is a left S-module, and for any left S-module N we have*

$$\mathrm{Hom}_S(S \otimes_R M, N) \cong \mathrm{Hom}_R(M, N) .$$

Proof. In the preceding proposition, take $_SU_R = {}_SS_R$. The result follows from the fact that $\operatorname{Hom}_R(S, N) \cong N$. $\square$

EXERCISES: SECTION 2.6

1. Show that although $\mathbf{Z}$ is a subgroup of $\mathbf{Q}$, $\mathbf{Z}_2 \otimes_Z \mathbf{Z}$ is not isomorphic to a subgroup of $\mathbf{Z}_2 \otimes_Z \mathbf{Q}$.

2. Show that if m, n are positive integers, then $\mathbf{Z}_m \otimes_Z \mathbf{Z}_n \cong \mathbf{Z}_d$, where d is the greatest common divisor of m and n.

3. Show that $\mathbf{Q} \otimes_Z \mathbf{Q} \cong \mathbf{Q}$.

4. Let R be a subring of S, and let U be an S-R-bimodule.
   (a) If M, N are left R-modules with $f \in \operatorname{Hom}_R(M, N)$, prove that the mapping $1 \otimes f : U \otimes_R M \to U \otimes_R N$ is an S-homomorphism.
   (b) If K, M, N are left R-modules with $f \in \operatorname{Hom}_R(K, M)$ and $g \in \operatorname{Hom}_R(M, N)$, show that $(1 \otimes g) \circ (1 \otimes f) = 1 \otimes gf$.

5. Let R be a subring of S, and let M be a left R-module. Prove that if M is projective as a left R-module, then $S \otimes_R M$ is projective as a left S-module.

6. Let I be an ideal of R, and let M be a left R-module. Prove that $(R/I) \otimes_R M \cong M/IM$, as left R-modules.

7. Let R be a commutative ring, with ideals I, J. Prove that $(R/I) \otimes_R (R/J) \cong R/(I + J)$, as abelian groups.

8. Let R and S be rings, let U be an S-R-bimodule, let M be a right S-module, and let N be a left R-module. Prove that $(M \otimes_S U) \otimes_R N \cong M \otimes_S (U \otimes_R N)$, as abelian groups.

9. Let F be a field. Prove that $F[x] \otimes_F F[y]$ is isomorphic to $F[x, y]$, as abelian groups. Use this isomorphism to define a multiplication on $F[x] \otimes_F F[y]$, and describe the necessary multiplication formula.

10. Let R be a commutative ring, and let M be an R-module. If S is a multiplicative set, introduce an equivalence relation on the set $M \times S$ as in Exercise 9 of Section 1.3. Show that the set M_S of equivalence classes can be given a module structure over the ring R_S, and that M_S is isomorphic to the R_S-module $R_S \otimes M$.

2.7 Modules over principal ideal domains

The fundamental theorem of finite abelian groups states that any finite abelian group is isomorphic to a direct product of cyclic groups of prime power order. From our current point of view, a finite abelian group is a module over the ring of integers $\mathbf{Z}$, and in this section we will show that we can extend the fundamental theorem to modules over any principal ideal domain. This includes the ring $\mathbf{Q}[x]$ of all polynomials with coefficients in the field $\mathbf{Q}$, and in this case all of the cyclic modules are infinite, so we cannot restrict ourselves to finite modules. The appropriate generalization is to consider finitely generated torsion modules, which we now define. We will also consider finitely generated torsionfree modules, which turn out to be free.

In this section all rings will be commutative, and so we simply refer to modules rather than left or right modules. It should be noted that part (c) of the next definition is consistent with Definition 2.4.8.

Definition 2.7.1 *Let D be an integral domain, and let M be a D-module.*

(a) *An element $m \in M$ is called a* torsion element *if $\mathrm{Ann}(m) \neq (0)$. The set of all torsion elements of M is denoted by $\mathrm{tor}(M)$.*

(b) *If $\mathrm{tor}(M) = (0)$, then M is said to be a* torsionfree *module.*

(c) *If $\mathrm{tor}(M) = M$, then M is said to be a* torsion *module.*

Proposition 2.7.2 *Let D be an integral domain, and let M be a D-module. Then $\mathrm{tor}(M)$ is a submodule of M, and $M/\mathrm{tor}(M)$ is a torsionfree module.*

Proof. If $x, y \in \mathrm{tor}(M)$, then by definition there exist nonzero elements $a, b \in D$ with $ax = 0$ and $by = 0$. Since D is an integral domain, we have $ab \neq 0$, and then $ab(x + y) = b(ax) + a(by) = 0$ shows that $x + y \in \mathrm{tor}(M)$. For any $d \in D$ we have $a(dx) = d(ax) = 0$, so $dx \in \mathrm{tor}(M)$. Thus $\mathrm{tor}(M)$ is a D-submodule of M.

To show that $M/\mathrm{tor}(M)$ is torsionfree, suppose that the coset $x + \mathrm{tor}(M)$ is a torsion element. Then there exists an element $b \in D$ with $b \neq 0$ and $b(x + \mathrm{tor}(M)) = 0$, so $bx \in \mathrm{tor}(M)$. By the definition of $\mathrm{tor}(M)$, there exists a nonzero element $a \in D$ with $a(bx) = 0$. Thus $ab \neq 0$ and $(ab)x = 0$, showing that $x \in \mathrm{tor}(M)$. We conclude that $M/\mathrm{tor}(M)$ has no nonzero torsion elements. $\square$

Let D be an integral domain, and let M be a finitely generated free module over D. By Theorem 2.2.5 we have $M \cong D^n$ for some n. If Q is the quotient field of D, then by Theorem 2.6.5 we have

$$Q \otimes_D M \cong Q \otimes_D D^n \cong (Q \otimes_D D)^n \cong Q^n \ .$$

Thus $Q \otimes_D M$ is a vector space over Q of dimension n, and the invariance of dimension for vector spaces implies that n is an invariant of M.

Definition 2.7.3 *Let D be an integral domain with quotient field Q, and let M be a finitely generated free D-module. The module M is said to be free of rank n, denoted by $\mathrm{rank}(M) = n$, if $\dim(Q \otimes_D M) = n$.*

Lemma 2.7.4 *Let D be a principal ideal domain with quotient field Q. Then any nonzero finitely generated submodule of Q is free of rank 1.*

Proof. Let M be a nonzero submodule of Q with generators $q_1, q_2, \ldots, q_n$, and let d be a common denominator for $q_1, q_2, \ldots, q_n$. Then $dq_i \in D \subseteq Q$, for $1 \leq i \leq n$, so for any $a \in D$ we have $aq_i = (aq_id)(1/d)$, showing that $Dq_i \subseteq D(1/d)$ for $1 \leq i \leq n$. It follows that $M \subseteq D(1/d)$, so M is isomorphic to an ideal of D because the mapping $f : D \to D(1/d)$ defined by $f(a) = a(1/d)$ for all $a \in D$ is a D-isomorphism. Thus M is a free module of rank 1 since D is a principal ideal domain. $\square$

Lemma 2.7.5 *Let D be a principal ideal domain, and let M be a finitely generated torsionfree D-module. If M contains a submodule N such that N is free of rank 1 and M/N is a torsion module, then M is free of rank 1.*

Proof. Let Q be the quotient field of D, and assume that M is a finitely generated D-module with a submodule N that is free of rank 1 such that M/N is a torsion module. Then there exists an isomorphism $f : N \to D$, and since Proposition 2.3.10 shows that Q is an injective D-module, there exists an extension $\widehat{f} : M \to Q$ of f. If $\widehat{f}(m) = 0$ for $m \in M$, then since M/N is a torsion module there exists $0 \neq d \in D$ with $dm \in N$. But then $f(dm) = \widehat{f}(dm) = d\widehat{f}(m) = 0$, so $dm = 0$ since f is one-to-one and therefore $m = 0$ since M is torsionfree. We conclude that $\widehat{f}$ is one-to-one, and so it suffices to show that $\widehat{f}(M)$ is free of rank 1. This follows immediately from the previous lemma. $\square$

Theorem 2.7.6 *If D is a principal ideal domain, then any nonzero finitely generated torsionfree D-module is free.*

Proof. Let D be a principal ideal domain, and let M be a nonzero finitely generated torsionfree D-module. If M is cyclic, say $M = Dm$ for some $m \in M$, then $Dm \cong D$ since by assumption $\mathrm{Ann}(m) = (0)$, and so M is free with $\mathrm{rank}(M) = 1$.

We now use induction on the number of generators of M. Assume that any finitely generated torsionfree D-module M with k generators is free with $\mathrm{rank}(M) \leq k$. Suppose that M is generated by $m_1, m_2, \ldots, m_{k+1} \in M$, with $m_{k+1} \neq 0$. Let

$$M_0 = \{x \in M \mid ax \in Dm_{k+1} \text{ for some } 0 \neq a \in D\} .$$

Thus $M_0/Dm_{k+1} = \text{tor}(M/Dm_{k+1})$, which implies that

$$M/M_0 \cong (M/Dm_{k+1})/\text{tor}(M/Dm_{k+1})$$

is torsionfree. Since M/M_0 is generated by the cosets determined by the elements $m_1, m_2, \ldots, m_k$, it follows from the induction hypothesis that M/M_0 is free of rank $\leq k$. Any free module is projective, so the natural projection of M onto M/M_0 splits, yielding a complement N of M_0 such that $M = N \oplus M_0$ and N is free of rank $\leq k$.

Since M is torsionfree, we have $\text{Ann}(m_{k+1}) = (0)$, so Dm_{k+1} is free of rank 1. Every submodule of M is finitely generated, since M is finitely generated and D is Noetherian, so M_0 satisfies the conditions of the preceding lemma, and hence is free of rank 1. Therefore $M = N \oplus M_0$ is free of rank at most $k + 1$, completing the proof. $\square$

Proposition 2.7.7 *Let D be a principal ideal domain, and let M be a finitely generated D-module. Then either M is a torsion module, or $\text{tor}(M)$ has a complement that is torsionfree. In the second case, $M = \text{tor}(M) \oplus N$ for a submodule $N \subseteq M$ such that N is free of finite rank.*

Proof. Consider the natural projection $p : M \to M/\text{tor}(M)$. If M is finitely generated, then $M/\text{tor}(M)$ is a finitely generated torsionfree module, and Theorem 2.7.6 shows that $M/\text{tor}(M)$ is a free module. Therefore p splits, and so $M = \text{tor}(M) \oplus N$, where $N \cong M/\text{tor}(M)$. $\square$

Proposition 2.7.7 shows that to complete the description of all finitely generated modules over a principal ideal domain we only need to characterize the finitely generated torsion modules. The first step is to show that any finitely generated torsion module can be written as a direct sum of finitely many indecomposable modules, and this is a consequence of the next propositions.

Proposition 2.7.8 *Let D be a principal ideal domain, and let a be a nonzero element of D. If $a = p_1^{\alpha_1} p_2^{\alpha_2} \cdots p_k^{\alpha_k}$ is the decomposition of a into a product of irreducible elements, then we have the following ring isomorphism.*

$$D/aD \cong (D/p_1^{\alpha_1} D) \oplus (D/p_2^{\alpha_2} D) \oplus \cdots \oplus (D/p_k^{\alpha_k} D)$$

Proof. Since $p_1^{\alpha_1}$ and $p_2^{\alpha_2} \cdots p_k^{\alpha_k}$ are relatively prime, there exist elements $b, c \in D$ such that

$$bp_1^{\alpha_1} + cp_2^{\alpha_2} \cdots p_k^{\alpha_k} = 1 \,.$$

If we let $I = p_1^{\alpha_1} D$ and $J = p_2^{\alpha_2} \cdots p_k^{\alpha_k} D$, then we have $I + J = D$ and $I \cap J = aD$. The Chinese remainder theorem (Theorem 1.2.12) implies that $D/(I \cap J) \cong (D/I) \oplus (D/J)$, so $D/aD \cong (D/I) \oplus (D/J)$, and then the desired result follows by induction. $\square$

Proposition 2.7.9 *Let D be a principal ideal domain, let p be an irreducible element of D, and let M be any indecomposable D-module with $\mathrm{Ann}(M) = p^k D$. Then M is a cyclic module isomorphic to $D/p^k D$.*

Proof. Since $p^k \in \mathrm{Ann}(M)$ but $p^{k-1} \notin \mathrm{Ann}(M)$, there exists $m \in M$ with $\mathrm{Ann}(m) = p^k D$, and then $Dm \cong D/p^k D$. Proposition 2.3.10 shows that Dm is injective as a $(D/p^k D)$-module, and therefore the inclusion mapping splits, yielding a complement N with $M = Dm \oplus N$. The indecomposability of M shows that $M = Dm$, and so M is in fact a cyclic module. $\square$

We recall that if D is a principal ideal domain, then any finitely generated torsion D-module has finite length.

Theorem 2.7.10 *Let D be a principal ideal domain, and let M be a finitely generated D-module. Then M is isomorphic to a finite direct sum of cyclic submodules each isomorphic to either D or $D/p^k D$, for some irreducible element p of D. Moreover, the decomposition is unique up to the order of the factors.*

Proof. The first decomposition is $M = T \oplus F$, where $T = \mathrm{tor}(M)$ is the torsion submodule of M and F is a free module of finite rank, say $F \cong D^m$ for some m. Since T is finitely generated, $\mathrm{Ann}(T) \neq (0)$, say $\mathrm{Ann}(T) = aD$, where a has the decomposition $a = p_1^{\alpha_1} p_2^{\alpha_2} \cdots p_k^{\alpha_k}$ into a product of irreducible elements. Then

$$D/aD \cong (D/p_1^{\alpha_1} D) \oplus (D/p_2^{\alpha_2} D) \oplus \cdots \oplus (D/p_k^{\alpha_k} D) \,,$$

and so part (c) of Proposition 2.2.7 shows that T has a unique decomposition into a direct sum of submodules M_{p_i} such that M_{p_i} is a module over $D/p_i^{\alpha_i} D$. Each submodule M_{p_i} has finite length, so it can be written as a direct sum of indecomposable submodules, each isomorphic to $D/p_i^{\beta} D$ for some $\beta \leq \alpha_i$.

Since each of the factors in the decomposition is indecomposable, uniqueness follows from the Krull–Schmidt theorem (Theorem 2.5.11). $\square$

Example 2.7.1 (Linear transformations)

Let V be an n-dimensional vector space over the field F, and let $T : V \to V$ be a linear transformation. Note that any power of T is a linear transformation of V. Since any linear combination of linear transformations is again a linear transformation, it follows that any polynomial $f(x)$ can be used to define a linear transformation $f(T)$ on V.

We can use the linear transformation T to define an $F[x]$-module structure on V. For any polynomial $f(x) \in F[x]$ and any vector

$v \in V$, define $f(x) \cdot v = [f(T)](v)$. Using this operation, it is clear that $K[x]$-submodules of V correspond to subspaces U of V such that $T(U) \subseteq U$. These are precisely the T-*invariant subspaces* of V. The cyclic submodule generated by a vector $v \in V$ is the subspace spanned by the vectors $\{T^i(v) \mid i \geq 0\}$. A subspace of this form is called a T-*cyclic subspace* of V.

The structure theorem for modules over a principal ideal domain can be used to analyze the action of T on V. This is outlined in the exercises.

EXERCISES: SECTION 2.7

1. Let D be an integral domain, and let M be a torsionfree D-module. Prove that M is injective if and only if it is divisible.

For the remaining exercises, let V be a fixed n-dimensional vector space over the field F, and let $T : V \to V$ be a fixed linear transformation.

2. (a) Show that there exists a monic polynomial $p(x) \in F[x]$ such that $p(T) = 0$ and $p(x)|f(x)$ for all $f(x) \in F[x]$ with $f(T) = 0$.

   (b) Show that if A is the matrix of T relative to any basis for V, then $p(A) = 0$.

   The polynomial $p(x)$ is called the *minimal polynomial* of T.

3. (a) Show that there exist monic polynomials $p_1(x), p_2(x), \ldots, p_r(x)$ in $F[x]$ of positive degree such that $p_1(x) \mid p_2(x) \mid \cdots \mid p_r(x)$, and corresponding T-cyclic subspaces $V_1, V_2, \ldots, V_r$ such that $V = V_1 \oplus V_2 \oplus \cdots \oplus V_r$.

   (b) Show that $p_i(x)$ is the minimal polynomial of the restriction of T to V_i, and that $p_r(x)$ is the minimal polynomial of T.

   The polynomials $p_1(x), p_2(x), \ldots, p_r(x)$ are called the *invariant factors* of T.

4. Let $p(x)$ be the minimal polynomial of T, and assume that $\deg(p(x)) = m$. Show that V is a cyclic $F[x]$-module if and only if $\dim(V) = m$ and V has an ordered basis $\mathcal{B}$ such that the matrix of T relative to $\mathcal{B}$ is the companion matrix of $p(x)$.

5. Show that there exists a basis $\mathcal{B}$ of V such that the matrix A of T relative to $\mathcal{B}$ consists of blocks on the main diagonal, where the blocks are the companion matrices of the invariant factors of T. The matrix A is said to be in *rational canonical form*.

2.8 *Modules over the Weyl algebras

We recall the construction in Definition 1.5.14 of the Weyl algebras over a field F. Let R be the polynomial ring $F[x_1, x_2, \ldots, x_n]$. Then the nth Weyl algebra is the iterated differential operator ring

$$\mathrm{A}_n(F) = R\left[y_1, \ldots, y_n; \frac{\partial}{\partial x_1}, \ldots, \frac{\partial}{\partial x_n}\right].$$

In this section we will consider some modules defined over $\mathrm{A}_n(F)$, assuming throughout that F is a field of characteristic 0.

Since the coefficient ring R is an integral domain, it follows from Proposition 1.5.13 (b) that $\mathrm{A}_n(F)$ is a domain. This result can be sharpened if F is a field of characteristic zero; in this case $\mathrm{A}_n(F)$ is a simple ring, as we will show in Corollary 2.8.4. We first show that the ring $\mathrm{A}_n(F)$ is both left and right Noetherian, by applying an extension of the Hilbert basis theorem. The proof of the generalized Hilbert basis theorem requires only the key assumption that the ring is generated by a left Noetherian subring R and an element x, subject to the condition that $xR + R = Rx + R$. Thus the theorem can be applied not only to rings of differential polynomials, but also to the ring $R[x; \sigma]$ of skew polynomials, where σ is an automorphism of R.

Theorem 2.8.1 (Generalized Hilbert basis theorem) *Let R be a subring of the ring S, such that S is generated as a ring by R and an element $x \in S$ for which $xR + R = Rx + R$. If R is left Noetherian, then so is S.*

Proof. By assumption, each element $s \in S$ has the form $s = \sum_{i=0}^{n} a_i x^i$, for some integer $n \geq 0$ and some elements $a_0, \ldots, a_n \in R$. Since $xR \subseteq Rx + R$, we can prove by induction that $x^n R \subseteq \sum_{i=0}^{n} Rx^i$. Thus we can also assume that each element $s \in S$ has the form $s = \sum_{i=0}^{n} x^i a_i$, for some integer $n \geq 0$ and some elements $a_0, \ldots, a_n \in R$. We say that this expression for s has degree n, and that its leading coefficient is a_n.

We will show that if S is not left Noetherian, then R is not left Noetherian. In particular, we suppose that there exists a left ideal I of S that is not finitely generated, and we use this left ideal to construct an infinite ascending chain of left ideals of R. Among nonzero elements of I, choose an element s_1 with an expression $s_1 = \sum_{i=0}^{n} x^i a_{1i}$ of minimal degree. Then $I \neq Ss_1$ since I is not finitely generated, so we can choose an element s_2 with an expression $s_2 = \sum_{i=0}^{m} x^i a_{2i}$ whose degree is minimal among the elements in $I \setminus Ss_1$. If such elements have already been chosen for $1 \leq i \leq k-1$, let s_k be an element whose degree is minimal among the elements in $I \setminus \left(\sum_{i=1}^{k-1} Ss_i\right)$.

Let $n(i)$ denote the degree of s_i. By our choice of the elements $\{s_i\}$, we see that if $i < j$, then $n(i) \leq n(j)$. Let a_i be the leading coefficient of s_i. We claim that the left ideals $Ra_1 \subseteq Ra_2 + Ra_2 \subseteq \cdots$ form a strictly ascending chain of left ideals of R. If $\sum_{i=1}^{k-1} Ra_i = \sum_{i=1}^{k} Ra_i$, then $a_k = \sum_{i=1}^{k-1} r_i a_i$, for some $r_1, \ldots, r_{k-1} \in R$. Since $x^{n(i)} R \subseteq \sum_{j=0}^{n(i)} Rx^j$, we can write $x^{n(i)} r_i = r_i' x^{n(i)} + t_i$, where t_i has an expression with degree less than $n(i)$.

Now consider the element $s = s_k - \sum_{i=1}^{k-1} x^{n(k)-n(i)} r_i' s_i$. To compute the leading coefficient of the term $x^{n(k)-n(i)} r_i' s_i$, we have

$$
\begin{aligned}
x^{n(k)-n(i)} r_i' s_i &= x^{n(k)-n(i)} r_i' x^{n(i)} a_i \\
&= x^{n(k)-n(i)} (x^{n(i)} r_i - t_i) a_i \\
&= x^{n(k)} r_i a_i + t_i' ,
\end{aligned}
$$

where the degree of t_i' is less than $n(k)$. It follows that the degree of s is less than $n(k)$, since the coefficient of $x^{n(k)}$ is $a_k - \sum_{i=1}^{k-1} r_i a_i = 0$. This contradicts the choice of s_k, since $s \in I \setminus \left(\sum_{i=1}^{k-1} Ss_i \right)$, but s has lower degree than s_k. $\square$

Corollary 2.8.2 *Let R be a left Noetherian ring. If δ is a derivation of R, then the differential polynomial ring $R[x; \delta]$ is left Noetherian.*

Recall that ring R is said to be *simple* if its only ideals are (0) and R (see Definition 1.2.10). It follows immediately from the definition that R is simple if and only if every nonzero R-module is faithful.

Our next proof is given in the more general context of differential operator rings, and we need to introduce some additional terminology. A derivation δ of the ring R is called an *inner derivation* if there exists an element $a \in R$ such that $\delta(r) = ar - ra$, for all $r \in R$.

Proposition 2.8.3 *Let R be a ring, and let δ be a derivation on R. Then the differential polynomial ring $R[x; \delta]$ is a simple ring, provided*
   (i) *the ring R is an algebra over $\mathbf{Q}$,*
   (ii) *there are no proper nontrivial ideals I of R with $\delta(I) \subseteq I$,*
   (iii) *the derivation δ is not an inner derivation of R.*

Proof. Let J be a nonzero ideal of $R[x; \delta]$, and let m be the minimal degree of the nonzero elements of J. Consider the set

$$
I = \{0\} \cup \left\{ a \in R \mid \text{there exist } a_0, \ldots, a_{m-1} \in R \text{ with } ax^m + \sum_{i=0}^{m-1} a_i x^i \in J \right\} .
$$

Since J is an ideal of $R[x; \delta]$, it is clear that I is a nonzero left ideal of R. Given $a \in I$ and $r \in R$, there is an element in J of the form $ax^m + \sum_{i=0}^{m-1} a_i x^i$. Multiplying on the right by r gives an element of the form $arx^m + \sum_{i=0}^{m-1} b_i x^i$, for some elements $b_0, \ldots, b_{m-1} \in R$, and so $ar \in I$, showing that I is a two-sided ideal.

We next show that $\delta(I) \subseteq I$. Given $a \in I$ there exists an element $s = ax^m + \sum_{i=0}^{m-1} a_i x^i$ in J, for some $a_0, \ldots, a_{m-1} \in R$. Since J is a two-sided ideal, it contains the element $xs - sx$. We have

$$
\begin{aligned}
xs - sx &= x\left(ax^m + \textstyle\sum_{i=0}^{m-1} a_i x^i\right) - \left(ax^m + \textstyle\sum_{i=0}^{m-1} a_i x^i\right) x \\
&= ax^{m+1} + \delta(a)x^m + a_{m-1}x^m + \textstyle\sum_{i=0}^{m-1} b_i x^i \\
&\quad - ax^{m+1} - a_{m-1}x^m - \textstyle\sum_{i=0}^{m-2} a_i x^{i+1} \\
&= \delta(a)x^m + \textstyle\sum_{i=0}^{m-1} c_i x^i \,,
\end{aligned}
$$

for some $b_0, c_0, \ldots, b_{m-1}, c_{m-1} \in R$. Thus $\delta(a) \in I$ since either $\delta(a) = 0$ or else $\delta(a)$ is the leading coefficient of an element in J.

It follows from condition (ii) that $I = R$, and so J must contain an element of degree m with leading coefficient 1. If $m \neq 0$, then J contains an element t of the form $t = x^m + \sum_{i=0}^{m-1} a_i x^i$, for some $a_0, \ldots, a_{m-1} \in R$. For any $r \in R$, we have $rt - tr \in J$, and as above we have

$$
\begin{aligned}
rt - tr &= r\left(x^m + \textstyle\sum_{i=0}^{m-1} a_i x^i\right) - \left(x^m + \textstyle\sum_{i=0}^{m-1} a_i x^i\right) r \\
&= rx^m + ra_{m-1}x^{m-1} + \textstyle\sum_{i=0}^{m-2} ra_i x^i \\
&\quad - rx^m - m\delta(r)x^{m-1} - a_{m-1}rx^{m-1} - \textstyle\sum_{i=0}^{m-2} b_i x^i \\
&= (ra_{m-1} - m\delta(r) - a_{m-1}r)x^{m-1} + \textstyle\sum_{i=0}^{m-2} c_i x^i \,,
\end{aligned}
$$

for some $b_0, c_0, \ldots, b_{m-2}, c_{m-2} \in R$. Since the degree of $rt - tr$ is less than m, the choice of m implies that $rt - tr$ must be identically zero, and so

$$
ra_{m-1} - m\delta(r) - a_{m-1}r = 0 \,.
$$

Now we need the assumption that R is an algebra over $\mathbf{Q}$, so that m is invertible in R. If we let $a = \dfrac{a_{m-1}}{m}$, then $\delta(r) = ra - ar$ for all $r \in R$, contradicting the fact that δ is not an inner automorphism.

We conclude that $m = 0$, so $1 \in J$, and $J = R[x; \delta]$, which completes the proof. $\square$

Corollary 2.8.4 *If F is a field of characteristic 0, then the Weyl algebra $A_n(F)$ is a simple Noetherian domain.*

Proof. We only need to show that $A_n(F)$ is simple. Given $S = A_{i-1}(F)$, we construct $A_i(F)$ as the differential polynomial ring $(S[x_i])\left[y_i; \dfrac{\partial}{\partial x_i}\right]$. By induction, it suffices to show that if S is a simple domain, then the differential polynomial ring $(S[x])[y; \delta]$ is a simple domain, where $\delta = \dfrac{d}{dx}$.

We will verify the conditions of Proposition 2.8.3. Since F has characteristic 0, the ring $S[x]$ is certainly a $\mathbf{Q}$-algebra, and since $\delta(x) = 1$, the derivation δ cannot be an inner derivation.

Suppose that I is a nonzero ideal of $S[x]$ with $\delta(I) \subseteq I$. We can choose a nonzero element $p(x)$ in I of minimal degree, and then since the degree of $\delta(p(x))$ is lower than that of $p(x)$, while $\delta(p(x)) \in I$, we must have $m a_m = 0$ for the leading coefficient a_m of $p(x)$. Because S is an algebra over $\mathbf{Q}$, we have $a_m = 0$, and thus $p(x)$ must be a constant in S. Since S is simple, it follows that $I = S[x]$, as required. $\square$

Proposition 2.8.5 *The polynomial ring $F[x_1, \ldots, x_n]$ has the structure of a simple left $\mathrm{A}_n(F)$-module.*

Proof. To define an action of $\mathrm{A}_n(F)$ on the polynomial ring $F[x_1, \ldots, x_n]$, for any polynomial $f \in R$, we let $x_i \cdot f = x_i f$ and $y_j \cdot f = \dfrac{\partial}{\partial x_j} f$. In effect, we can think of x_i and y_j as operators in the vector space $\mathrm{End}_F(F[x_1, \ldots, x_n])$ over F, and so this action represents $\mathrm{A}_n(F)$ as a ring of differential operators with polynomial coefficients. (Compare the original construction of the differential polynomial ring $\mathbf{C}[z; \partial]$ in Example 1.1.7.)

As a left module, $F[x_1, \ldots, x_n]$ is cyclic, generated by the element 1. For any nonzero polynomial $f \in F[x_1, \ldots, x_n]$, repeated differentiation will eventually lead to a nonzero constant. Thus multiplying by the appropriate element of $\mathrm{A}_n(F)$ yields 1, and so $F[x_1, \ldots, x_n]$ is a simple $\mathrm{A}_n(F)$-module since each nonzero element is a generator. $\square$

We recall the correspondence between left R-modules M and ring homomorphisms ρ from R into $\mathrm{End}_Z(M)$. If α is any automorphism of R, then the composition $\rho\alpha$ defines another R-module structure on M. This technique can be used to define additional $\mathrm{A}_n(F)$-modules.

Example 2.8.1 (Some simple modules over $\mathrm{A}_n(F)$)

Using appropriate automorphisms of $\mathrm{A}_n(F)$, we can define an infinite family of pairwise nonisomorphic simple modules.

For each positive integer k, define an automorphism α_k of $\mathrm{A}_n(F)$ by setting $\alpha_k(x_i) = x_i$ and $\alpha_k(y_j) = y_j - x_j^k$. Using the module structure on $F[x_1, \ldots, x_n]$ that is defined in Proposition 2.8.5, we can 'twist' the structure using α_k, to define a module S_k. Each of the new modules is still simple, and generated by 1. Suppose that there exists a homomorphism $h : S_k \to S_t$, where $k < t$, with $h(1) = f$. We must have

$$h(y_j \cdot 1) = y_j \cdot h(1) = y_j \cdot f .$$

Since $y_j \cdot 1 = \dfrac{\partial}{\partial x_j}(1) - x_j^k(1)$, this translates into the differential equation

$$-x_j^k f = \frac{\partial}{\partial x_j}(f) - x_j^t f$$

or

$$\frac{\partial f}{\partial x_j} = \left(x_j^t - x_j^k \right) f .$$

The left hand side of the equation has degree $\leq \deg(f) - 1$, while the right hand side has degree $\deg(f) + t$. This is a contradiction, so there are no nonzero module homomorphisms from S_k into S_t.

As we have seen in Proposition 2.8.5, we have a natural action of $A_n(F)$ on the polynomial ring $F[x_1, \dots, x_n]$. Homogeneous linear partial differential equations, with polynomial solutions, can be expressed in the form

$$L_1(f) = 0, \quad L_2(f) = 0, \quad \dots, \quad L_m(f) = 0 ,$$

for differential operators $L_1, L_2, \dots, L_m \in A_n(F)$. If I is the left ideal of $A_n(F)$ generated by the elements $L_1, \dots, L_m$, then $M = A_n(F)/I$ is called the $A_n(F)$-module associated with the system of differential equations. This terminology is justified by the fact that the vector space of solutions to the given system of differential equations can be shown to be isomorphic to

$$\text{Hom}_{A_n}(M, F[x_1, \dots, x_n]) .$$

This module theoretic approach also leads to the definition of more general kinds of solutions.

Our final example of the construction of modules over the Weyl algebras follows Chapter 6 of [7]. The example requires the use of direct limits, which we will discuss below. For the most part, we will use the notation and terminology of Coutinho in [7].

From this point on, we specialize to the ring $A_1(\mathbf{C})$, where $\mathbf{C}$ is the field of complex numbers. We use the notation of Example 1.1.7, and view $A_1(\mathbf{C})$ as the ring $\mathbf{C}[z; \partial]$, where $\partial = \dfrac{d}{dz}$.

For the positive real number ϵ, let $D(\epsilon)$ be the open disk with center at the origin and radius ϵ. Let $\mathcal{H}(D(\epsilon)) = \mathcal{H}(\epsilon)$ be the set of all holomorphic functions defined on $D(\epsilon)$. (Recall that a complex valued function f defined on an open set $\mathcal{U}$ is said to be holomorphic on $\mathcal{U}$ if f has derivatives of all orders on $\mathcal{U}$.) The vector space $\mathcal{H}(\epsilon)$ can be given a left $A_1(\mathbf{C})$-module structure by letting z act via left multiplication, and letting ∂ act as d/dz. If $0 < \epsilon < \delta$, then we can restrict a function in $\mathcal{H}(\delta)$ to the smaller disk with radius ϵ, and this yields a homomorphism $f_\delta^\epsilon : \mathcal{H}(\delta) \to \mathcal{H}(\epsilon)$. If $\delta < \gamma$, then we have $f_\delta^\epsilon f_\gamma^\delta = f_\gamma^\epsilon$; of course, f_ϵ^ϵ is the identity mapping. This provides some motivation for the next definitions.

Definition 2.8.6 *A set I is said to be a* directed set *if I has a relation $\alpha \leq \beta$ satisfying the following conditions, for all $\alpha, \beta, \gamma \in I$:*

(i) $\alpha \leq \alpha$;

(ii) $\alpha \leq \beta$ and $\beta \leq \gamma$ imply $\alpha \leq \gamma$;

(iii) $\alpha \leq \beta$ and $\beta \leq \alpha$ imply $\alpha = \beta$;

(iv) *given α, β, there exists $\gamma \in I$ such that $\alpha \leq \gamma$ and $\beta \leq \gamma$.*

Definition 2.8.7 *Let I be a directed set. A* directed family of homomorphisms *is a set $\{M_\alpha\}_{\alpha \in I}$ of left R-modules, and a set $\mathcal{F} = \{f_\alpha^\beta\}_{\alpha \in I}$ of homomorphisms, with $f_\alpha^\beta : M_\alpha \to M_\beta$, such that*

(i) *if $\alpha, \beta \in I$ with $\alpha \leq \beta$, then there is a unique element $f_\alpha^\beta \in \mathcal{F}$,*

(ii) *if $\alpha = \beta$, then f_α^β is the identity,*

(iii) *if $\alpha \leq \beta \leq \gamma$, then $f_\delta^\epsilon f_\gamma^\delta = f_\gamma^\epsilon$.*

Given a directed family of homomorphisms, we are interested in constructing a module (together with appropriate homomorphisms) that forms a 'limit' for the directed family.

To help understand the definition of a direct limit we need an example of a familiar module that can be constructed as a direct limit. In the next example we will show that the field of rational numbers is a direct limit in a natural way. In fact, direct limits can be used to construct rings of fractions in much more general settings.

Example 2.8.2 (The field of rational numbers as a direct limit)

Since the group $\mathbf{Z}$ of integers is cyclic, its endomorphism ring is isomorphic to $\mathbf{Z}$ itself. Every endomorphism $\phi : \mathbf{Z} \to \mathbf{Z}$ is given by multiplication, since $\phi(x) = kx$, for all $x \in \mathbf{Z}$, where $k = \phi(x)$.

To construct fractions, we can consider 'partial endomorphisms' $\phi : n\mathbf{Z} \to \mathbf{Z}$. If $nq \in n\mathbf{Z}$, then

$$\phi(nq) = q \cdot \phi(n) = \frac{\phi(n)}{n} nq \, ,$$

and thus for all $x \in n\mathbf{Z}$ we have $\phi(x) = \dfrac{k}{n}x$, for $k = \phi(n)$. The fractions with denominator n can be identified with the elements of $\mathrm{Hom}_Z(n\mathbf{Z}, \mathbf{Z})$.

We need a construction that puts these sets together to give the field $\mathbf{Q}$ of rational numbers. Note that if $n \mid m$ for $n, m \in \mathbf{Z}$, then $m\mathbf{Z} \subseteq n\mathbf{Z}$, and so there is a natural mapping

$$f_n^m : \mathrm{Hom}_Z(n\mathbf{Z}, \mathbf{Z}) \to \mathrm{Hom}_Z(m\mathbf{Z}, \mathbf{Z})$$

given by restricting elements of $\mathrm{Hom}_Z(n\mathbf{Z}, \mathbf{Z})$ to $m\mathbf{Z}$. This mapping respects the identifications we have made, since if $m = nq$ and $\phi \in \mathrm{Hom}_Z(n\mathbf{Z}, \mathbf{Z})$, then ϕ is identified with the fraction $\dfrac{\phi(n)}{n}$, while $f_n^m(\phi)$ is identified with $\dfrac{\phi(m)}{m}$. These fractions are equal since $m = nq$ and $\phi(m) = q\phi(n)$.

The construction of the direct limit of the system of groups $\mathrm{Hom}_Z(n\mathbf{Z}, \mathbf{Z})$ and homomorphisms

$$f_n^m : \mathrm{Hom}_Z(n\mathbf{Z}, \mathbf{Z}) \to \mathrm{Hom}_Z(m\mathbf{Z}, \mathbf{Z})$$

is given by first taking the direct sum of the groups. Then in the direct sum it is necessary to identify certain elements, so that we introduce the usual equivalence relation for fractions. This is done by factoring out the submodule generated by the necessary relations, just as in the construction of the tensor product of two modules.

This construction of $\mathbf{Q}$ can be generalized to construct the ring $\mathbf{Z}_{(p)}$ of all fractions whose denominator is relatively prime to p. To do so we use the subgroups $n\mathbf{Z}$ of $\mathbf{Z}$ that do not contain p, ordered by reverse inclusion (as before).

We are now ready to give the definition of a direct limit. In the definition, this limit is specified by requiring the existence of a unique homomorphism. The fact that the homomorphism is unique makes it easy to prove the the direct limit of a directed family of homomorphisms is unique (up to isomorphism), provided it can be shown to exist. In Theorem 2.8.9 we will prove that direct limits exist, but the uniqueness is left as an exercise for the reader.

Definition 2.8.8 *Let I be a directed set, let $\{M_\alpha\}_{\alpha \in I}$ be a set of left R-modules, and let $\mathcal{F} = \{f_\alpha^\beta\}_{\alpha \in I}$ be a directed family of homomorphisms, with $f_\alpha^\beta : M_\alpha \to M_\beta$. A direct limit of $\mathcal{F}$ is a left R-module $\varinjlim_I M_\alpha$, together with homomorphisms $f_\alpha : M_\alpha \to \varinjlim_I M_\alpha$ with $f_\beta f_\alpha^\beta = f_\alpha$ for all $\alpha \leq \beta$ in I*

$$
\cdots \longrightarrow M_\alpha \xrightarrow{\ f_\alpha^\beta\ } M_\beta \longrightarrow \cdots
$$

$$
f_\alpha \searrow \qquad \Big\downarrow f_\beta
$$

$$
\varinjlim_I M_\alpha
$$

such that the following condition is satisfied for all $\alpha \leq \beta$ in I:

for every left R-module X and every set $\{g_\alpha\}_{\alpha \in I}$ of homomorphisms $g_\alpha : M_\alpha \to X$ satisfying $g_\beta f_\alpha^\beta = g_\alpha$,

$$
\cdots \longrightarrow M_\alpha \xrightarrow{\ f_\alpha^\beta\ } M_\beta \longrightarrow \cdots
$$

$$
g_\alpha \searrow \qquad \Big\downarrow g_\beta
$$

$$
X
$$

there exists a unique homomorphism $g : \varinjlim_I M_\alpha \to X$ with $gf_\beta = g_\beta$ for all $\beta \in I$.

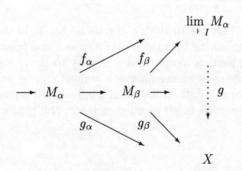

Theorem 2.8.9 *Let I be a directed set, let $\{M_\alpha\}_{\alpha \in I}$ be a set of left R-modules, and let $\mathcal{F} = \{f_\alpha^\beta\}$ be a directed family of homomorphisms, with f_α^β : $M_\alpha \to M_\beta$. For the direct sum $\bigoplus_{\alpha \in I} M_\alpha$, let $i_\alpha : M_\alpha \to \bigoplus_{\alpha \in I} M_\alpha$ denote the inclusion mapping.*

(a) *The direct limit $\varinjlim_I M_\alpha$ can be constructed as $\bigoplus_{\alpha \in I} M_\alpha / N$, where N is the submodule generated by all elements of the form $i_\alpha(m_\alpha) - i_\beta f_\alpha^\beta(m_\alpha)$.*

(b) *Each element of $\varinjlim_I M_\alpha$ has the form $i_\alpha(m_\alpha) + N$, for some $m_\alpha \in M_\alpha$. With this representation, $i_\alpha(m_\alpha) + N = 0$ if and only if $f_\alpha^\beta(m_\alpha) = 0$ for some $\beta \in I$ with $\alpha \le \beta$.*

Proof. (a) Given $_R X$ and a set of homomorphisms $g_\alpha : M_\alpha \to X$ satisfying $g_\beta f_\alpha^\beta = g_\alpha$, the definition of the direct sum $\bigoplus_{\alpha \in I} M_\alpha$ guarantees the existence of a homomorphism $g' : \bigoplus_{\alpha \in I} M_\alpha \to X$. The conditions on the homomorphisms g_α imply that g' factors through the submodule N. This yields the required homomorphism $g : \bigoplus_{\alpha \in I} M_\alpha / N \to X$, and since g is unique, we see that $\bigoplus_{\alpha \in I} M_\alpha / N$ satisfies the definition of the direct limit $\varinjlim_I M_\alpha$.

(b) We first need to find a 'standard' form for the elements of N, which are linear combinations of generators. Since the mappings i_α and f_α^β are R-homomorphisms, we only need to consider sums of generators. If the element $x = i_\alpha(m_\alpha) - i_\beta f_\alpha^\beta(m_\alpha)$ is a generator, with $\alpha \le \beta$, suppose that $\beta \le \gamma$. Then

$$
\begin{aligned}
i_\alpha(m_\alpha) - i_\beta f_\alpha^\beta(m_\alpha) &= i_\alpha(m_\alpha) - i_\gamma f_\beta^\gamma f_\alpha^\beta(m_\alpha) + i_\gamma f_\beta^\gamma f_\alpha^\beta(m_\alpha) - i_\beta f_\alpha^\beta(m_\alpha) \\
&= [i_\alpha(m_\alpha) - i_\gamma f_\alpha^\gamma(m_\alpha)] - [i_\beta(f_\alpha^\beta(m_\alpha)) - i_\gamma f_\beta^\gamma(f_\alpha^\beta(m_\alpha))]
\end{aligned}
$$

which shows that we can write the generator x as a sum of two generators whose last nonzero entry occurs at γ. Given any sum of generators of N, since I is a directed set we can choose an element $\gamma \in I$ such that $\alpha \le \gamma$ for all indices α that occur in the sum. Then we can expand the sum of generators to obtain one of the form

$$
\sum_{\alpha \in F} (i_\alpha(m_\alpha) - i_\gamma f_\alpha^\gamma(m_\alpha)) ,
$$

where F is a finite subset of I, and the index γ is fixed. In this sum, we can assume that each index α occurs only once (the mappings i_α and f_α^γ are additive, allowing us to combine terms with the same indices) and in each case we have $\alpha < \gamma$, since by assumption $f_\alpha^\alpha = 1$. $\square$

We now return to the construction of the space of 'microsolutions'.

Example 2.8.3 (The module of microfunctions)

As in the remarks preceding Definition 2.8.6, for the positive real number ϵ let $D(\epsilon)$ be the open disk with center at the origin and radius ϵ, and let $D^*(\epsilon) = D(\epsilon) \setminus \{0\}$. The universal cover of $D^*(\epsilon)$ is the set

$$\widetilde{D}(\epsilon) = \{z \in \mathbf{C} \mid Re(z) < \log(\epsilon)\},$$

with the projection $\pi : \widetilde{D}(\epsilon) \to D^*(\epsilon)$ defined by $\pi(z) = e^z$, for all $z \in \widetilde{D}(\epsilon)$. Note that π maps $\widetilde{D}(\epsilon)$ onto $D^*(\epsilon)$, since if $r < \epsilon$ is a positive real number, then $re^{i\theta} = \pi(\log(r) + i\theta)$ for the element $\log(r) + i\theta \in \widetilde{D}(\epsilon)$.

The set $\mathcal{H}(\widetilde{D}(\epsilon))$ of functions holomorphic in $\widetilde{D}(\epsilon)$ has an $A_1(\mathbf{C})$-module structure defined as follows. For $h \in \mathcal{H}(\widetilde{D}(\epsilon))$, instead of letting $z \in \mathbf{C}[z; \partial]$ act via multiplication, we define $[z \cdot h](z) = e^z h(z)$, for all $z \in \mathbf{C}$. In addition, we define the action of ∂ by setting $[\partial \cdot h](z) = h'(z)e^{-z}$.

The mapping $\pi^* : \mathcal{H}(D^*(\epsilon)) \to \mathcal{H}(\widetilde{D}(\epsilon))$ defined by $[\pi^*(h)](z) = h(\pi(z))$, for all $h \in \mathcal{H}(D^*(\epsilon))$ and all $z \in \mathbf{C}$, can be shown to be a one-to-one $A_1(\mathbf{C})$-homomorphism. Therefore we can consider the factor module $\mathcal{H}(\widetilde{D}(\epsilon))/\pi^*(\mathcal{H}(D^*(\epsilon)))$, which we denote by $\mathcal{M}_\epsilon$. If $\delta \leq \epsilon$, then $\mathcal{H}(\widetilde{D}(\epsilon)) \subseteq \mathcal{H}(\widetilde{D}(\delta))$, and this induces an $A_1(\mathbf{C})$-module homomorphism $f_\epsilon^\delta : \mathcal{M}_\epsilon \to \mathcal{M}_\delta$. The set $\{f_\epsilon^\delta \mid \delta, \epsilon \in \mathbf{R}^+\}$ is a directed family of homomorphisms.

The direct limit of this family is denoted by $\mathcal{M}$, and is called the *module of microfunctions*.

EXERCISES: SECTION 2.8

1. Let $R[x; \delta]$ be a differential polynomial ring. Show that if δ is an inner derivation, with $\delta(r) = ar - ra$ for all $r \in R$, then $R[x; \delta]$ is isomorphic to the polynomial ring $R[t]$, where $t = x - a$.

2. Let $R[x; \delta]$ be a differential polynomial ring. Show that if I is any ideal of R such that $\delta(I) \subseteq I$, then $R[x; \delta] \cdot I$ is a two-sided ideal of $R[x; \delta]$.

3. Prove that the direct limit of a directed family of homomorphisms is unique up to isomorphism.

4. Verify that the action of $A_1(\mathbf{C})$ on $\mathcal{H}(\widetilde{D}(\epsilon))$, as given in Example 2.8.3, does in fact define an $A_1(\mathbf{C})$-module structure.

5. Verify that the set $\{f_\epsilon^\delta \mid \delta, \epsilon \in \mathbf{R}^+\}$ defined in Example 2.8.3 is a directed family of homomorphisms.

Chapter 3

STRUCTURE OF NONCOMMUTATIVE RINGS

If we use the standard identification of the complex number $a + b\mathrm{i}$ with the ordered pair (a, b) in the Euclidean plane $\mathbf{R}^2$, then multiplication in $\mathbf{C}$ allows us to multiply ordered pairs. This multiplication respects the structure of $\mathbf{R}^2$ as a vector space over $\mathbf{R}$, where $r(a, b) = (ra, rb)$ for $r \in \mathbf{R}$, since if $x = (a, b)$ and $y = (c, d)$, then $r(xy) = (rx)y = x(ry)$. Hamilton discovered the quaternions $\mathbf{H}$ (recall Example 1.1.6) in 1843 after all of his efforts to find such a multiplication on $\mathbf{R}^3$ ended in failure. He finally realized that it was necessary to work in $\mathbf{R}^4$ and that he had to drop the requirement of commutativity. In our current terminology, Hamilton had found the most elementary example of a division ring. (Recall Definition 1.1.6: a division ring satisfies all of the axioms of a field, with the possible exception of the commutative law for multiplication.) We note that Wedderburn showed in 1909 that it is impossible to give a finite example of a division ring that is not a field.

An *algebra* over a field F is a ring A that is also a vector space over F, in which $c(xy) = (cx)y = x(cy)$ for all $x, y \in A$ and $c \in F$. The algebra A is called *finite dimensional* if A is a finite dimensional vector space over F, and it is called a *division algebra* if it is a division ring. (Refer to Definition 1.5.7.) With this terminology, Hamilton's discovery of the quaternions represents the first step in the study of finite dimensional division algebras. In 1878, Frobenius classified all finite dimensional division algebras over the field of real numbers: any such algebra must be isomorphic to $\mathbf{R}$, $\mathbf{C}$, or $\mathbf{H}$.

In 1908, Wedderburn gave a structure theorem for simple algebras, showing that any simple finite dimensional algebra can be described as $\mathrm{M}_n(D)$, for some division ring D and some positive integer n. He also showed that in any finite dimensional algebra A there is an ideal N maximal with respect to the property $N^k = (0)$, and that A/N is isomorphic to a finite direct sum of simple algebras. If A is a finite dimensional algebra over the field F, then any left ideal of A must be a subspace of A, and so it is easy to see that the lattice of left ideals of A must satisfy both the ascending and descending chain conditions. The modern theory of Artinian rings thus includes the classical theory of finite dimensional algebras.

Wedderburn's results were extended by Artin in 1927 to rings satisfying both chain conditions on left ideals. The Artin–Wedderburn theorem is the main goal of this chapter, and it gives the structure of what are now known as semisimple Artinian rings. It should be pointed out that it was Emmy Noether who introduced the chain conditions, and who recognized that many proofs depended only on the chain conditions. We will also be able to give several module theoretic characterizations of semisimple Artinian rings, making use of our study in Chapter 2 of semisimple, completely reducible, projective, and injective modules.

The final section in this chapter studies one possible noncommutative analog of the embedding of an integral domain in its quotient field. Instead of an embedding in a field, we consider an embedding in a simple Artinian ring. The characterization of rings that have a simple Artinian ring of fractions dates from 1958, and has played a major role in the study of noncommutative Noetherian rings.

3.1   Prime and primitive ideals

We begin by recalling Definition 1.3.1: an ideal P of a commutative ring R is said to be a prime ideal if $ab \in P$ implies $a \in P$ or $b \in P$, for all $a, b \in R$. We know, of course, that the zero ideal of any field F is a maximal ideal, and so it is certainly a prime ideal. When looking for an appropriate general definition of a prime ideal, it is natural to consider the zero ideal in the ring $\mathrm{M}_2(F)$ of 2×2 matrices over F. Since F is a simple ring, so is $\mathrm{M}_2(F)$ (see Proposition 1.5.4), so the zero ideal of $\mathrm{M}_2(F)$ is maximal, and thus it should satisfy the general definition of a prime ideal. The complexities of the noncommutative situation are already evident in this basic example, since it is easy to find nonzero matrices in $M_2(F)$ whose product is zero. Thus the elementwise condition does not give the expected result in the noncommutative setting.

The zero ideal of a division ring does satisfy the elementwise definition of a prime ideal. Another situation in which the elementwise condition holds is if P is an ideal for which the factor ring R/P is an integral domain. The ideal $P = \begin{bmatrix} 0 & 0 \\ \mathbf{Z} & \mathbf{Z} \end{bmatrix}$ of the triangular matrix ring $R = \begin{bmatrix} \mathbf{Z} & 0 \\ \mathbf{Z} & \mathbf{Z} \end{bmatrix}$ provides such an example.

In the general (noncommutative) situation we end up with two different definitions, each of which generalizes the definition used for commutative rings. In the first of the two definitions (which we will rarely use) we say that an ideal P of the ring R is *completely prime* if $ab \in P$ implies $a \in P$ or $b \in P$, for all $a, b \in R$. The ring R is called a *domain* if (0) is a completely prime ideal (see Definition 1.1.6). Since an element c of R is *regular* if $xc = 0$ or $cx = 0$ implies $x = 0$, for all $x \in R$, it follows that the ring R is a domain if and only if every nonzero element of R is regular.

For the second definition of a prime ideal we turn to the ideal theoretic characterization of a prime ideal given in Proposition 1.3.2 (a). Using the definition $AB = \left\{ \sum_{j=1}^{n} a_j b_j \mid a_j \in A,\ b_j \in B \right\}$ of the product of two ideals A and B, an ideal P of a commutative ring R is prime if and only if $AB \subseteq P$ implies $A \subseteq P$ or $B \subseteq P$, for all ideals A and B of R. Consider the ring $R = \begin{bmatrix} \mathbf{Z} & \mathbf{Z} \\ \mathbf{Z} & \mathbf{Z} \end{bmatrix} = M_2(\mathbf{Z})$. If p is a prime number, then the ideal $P = M_2(p\mathbf{Z})$ is not completely prime since for $a = \begin{bmatrix} 1 & 0 \\ 0 & 0 \end{bmatrix}$ and $b = \begin{bmatrix} 0 & 0 \\ 0 & 1 \end{bmatrix}$ we have $ab \in P$, even though $a \notin P$ and $b \notin P$. On the other hand, if A and B are ideals of R with $AB \subseteq P$, then Proposition 1.5.4 shows that $A = M_2(n\mathbf{Z})$ and $B = M_2(m\mathbf{Z})$ for some $n, m \in \mathbf{Z}$. Since $AB \subseteq P$, we have $(n\mathbf{Z})(m\mathbf{Z}) \subseteq p\mathbf{Z}$, so either $n\mathbf{Z} \subseteq p\mathbf{Z}$ or $m\mathbf{Z} \subseteq p\mathbf{Z}$. Thus either $A \subseteq P$ or $B \subseteq P$, and so P does satisfy the ideal theoretic characterization of a prime ideal. In the noncommutative setting we will take this as the definition.

In addition to prime ideals, we also consider ideals that are intersections of prime ideals. If N is an intersection of prime ideals, we have the following condition: for any ideal I, if $I^k \subseteq N$, then $I \subseteq N$. At this point we need the following definition. We say that an ideal I of the ring R is a *nilpotent ideal* if $I^k = (0)$ for some positive integer k. Proposition 3.1.7 will show that if N is an intersection of prime ideals, then the factor ring R/N has no nontrivial nilpotent ideals.

For commutative rings we considered maximal ideals, as well as prime ideals, and showed (in Proposition 1.3.2 (c)) that any maximal ideal is prime. In the general case there is a useful intermediate notion, that of a primitive ideal.

There is also a corresponding definition of a primitive ring, and these rings will be important in later work.

Definition 3.1.1 *Let R be a ring.*

(a) *A proper ideal P of R is called a* prime ideal *if $AB \subseteq P$ implies $A \subseteq P$ or $B \subseteq P$, for any ideals A, B of R.*

(b) *A proper ideal I of R is called a* semiprime ideal *if it is an intersection of prime ideals of R.*

(c) *A proper ideal I of R is called a* left primitive ideal *if it is the annihilator of a simple left R-module.*

Recall from Definition 1.2.10 (b) that the ring R is said to be a *simple ring* if (0) is a maximal ideal of R. In a similar way we identify rings for which the zero ideal has the properties defined above.

Definition 3.1.2 *Let R be a ring.*

(a) *The ring R is called a* prime ring *if (0) is a prime ideal.*

(b) *The ring R is called a* semiprime ring *if (0) is a semiprime ideal.*

(c) *The ring R is said to be a (left)* primitive ring *if (0) is a primitive ideal.*

Example 2.1.7 showed that if V is a vector space over the field F, then V is a faithful simple module when considered as a left module over $\mathrm{End}_F(V)$. Thus the ring $\mathrm{End}_F(V)$ of all linear transformations of V is a left-primitive ring.

We will show in Proposition 3.1.6 (b) that any primitive ideal is prime. The converse is false, as is shown by considering the ring $\mathbf{Z}$ of integers. This is certainly a prime ring, but since the simple $\mathbf{Z}$-modules are the abelian groups $\mathbf{Z}_p$, where p is a prime number, there does not exist a faithful simple $\mathbf{Z}$-module. Primitive rings will be studied in more detail in Section 3.3.

The next proposition gives several different ways to interpret the notion of a prime ideal.

Proposition 3.1.3 *The following conditions are equivalent for the proper ideal P of the ring R:*

(1) *P is a prime ideal;*

(2) *$AB \subseteq P$ implies $A = P$ or $B = P$, for any ideals A, B of R with $P \subseteq A$ and $P \subseteq B$;*

(3) *$AB \subseteq P$ implies $A = P$ or $B = P$, for any left ideals A, B of R with $P \subseteq A$ and $P \subseteq B$;*

(4) *$aRb \subseteq P$ implies $a \in P$ or $b \in P$, for any $a, b \in R$.*

Proof. First, it is clear that (2) is a special case of (1). To show that (2) implies (3), we only need to observe that if A and B are left ideals which contain P and satisfy $AB \subseteq P$, then the same condition holds for the two-sided ideals AR and BR. To show that (3) implies (4), note that if $aRb \subseteq P$, then $(Ra + P)(Rb + P) \subseteq P$, for the left ideals $Ra + P$ and $Rb + P$. Finally, to show that (4) implies (1), suppose that A, B are ideals for which neither A nor B is contained in P. Let $a \in A \setminus P$ and $b \in B \setminus P$. Then by assumption aRb is not contained in P, so certainly AB is not contained in P. $\square$

Example 3.1.1 (Commutative rings)

Let R be a commutative ring. The general definition of a prime ideal (in the noncommutative case) reduces to the familiar one for commutative rings. Condition (4) of Proposition 3.1.3 shows that for any ideal P we have $aRb = Rab \subseteq P$ if and only if $ab \in P$.

Next consider a primitive ideal P of R. By definition there exists a simple R-module S with $P = \text{Ann}(S)$, and if x is a nonzero element of S, then $Rx = S$ since S is simple. If $a \in \text{Ann}(x)$, then $a(rx) = r(ax) = 0$ for all $r \in R$, since R is commutative, showing that $\text{Ann}(x) = \text{Ann}(S)$. Thus P is a maximal ideal of R since $R/P = R/\text{Ann}(x) \cong Rx = S$.

We now show that prime ideals correspond to the annihilators of certain modules, just as primitive ideals correspond to annihilators of simple modules.

Definition 3.1.4 *The nonzero module $_RM$ is called a* prime module *if*

$$AN = (0) \text{ implies } N = (0) \text{ or } AM = (0),$$

for any ideal A of R and any submodule N of M.

It follows immediately from the definition that any simple left R-module is prime. It follows from condition (3) of Proposition 3.1.3 that an ideal P is prime if and only if the left R-module R/P is a prime module. This gives us two standard examples of prime modules.

Proposition 3.1.5 *The annihilator of a prime module is a prime ideal.*

Proof. Let $_RM$ be a prime module, and suppose that A, B are ideals with $AB \subseteq \text{Ann}(M)$. If $BM = (0)$, then $B \subseteq \text{Ann}(M)$. On the other hand, suppose that $BM \neq (0)$. Then since M is prime and $A(BM) = (0)$, it follows from the definition of a prime module that $AM = (0)$. Thus $A \subseteq \text{Ann}(M)$. $\square$

Since the annihilator of a prime module is a prime ideal, it follows that R is a prime ring if and only if there exists a faithful prime left R-module.

Proposition 3.1.6 *Let R be any ring.*

(a) *Any maximal ideal of R is (left) primitive.*

(b) *Any (left) primitive ideal of R is prime.*

(c) *If R is a left Artinian ring, then the notions of maximal ideal, (left) primitive ideal, and prime ideal coincide.*

Proof. (a) Let I be a maximal ideal of the ring R. Then I is contained in a maximal left ideal A of R, by Corollary 2.1.15, and $I \subseteq \text{Ann}(R/A)$. Since I is a maximal ideal of R, we must have $I = \text{Ann}(R/A)$, and thus I is the annihilator of a simple left R-module.

(b) If P is a primitive ideal of R, then it is the annihilator of a simple left R-module, which is a prime module. It follows from Proposition 3.1.5 that P is a prime ideal.

(c) Assume that R is a left Artinian ring and P is a prime ideal of R. Since P is a maximal ideal if and only if R/P is a simple ring, it suffices to show that any prime Artinian ring is simple, so we may assume that R is a prime ring.

Since R is left Artinian, it contains a minimal left ideal M, and $\text{Ann}(M) = (0)$ since R is a prime ring. The set of all left annihilators of finite subsets of M must have a minimal element A, since R satisfies the minimum condition for left ideals. Thus $A = \text{Ann}(m_1, \ldots, m_n)$, for elements $m_1, \ldots, m_n \in M$, and we must have $A = \text{Ann}(M)$ since otherwise we could construct a smaller left annihilator of a finite subset of M.

Define the R-homomorphism $f : R \to M^n$ by $f(r) = (rm_1, \ldots, rm_n)$, for all $r \in R$. Then $\ker(f) = \text{Ann}(m_1, \ldots, m_n) = \text{Ann}(M)$, so f must be one-to-one since $\text{Ann}(M) = (0)$. Since M is simple, this shows that R is isomorphic to a submodule of a semisimple left R-module. Thus $_RR$ is semisimple by Theorem 2.3.2, and hence completely reducible by Corollary 2.3.3.

If I is a proper nonzero ideal of R, then since $_RR$ is completely reducible there exists a nonzero left ideal A of R with $R = I \oplus A$. This contradicts the assumption that R is prime, since $IA \subseteq I \cap A = (0)$. We conclude that R cannot have a proper nonzero ideal, so it is a simple ring. $\square$

We next characterize semiprime ideals. Condition (5) of the next proposition implies that an ideal I of a commutative ring R is semiprime if and only if $a^n \in I$ implies $a \in I$, for all $a \in R$.

Proposition 3.1.7 *The following conditions are equivalent for the proper ideal I of the ring R:*

(1) *I is a semiprime ideal;*

(2) *the ring R/I has no nonzero nilpotent ideals;*

(3) *$AB \subseteq I$ implies $A \cap B \subseteq I$, for any ideals A, B of R;*

(4) *$AB \subseteq I$ implies $A \cap B = I$, for any left ideals A, B of R with $I \subseteq A$ and $I \subseteq B$;*

(5) *$aRa \subseteq I$ implies $a \in I$, for all $a \in R$.*

Proof. (1) $\Rightarrow$ (2) If N/I is a nilpotent ideal of R/I, then $N^k \subseteq I$ for some positive integer k, and this shows that N is contained in every prime ideal that contains I. Since I is an intersection of prime ideals, this forces $N \subseteq I$.

(2) $\Rightarrow$ (3) If $AB \subseteq I$, then $(A \cap B)^2 \subseteq I$, since $A \cap B \subseteq A$ and $A \cap B \subseteq B$, and so $A \cap B \subseteq I$.

(3) $\Rightarrow$ (4) If $AB \subseteq I$ for left ideals A, B of R, then $(AR)(BR) \subseteq I$ for the ideals AR and BR. Therefore $A \cap B \subseteq AR \cap BR \subseteq I$.

(4) $\Rightarrow$ (5) If $aRa \subseteq I$, then $(Ra + I)(Ra + I) \subseteq I$, and so $Ra \subseteq I$.

(5) $\Rightarrow$ (1) Suppose that $a \in R \setminus I$. Then there exists $r \in R$ such that $ara \notin I$, and if we let $a_1 = ara$, then there exists $r_1 \in R$ such that $a_1 r_1 a_1 \notin I$. Continuing by induction we obtain a sequence $\{a, a_1, a_2, \ldots\} = \mathcal{A}$ of elements that lie outside I. An easy application of Zorn's lemma shows that there exists an ideal P maximal with respect to $P \cap \mathcal{A} = \emptyset$.

Let A and B be ideals properly containing P. Then there exist $a_m, a_k \in \mathcal{A}$ with $a_m \in A$ and $a_k \in B$. If n is the maximum of m and k, then $a_n \in A \cap B$, so $a_{n+1} = a_n r_n a_n \in AB \setminus P$. Thus AB properly contains P, showing that P is prime. Since $a \notin P$, we have shown that each element not in I belongs to some prime ideal, and so I is the intersection of all prime ideals of R that contain it. $\square$

Our next task is to provide a setting in which we can exploit the fact that a primitive ideal is the annihilator of a simple module. We will make crucial use of the fact (proved in Lemma 2.1.10) that the endomorphism ring of a simple module is a division ring.

Recall that $\mathrm{End}_R(M)$ is used to denote the set of R-endomorphisms of the module $_R M$ (see Definition 2.1.6). If $\rho : R \to \mathrm{End}_Z(M)$ is the representation of R that defines the left R-module structure on M, then $\mathrm{End}_R(M)$ is the subring of $\mathrm{End}_Z(M)$ defined as follows:

$$\begin{aligned} \mathrm{End}_R(M) &= \{f \in \mathrm{End}_Z(M) \mid f(rm) = rf(m) \text{ for all } r \in R \text{ and } m \in M\} \\ &= \{f \in \mathrm{End}_Z(M) \mid f \cdot \rho(r) = \rho(r) \cdot f \text{ for all } r \in R\}. \end{aligned}$$

Thus $\operatorname{End}_R(M)$ is the centralizer in $\operatorname{End}_Z(M)$ of the image of R under the representation $\rho : R \to \operatorname{End}_Z(M)$. Since $\operatorname{End}_R(M)$ is a subring of $\operatorname{End}_Z(M)$, the group M also defines a representation of $\operatorname{End}_R(M)$, and so M has the structure of a left module over $\operatorname{End}_R(M)$.

If we repeat the above procedure and consider the centralizer of $\operatorname{End}_R(M)$ in $\operatorname{End}_Z(M)$, we obtain the double centralizer of $\rho(R)$ in $\operatorname{End}_Z(M)$, which is also the endomorphism ring of M considered as a left $\operatorname{End}_R(M)$-module. We use the following terminology and notation.

Definition 3.1.8 *Let $_R M$ be a left R-module. The* biendomorphism ring *of M is the subring of $\operatorname{End}_Z(M)$ defined by*

$$\operatorname{BiEnd}_R(M) = \{\alpha \in \operatorname{End}_Z(M) \mid f\alpha = \alpha f \text{ for all } f \in \operatorname{End}_R(M)\}.$$

Lemma 3.1.9 *If M is any left R-module, then $R/\operatorname{Ann}(M)$ is isomorphic to a subring of $\operatorname{BiEnd}_R(M)$.*

Proof. Let $\rho : R \to \operatorname{End}_Z(M)$ be the representation that defines the action of R on M. By definition, for any $r \in R$ the endomorphism $\rho(r)$ must commute with all elements in the centralizer of $\rho(R)$, and so $\rho(r)$ belongs to $\operatorname{BiEnd}_R(M)$. Thus $\rho : R \to \operatorname{BiEnd}_R(M)$, and since $\ker(\rho) = \operatorname{Ann}(M)$, it follows from Theorem 1.2.7 that $R/\operatorname{Ann}(M)$ is isomorphic to a subring of $\operatorname{BiEnd}_R(M)$. $\square$

Proposition 3.1.10 *For any ring R the following conditions hold.*

(a) $\operatorname{End}_R(R) \cong R^{op}$.

(b) $\operatorname{BiEnd}_R(R) = R$.

Proof. Let ρ be the regular representation $\rho : R \to \operatorname{End}_Z(R)$, defined by $[\rho(r)](x) = rx$, for all $r, x \in R$. If $r \neq 0$, then $\rho(r) \neq 0$, since $[\rho(r)](1) = r$, and so ρ is one-to-one. Thus R is isomorphic to the subring $\rho(R)$ of $\operatorname{End}_Z(R)$.

(a) For any $a \in R$, define $\mu_a : R \to R$ by $\mu_a(x) = xa$, for all $x \in R$. Then $\mu_a \in \operatorname{End}_R(R)$, since

$$\mu_a(rx) = (rx)a = r(xa) = r\mu_a(x)$$

for all $r, x \in R$. If f is any R-endomorphism in $\operatorname{End}_R(R)$, then for all $x \in R$ we have $f(x) = xf(1)$, showing that $f = \mu_a$, for $a = f(1)$. Thus there is a one-to-one correspondence between elements of R and elements of $\operatorname{End}_R(R)$. We note that $\mu_a \mu_b = \mu_{ba}$, since $\mu_a(\mu_b(1)) = ba = \mu_{ba}(1)$.

The function $\phi : R \to \operatorname{End}_R(R)$ defined by $\phi(a) = \mu_a$, for all $a \in R$, is an anti-isomorphism since $\phi(ab) = \mu_{ab}$, and $\phi(a)\phi(b) = \mu_a \mu_b = \mu_{ba} = \phi(ba)$. It follows that $\operatorname{End}_R(R)$ is isomorphic to the opposite ring R^{op}.

(b) Given $a \in R$, define μ_a as in part (a). If $\alpha \in \mathrm{BiEnd}_R(R)$, then we have $\mu_a \alpha = \alpha \mu_a$ for all $a \in R$. Thus

$$\alpha(a) = \alpha(\mu_a(1)) = \mu_a(\alpha(1)) = \alpha(1)a \;,$$

and so $\alpha = \rho(\alpha(1))$. Therefore $\mathrm{BiEnd}_R(R) = R$, completing the proof. $\square$

Proposition 3.1.11 *If P is a primitive ideal of the ring R, then there exist a division ring D and a vector space V over D for which R/P is isomorphic to a subring of the ring of all linear transformations from V into V.*

Proof. If P is a primitive ideal, then there exists a simple left R-module S with $P = \mathrm{Ann}(S)$. By Schur's lemma, the endomorphism ring $\mathrm{End}_R(S)$ is a division ring, which we denote by D. Then S is naturally a left D-module, since $D = \mathrm{End}_R(S)$ is a subring of $\mathrm{End}_Z(S)$. Thus S is a vector space over D, and $\mathrm{BiEnd}_R(S) = \mathrm{End}_D(S)$ is the ring of all D-linear transformations from V into V. Lemma 3.1.9 shows that R/P is isomorphic to a subring of $\mathrm{End}_D(S) = \mathrm{BiEnd}_R(S)$. $\square$

The next proposition will be used in the proof of the Artin–Wedderburn theorem.

Proposition 3.1.12 *Let M be a left R-module.*

(a) *The endomorphism ring $\mathrm{End}_R(M^n)$ is isomorphic to the ring of $n \times n$ matrices with entries in $\mathrm{End}_R(M)$.*

(b) *The biendomorphism ring $\mathrm{BiEnd}_R(M^n)$ is isomorphic to $\mathrm{BiEnd}_R(M)$.*

Proof. (a) For $1 \le j \le n$ we have the inclusion mapping $i_j : M \to M^n$ that takes M into the jth component of M^n and the projection mapping $p_j : M^n \to M$ defined by $p_j(m_1, \ldots, m_j, \ldots, m_n) = m_j$. For any endomorphism $f \in \mathrm{End}_R(M^n)$, we define $f_{jk} = p_j f i_k \in \mathrm{End}_R(M)$. By Proposition 2.2.6, for any element $x \in M^n$ we have $x = \sum_{k=1}^n i_k p_k(x)$ and $f(x) = \sum_{j=1}^n i_j p_j f(x)$. Substituting for x in the second expression yields

$$
\begin{aligned}
f(x) &= \textstyle\sum_{j=1}^n i_j \left(\sum_{k=1}^n p_j f i_k (p_k(x)) \right) \\
&= \textstyle\sum_{j=1}^n i_j \left(\sum_{k=1}^n f_{jk}(p_k(x)) \right) \;.
\end{aligned}
$$

This can be written in matrix format by representing x as the column vector that is the transpose of $(p_1(x), \ldots, p_n(x))$ and multiplying by the $n \times n$ matrix with entries f_{jk}. Then $\sum_{k=1}^n f_{jk}(p_k(x))$ is the jth component of the column vector that represents $f(x)$.

(b) Let ρ be the representation that defines the module structure on M, so that we have $\rho : R \to \mathrm{End}_Z(M)$. Then the module structure on M^n is given by $\lambda : R \to \text{End}_Z(M^n)$, where $[\lambda(r)](m_1, \ldots, m_n) = (rm_1, \ldots, rm_n)$. We can view $\text{End}_Z(M^n)$ as a ring of $n \times n$ matrices over $\text{End}_Z(M)$, and so for $r \in R$, the endomorphism $\lambda(r)$ corresponds to a 'scalar' matrix with entries $\rho(r)$.

The subring $\text{End}_R(M^n)$ of $\text{End}_Z(M^n)$ consists of the matrices with entries in $\text{End}_R(M)$. This includes all of the matrix units e_{ij}, in which the (i, j)-entry is the identity function $1_M : M \to M$ and all other entries are the zero function. If $[f_{ij}] \in \text{End}_Z(M^n)$ commutes with all matrix units, then $e_{kk}[f_{ij}] = [f_{ij}]e_{kk}$ implies $a_{ik} = 0$ for $i \neq k$ and $a_{kj} = 0$ for $j \neq k$, and $e_{1k}[f_{ij}] = [f_{ij}]e_{1k}$ implies that $f_{kk} = f_{11}$. Thus $[f_{ij}]$ is a 'scalar' matrix, with the same entry f_{11} on the main diagonal and zeros elsewhere. It follows that the centralizer of $\text{End}_R(M^n)$ in $\text{End}_Z(M^n)$ is the set of all 'scalar' matrices whose entries belong to the centralizer of $\text{End}_R(M)$. This implies that $\text{BiEnd}_R(M^n)$ can be described as all 'scalar' matrices with entries in $\text{BiEnd}_R(M)$, and this subring of $\text{End}_Z(M^n)$ is isomorphic to $\text{BiEnd}_R(M)$. $\square$

EXERCISES: SECTION 3.1

1. Show that the intersection of any collection of semiprime ideals is a semiprime ideal.

2. Let R be the ring $\begin{bmatrix} \mathbf{Z} & 0 \\ \mathbf{Z} & \mathbf{Z} \end{bmatrix}$ of lower triangular matrices with integer entries. Show that each prime ideal of R has the form $\begin{bmatrix} p\mathbf{Z} & 0 \\ \mathbf{Z} & \mathbf{Z} \end{bmatrix}$ or $\begin{bmatrix} \mathbf{Z} & 0 \\ \mathbf{Z} & q\mathbf{Z} \end{bmatrix}$, for primes $p, q \in \mathbf{Z}$.

3. Prove that in a commutative ring every primitive ideal is maximal.

4. Prove that the center of a simple ring is a field.

5. Prove that a prime ring with a minimal left ideal must be (left) primitive.

6. Let S be a simple left R-module. Let M be the set of all n-tuples of elements of S, considered as column vectors. Show that the usual matrix multiplication defines on M the structure of a simple left $\text{M}_n(R)$-module.

7. Let A be a left ideal of the ring R. In $\text{M}_n(R)$, let I_k be the set of all matrices whose entries in column k come from A, while all other entries are zero. Let J_k be the set of all matrices whose entries in column k come from A, while all other entries are unrestricted.
   (a) Show that I_k and J_k are left ideals of $\text{M}_n(R)$.
   (b) Show that if A is a minimal left ideal, then so is I_k.
   (c) Show that if A is a maximal left ideal, then so is J_k.

8. Prove that if the ring R is prime, then so is the $n \times n$ matrix ring $\text{M}_n(R)$.

9. Let e be a nonzero idempotent element of the ring R (that is, $e \neq 0$ and $e^2 = e$). Prove that if R is a primitive ring, then so is the ring eRe.

10. Let R be a set that satisfies all of the axioms of a ring, with the possible exception of the existence of a multiplicative identity element. Prove that R must have an identity element if it satisfies the following conditions: (i) every nonempty set of left ideals of R contains a minimal element; (ii) if A is a nonzero left ideal, or a nonzero right ideal, then $A^2 \neq (0)$.

3.2 The Jacobson radical

Wedderburn showed that any finite dimensional algebra over a field has a unique maximal nilpotent ideal, which is referred to as the *nilpotent radical* of the algebra. He then proved that any finite dimensional algebra over a perfect field is equal to the direct sum (as a vector space) of its nilpotent radical and a semisimple algebra.

In 1943 Jacobson developed a structure theory for rings without finiteness assumptions. Considering a simple module over the ring R as a representation of the ring, he defined what is now known as the Jacobson radical to be the intersection of the kernels of all representations of R by simple left R-modules. In other words, he defined the radical of R to be the intersection of all (left) primitive ideals of R.

We present a somewhat more general point of view, by considering radicals for modules. In particular, we are interested in how closely a given module can be related to simple modules. A submodule is minimal if and only if it is simple; it is maximal if and only if the corresponding factor module is simple. For a given module M we therefore consider both simple submodules and simple factor modules (working both from the bottom up and from the top down).

Recall from Definition 2.3.1 that the sum of all minimal submodules of M is the *socle* of M, denoted by $\mathrm{Soc}(M)$. The module M is called *semisimple* if it can be expressed as a sum of minimal submodules. If $_R M$ is any module, and $_R S$ is any simple module, we can consider the sum in M of all submodules isomorphic to S. This submodule, if it is nonzero, is called a *homogeneous component* of the socle of M. It is left as an exercise to show that $\mathrm{Soc}(M)$ is a direct sum of its homogeneous components.

Definition 3.2.1 *Let M be a left R-module. The intersection of all maximal submodules of M is called the* Jacobson radical *of M, and is denoted by* $\mathrm{J}(M)$. *If M has no maximal submodules, then* $\mathrm{J}(M) = M$.

The socle and Jacobson radical of a module can be characterized in a variety of ways. To do so we need some additional definitions.

Definition 3.2.2 *Let M be a left R-module.*

(a) *The submodule N of M is called* essential *or* large *in M if $N \cap K \neq (0)$ for all nonzero submodules K of M.*

(b) *The submodule N is called* superfluous *or* small *in M if $N + K \neq M$ for all proper submodules K of M.*

Example 3.2.1 (Essential ideals in prime rings)

> We will show that an ideal of a prime ring is essential as a left ideal if and only if it is nonzero. In particular, any nonzero ideal of an integral domain must be essential. Let I be any nonzero two-sided ideal of the prime ring R. If A is any nonzero left ideal of R, then $IA \neq (0)$, and so this implies that $I \cap A \neq (0)$. On the other hand, any essential ideal must be nonzero.

It is clear that a submodule $N \subseteq M$ is essential if and only if it has nontrivial intersection with every nontrivial cyclic submodule of M. This gives a condition on elements: N is an essential submodule of M if and only if for each nonzero element $m \in M$ there exists $r \in R$ such that $rm \in N$ and $rm \neq 0$. It then follows that for any R-homomorphism, the inverse image of an essential submodule is essential. Furthermore, if $M_1 \subseteq M_2 \subseteq M$ are submodules, then M_1 is essential in M if and only if M_1 is essential in M_2 and M_2 is essential in M.

Proposition 3.2.3 *Let N be a submodule of $_R M$. If K is maximal in the set of all submodules of M that have trivial intersection with N, then $N + K$ is essential in M, and $(N + K)/K$ is essential in M/K.*

Proof. Let K be maximal such that $K \cap N = (0)$. We will first show that $(N + K)/K$ is essential in M/K. If $L \supset K$ with $(L/K) \cap (N + K)/K = (0)$, then we have $L \cap (N + K) = K$. By Theorem 2.1.12, we have $(L \cap N) + K = K$, so $L \cap N \subseteq K$ and therefore $L \cap N = 0$. This contradicts the choice of K, and so we have shown that $(N + K)/K$ has nonzero intersection with each nonzero submodule of M/K.

Finally, $N + K$ is the inverse image of $(N + K)/K$ under the projection of M onto M/K, and therefore it is essential in M.   □

Proposition 3.2.4 *The socle of any module is the intersection of its essential submodules.*

Proof. Let $_RM$ be any module. If S is a simple submodule of M and N is an essential submodule of M, then $N \cap S \neq (0)$ and therefore $S \subseteq N$. It follows that the intersection of all essential submodules contains all simple submodules and hence it contains their sum. Thus $\mathrm{Soc}(M)$ is contained in the intersection of all essential submodules of M.

Suppose that $m \in M \setminus \mathrm{Soc}(M)$. By Proposition 2.1.14 there exists a submodule N maximal with respect to $N \supseteq \mathrm{Soc}(M)$ and $m \notin N$. If we can show that N is an essential submodule, then m lies outside an essential submodule, and so $\mathrm{Soc}(M)$ is equal to the intersection of all essential submodules of M.

To show that N is an essential submodule of M, suppose that $K \cap N = (0)$ for some nonzero submodule K. By Proposition 2.1.14 the image of K in M/N contains a minimal submodule, and since this image is isomorphic to K, it follows that $K \cap \mathrm{Soc}(M) \neq (0)$. This contradicts the choice of N, and completes the proof. $\square$

Before studying the Jacobson radical, it is convenient to study radicals from a general point of view. Of course, these are radicals of modules. There is also an extensive (but quite different) theory of radicals of rings. The general idea is to factor out a 'bad' part, in some sense. In the case of the Jacobson radical, it is hoped that the part that is left, $M/J(M)$, is as close as possible to a semisimple module.

Definition 3.2.5 *A radical for the class of left R-modules is a function that assigns to each module $_RM$ a submodule $\tau(M)$ such that*

(i) $f(\tau(M)) \subseteq \tau(N)$, *for all modules $_RN$ and all $f \in \mathrm{Hom}_R(M,N)$,*

(ii) $\tau(M/\tau(M)) = (0)$.

We note that the socle does not define a radical, since it does not in general satisfy condition (ii) of the above definition. To see this, just note that the socle of the **Z**-module **Z**$_4$ is $2\mathbf{Z}_4$, and that $\mathbf{Z}_4/\mathrm{Soc}(\mathbf{Z}_4)$ is a simple **Z**-module.

Example 3.2.2 (The radical of a direct sum)

Let $M = M_1 \oplus \cdots \oplus M_n$ be a direct sum of left R-modules, with inclusion and projection mappings $i_j : M_j \to M$ and $p_j : M \to M_j$. We will show that if τ is a radical, then

$$\tau(M_1 \oplus \cdots \oplus M_n) = \tau(M_1) \oplus \cdots \oplus \tau(M_n).$$

For each index j we have $i_j(\tau(M_j)) \subseteq \tau(M)$, and so

$$\tau(M_1) \oplus \cdots \oplus \tau(M_n) \subseteq \tau(M).$$

We also have $p_j(\tau(M)) \subseteq \tau(M_j)$, and so

$$\tau(M) \subseteq p_j^{-1}(\tau(M_j)) = M_1 \oplus \cdots \oplus \tau(M_j) \oplus \cdots \oplus M_n.$$

Therefore $\tau(M) \subseteq \tau(M_1) \oplus \cdots \oplus \tau(M_n)$, which completes the proof.

The next definition gives a general method for constructing a radical, given any class of modules. Proposition 3.2.7 will show that every radical is of this form. Our primary interest lies in the fact that the Jacobson radical can be constructed in this way, from the class of simple modules.

Definition 3.2.6 *Let C be any class of left R-modules. For any module $_RM$ we make the following definition:*

$$\text{rad}_C(M) = \bigcap\nolimits_{f:M \to X,\; X \in C} \ker(f) \;.$$

Proposition 3.2.7 *Let τ be a radical for the class of left R-modules, and let $\mathcal{F}$ be the class of left R-modules X for which $\tau(X) = (0)$.*

(a) *The subset $\tau(R)$ is a two-sided ideal of R.*

(b) *For all modules $_RM$, we have $\tau(R)M \subseteq \tau(M)$.*

(c) *The construction $\text{rad}_{\mathcal{F}}$ defines a radical, and $\tau = \text{rad}_{\mathcal{F}}$.*

(d) *The radical of R is $\tau(R) = \bigcap_{X \in \mathcal{F}} \text{Ann}(X)$.*

Proof. (a) Since τ is a radical, we have $f(\tau(R)) \subseteq \tau(R)$ for all $f \in \text{End}_R(R)$. But the endomorphisms of $_RR$ coincide with the functions determined by right multiplication by elements of R, so $\tau(R)$ is invariant under right multiplication.

(b) For any $m \in M$, we have $\tau(R)m = f(\tau(R))$ for the R-homomorphism $f_m : R \to M$ defined by $f_m(r) = rm$ for all $r \in R$. Since τ is a radical, this implies that $\tau(R)m \subseteq \tau(M)$.

(c) Let $m \in \text{rad}_{\mathcal{F}}(M)$ and $g \in \text{Hom}_R(M, N)$. Then for any $f \in \text{Hom}_R(N, X)$ with $X \in \mathcal{F}$ we have $fg(m) = 0$, so $g(m) \in \ker(f)$. Thus $g(\text{rad}_{\mathcal{F}}(M)) \subseteq \text{rad}_{\mathcal{F}}(N)$. If $m \notin \text{rad}_{\mathcal{F}}(M)$, then there exists $f \in \text{Hom}_R(M, X)$ with $f(m) \neq 0$ for some $X \in \mathcal{F}$. But $\text{rad}_{\mathcal{F}}(M) \subseteq \ker(f)$, and so f is defined on $M/\text{rad}_{\mathcal{F}}(M)$ and $f(\overline{m}) \neq 0$ for the coset $\overline{m}$ containing m. Thus $\overline{m} \notin \text{rad}_{\mathcal{F}}(M/\text{rad}_{\mathcal{F}}(M))$, showing that $\text{rad}_{\mathcal{F}}(M/\text{rad}_{\mathcal{F}}(M)) = (0)$. This completes the proof that $\text{rad}_{\mathcal{F}}$ is a radical.

If $f \in \text{Hom}_R(M, X)$, with $X \in \mathcal{F}$, then $f(\tau(M)) \subseteq \tau(X) = (0)$, and so we have $\tau(M) \subseteq \ker(f)$. This implies that $\tau(M) \subseteq \text{rad}_{\mathcal{F}}(M)$. On the other hand, $\tau(M)$ is the kernel of the projection onto $M/\tau(M)$, which belongs to $\mathcal{F}$ since τ is a radical. Thus $\text{rad}_{\mathcal{F}}(M) \subseteq \tau(M)$, and so $\tau = \text{rad}_{\mathcal{F}}$.

(d) Since $1 \in R$, any R-homomorphism $f : R \to X$ has the form $f(r) = rf(1)$, for all $r \in R$, so $\ker(f) = \mathrm{Ann}(f(1))$. Thus $\mathrm{Ann}(X) = \bigcap_{f:R \to X} \ker(f)$, and so $\mathrm{rad}_{\mathcal{F}}(R) = \bigcap_{X \in \mathcal{F}} \mathrm{Ann}(X)$. $\square$

We now observe that the Jacobson radical is just $\mathrm{rad}_{\mathcal{S}}$, where $\mathcal{S}$ is the class of all simple left R-modules. Since an ideal is primitive if it is the annihilator of a simple left R-module, it follows from the previous proposition that $J(R)$ is the intersection of all (left) primitive ideals of R.

Example 3.2.3 (The Jacobson radical of a finite abelian group)

In a cyclic group of order p^k, where p is a prime number, the subgroups are linearly ordered. Thus the group has a unique maximal subgroup, and this is the Jacobson radical of the group. Note that the corresponding factor group is isomorphic to the simple abelian group $\mathbf{Z}_p$. In the general case, any finite abelian group is isomorphic to a direct sum of cyclic groups of prime power order (recall Proposition 2.7.8). The Jacobson radical of the group can be computed componentwise, using Example 3.2.2.

The next result is usually referred to as Nakayama's lemma. Nakayama himself credited Jacobson for the special case of a left ideal, and Azumaya for the general case of a finitely generated module.

Lemma 3.2.8 (Nakayama) *If $_R M$ is finitely generated and $J(R)M = M$, then $M = (0)$.*

Proof. If M is finitely generated and nonzero, then by Corollary 2.1.15 it has a proper maximal submodule, and so it follows from Proposition 3.2.7 (b) that $J(R)M \subseteq J(M) \neq M$. $\square$

Proposition 3.2.9 *Let M be a left R-module.*

(a) $J(M) = \{m \in M \mid Rm$ *is small in* $M\}$.

(b) $J(M)$ *is the sum of all small submodules of* M.

(c) *If M is finitely generated, then $J(M)$ is a small submodule.*

(d) *If M is finitely generated, then $M/J(M)$ is semisimple if and only if it is Artinian.*

Proof. (a) If $m \in M \setminus J(M)$, then $m \notin N$ for some maximal submodule N of M. It follows that Rm is not small, since $Rm + N = M$ but $N \neq M$.

Conversely, if Rm is not small in M, then there exists $N \neq M$ such that $Rm + N = M$. By Proposition 2.1.14 there exists a submodule $K \supseteq N$ maximal with respect to missing m. Then K is maximal in M since any larger submodule contains m and hence is equal to $M = Rm + N$. Thus $m \notin J(M)$.

(b) It suffices to show that $J(M)$ contains all small submodules. If N is small, then so is Rm, for all $m \in N$. It follows from part (a) that $m \in J(M)$ for all $m \in N$.

(c) Assume that M is finitely generated and $N + J(M) = M$ for some submodule $N \subseteq M$. The natural projection $p : M \to M/N$ maps $J(M)$ into $J(M/N)$, and so $J(M/N) = M/N$. Since M is finitely generated, so is M/N, and so Nakayama's lemma implies that $M/N = (0)$. Thus $N = M$ and $J(M)$ is a small submodule.

(d) If $M/J(M)$ is semisimple, then it is a direct sum of simple modules. Since it is finitely generated, this must be a finite direct sum, and then it obviously has a composition series.

Conversely, if $M/J(M)$ is Artinian, then $J(M)$ is a finite intersection of kernels of R-homomorphisms into simple modules, say $J(M) = \bigcap_{j=1}^{n} \ker(f_j)$, where $f_j : M \to S_j$ and S_j is simple. The mappings $\{f_j\}_{j=1}^{n}$ define an embedding of $M/J(M)$ into $\bigoplus_{j=1}^{n} S_j$, and so $M/J(M)$ is semisimple. $\square$

For any module $_RM$ we have $J(R)M \subseteq J(M)$, and so the preceding result implies that if M is finitely generated, then $J(R)M$ is a small submodule of M. This proves the following condition, which is sometimes stated as part of Nakayama's lemma: if M is finitely generated, then for any submodule N the condition $N + J(R)M = M$ implies $N = M$.

Theorem 3.2.10 *The Jacobson radical $J(R)$ of the ring R is equal to each of the following sets:*

(1) *the intersection of all maximal left ideals of R;*

(2) *the intersection of all maximal right ideals of R;*

(3) *the intersection of all left primitive ideals of R;*

(4) *the intersection of all right primitive ideals of R;*

(5) $\{x \in R \mid 1 - ax$ *is left invertible for all* $a \in R\}$;

(6) $\{x \in R \mid 1 - xa$ *is right invertible for all* $a \in R\}$;

(7) *the largest ideal J of R such that $1 - x$ is invertible in R for all $x \in J$.*

Proof. Condition (1) is just our original definition of the Jacobson radical of R, considered as a left R-module.

(1) $\Leftrightarrow$ (3) We have already observed that $J(R) = \text{rad}_{\mathcal{S}}$, where $\mathcal{S}$ is the class of all simple left R-modules. It follows from Proposition 3.2.7 that $J(R)$ is the intersection of the left annihilators of all simple left R-modules.

(1) $\Leftrightarrow$ (5) By Proposition 3.2.9, we know that $J(R) = \{x \in R \mid Rx \text{ is small}\}$. We will show that Rx is small in R if and only if $1 - ax$ is left invertible, for all $a \in R$. First suppose that Rx is small in R. For any nonzero element $a \in R$ we have $(1 - ax) + (ax) = 1$, which shows that $R(1 - ax) + Rx = R$. It follows that $R(1 - ax) = R$, and so $1 - ax$ has a left inverse.

Conversely, suppose that $1 - ax$ has a left inverse, for all $a \in R$, and that $I + Rx = R$ for some left ideal I. Then $y + ax = 1$ for some $y \in I$ and some $a \in R$, and so $R = Ry \subseteq I$ since $y = 1 - ax$ has a left inverse. Thus Rx is small in R.

(5) $\Leftrightarrow$ (7) If $1 - ax$ has a left inverse for all $a \in R$, then for $a = 1$ there exists $r \in R$ such that $r(1 - x) = 1$. Now $1 - r = -rx \in J(R)$, so as before there exists $s \in R$ such that $s(1 - (1 - r)) = 1$. Thus $sr = 1$, and so $s = s1 = sr(1 - x) = 1 - x$. Finally, $(1 - x)r = sr = 1$, and so $1 - x$ is invertible.

Let J be an ideal such that $1 - x$ is invertible for all $x \in J$. If $Rx + I = R$ for some $x \in J$ and some left ideal I, then $1 = rx + y$ for some $r \in R$ and some $y \in I$. Then $x \in J$ implies that $y = 1 - rx$ is invertible, so $I = R$. This shows that Rx is small in R, and thus $J \subseteq J(R)$.

(2) $\Leftrightarrow$ (4) $\Leftrightarrow$ (6) $\Leftrightarrow$ (7) The equivalence of these conditions follows as above, using the fact that condition (7) is left–right symmetric. $\square$

Example 3.2.4 (The Jacobson radical of a triangular matrix ring)

Let F be a field, and let R be the ring $\begin{bmatrix} F & 0 \\ F & F \end{bmatrix}$ of lower triangular matrices over F. If $x = \begin{bmatrix} 0 & 0 \\ a & 0 \end{bmatrix}$ is any element of R, then the element $1 - x$ is invertible, and so $J(R)$ contains the ideal $\begin{bmatrix} 0 & 0 \\ F & 0 \end{bmatrix}$. On the other hand, this ideal contains $J(R)$ since it is the intersection of two maximal left ideals: $\begin{bmatrix} 0 & 0 \\ F & F \end{bmatrix}$ and $\begin{bmatrix} F & 0 \\ F & 0 \end{bmatrix}$.

The fact that the intersection $\bigcap_p p\mathbf{Z}$ of all maximal ideals of $\mathbf{Z}$ is zero shows that $J(\mathbf{Z}) = (0)$. If F is any field, then we also have $J(F[x]) = (0)$. This can be seen by noting that only the zero ideal of $F[x]$ satisfies condition (7) of Theorem 3.2.10.

The role played by primitive ideals in the preceding theorem provides the motivation for naming rings with zero Jacobson radical. We should note that such rings are also called Jacobson semisimple by some authors.

Definition 3.2.11 *The ring R is said to be* semiprimitive *if* $J(R) = (0)$.

A very useful result is that any element that is invertible modulo the Jacobson radical is invertible. To see this, assume that $a + J(R)$ is invertible in $R/J(R)$. Then there exist $b \in R$ and $x, y \in J(R)$ with $ab = 1 - x$ and $ba = 1 - y$. Since $1 - x$ is invertible, it follows that ab has a right inverse, and so a has a right inverse. Similarly, a has a left inverse since $1 - y$ is invertible. Therefore a is invertible in R.

An ideal N of R is said to be *nil* if each element of N is nilpotent. As the final result of the section, we show that the Jacobson radical contains all nil ideals. For left Artinian rings, we show that the Jacobson radical is nilpotent, and so in this case the Jacobson radical is the largest nil ideal.

Proposition 3.2.12 *Let R be any ring.*

(a) *The Jacobson radical of R contains every nil ideal of R.*

(b) *If R is left Artinian, then the Jacobson radical of R is nilpotent.*

Proof. (a) Let N be a nil ideal of R, and let $x \in N$. Then there exists $n > 0$ with $x^n = 0$, so $(1-x)(1+x+\cdots+x^{n-1}) = (1+x+\cdots+x^{n-1})(1-x) = 1-x^n = 1$, and thus $1 - x$ is invertible. It follows that $N \subseteq J(R)$.

(b) If R is left Artinian, then the set of powers $J(R)^k$ of the Jacobson radical contains a minimal element, say $J(R)^n = I$, and we must have $I^2 = I$. If I is nonzero, then the set $\mathcal{K} = \{K \subseteq R \mid K$ is a left ideal and $IK \neq (0)\}$ is nonempty, so by assumption it contains a minimal element, say A, with $IA \neq (0)$. Then $Ia \neq (0)$ for some $a \in A$, and it follows from the minimality of A that $Ia = A$, since $I \cdot Ia \neq (0)$. Therefore $a = xa$ for some $x \in I$, and so $(1 - x)a = 0$. Since $x \in J(R)$, the element $1 - x$ is invertible, which implies that $a = 0$, a contradiction, We conclude that $J(R)^n = (0)$. □

EXERCISES: SECTION 3.2

1. Prove that if the ring R is primitive, then so is the $n \times n$ matrix ring $M_n(R)$.

2. For any positive integer n, compute the Jacobson radical $J(\mathbf{Z}_n)$ of the ring $\mathbf{Z}_n$.

3. Find the Jacobson radical of the ring of lower triangular $n \times n$ matrices over $\mathbf{Z}$.

4. Prove that if $_R M$ is a semisimple module, then $J(M) = (0)$.

5. Prove that for any module $_R M$, the socle $\mathrm{Soc}(M)$ can be written as a direct sum of its homogeneous components.

6. Let $M_1 \subseteq M_2 \subseteq M$ be R submodules of the module $_R M$. Show that M_1 is an essential submodule of M if and only if M_1 is an essential submodule of M_2 and M_2 is an essential submodule of M.

7. Let $\{M_\alpha\}_{\alpha \in I}$ be a collection of left R-modules, with submodules $\{N_\alpha\}_{\alpha \in I}$, where $N_\alpha \subseteq M_\alpha$. Prove that $\bigoplus_{\alpha \in I} N_\alpha$ is an essential submodule of $\bigoplus_{\alpha \in I} M_\alpha$.

8. For any ring R, prove that $J(M_n(R)) = M_n(J(R))$.

9. Prove that if e is a nonzero idempotent in the ring R, then $J(eRe) = eJ(R)e$.

10. Let V be a left vector space over a division ring D. A subring R of $\text{End}_D(V)$ is called a *dense subring* if for each $n > 0$, each linearly independent subset $\{u_1, u_2, \ldots, u_n\}$ of V, and each arbitrary subset $\{v_1, v_2, \ldots, v_n\}$ of V there exists an element $\theta \in R$ such that $\theta(u_i) = v_i$ for all $i = 1, \ldots, n$. Prove that any such subring must be a primitive ring.

11. Let R be a commutative ring. Prove the following extension of Nakayama's lemma: if M is a finitely generated R-module and I is an ideal of R with $IM = M$, then there exists $a \in R$ with $aM = (0)$ and $a + I = 1 + I$. Also prove that if $I \subseteq J(R)$, then $M = (0)$.

12. Let R be a commutative ring, let M be a finitely generated R-module, and let $f : M \to M$ be an R-homomorphism. Use Exercise 11 to prove that if f is onto, then it must also be one-to-one.

3.3 Semisimple Artinian rings

Recall that a ring R is left primitive if there exists a faithful simple left R-module. The most elementary example is the ring of all linear transformations of a vector space over a field. We have two goals. The first is to show that any left Artinian, left primitive ring is isomorphic to the ring of all linear transformations of a finite dimensional vector space over a division ring (the Artin–Wedderburn theorem). The second is to show, more generally, that any primitive ring is isomorphic to a dense subring of a ring of linear transformations of a vector space over a division ring (the Jacobson density theorem).

Theorem 3.3.1 *Any simple ring with a minimal left ideal is isomorphic to a ring of $n \times n$ matrices over a division ring.*

Proof. Let R be a simple ring with minimal left ideal S. Then S is simple as a left R-module, and so the homogeneous component of $\text{Soc}(R)$ determined by S is nonzero. Since this component of the socle is a two-sided ideal of R, it must be equal to R since R is simple, and so $_R R$ is a direct sum of minimal left ideals, each isomorphic to S. Because R is a ring with identity, this is a finite direct sum, say $R \cong S^n$. Proposition 3.1.10 (b) shows that $R \cong \text{BiEnd}_R(R)$, and so we have $R \cong \text{BiEnd}_R(S^n)$. Proposition 3.1.12 (b) implies that $R \cong \text{BiEnd}_R(S)$.

By Schur's lemma (Lemma 2.1.10), $\text{End}_R(S)$ is a division ring, which we will denote by D. We next show that S is a finite dimensional vector space over D. In the isomorphism $R \cong S^n$, let $(x_1, x_2, \ldots, x_n) \in S^n$ correspond to $1 \in R$.

We claim that $\{x_1, x_2, \ldots, x_n\}$ is a basis for $_DS$. Given $x \in S$, the mapping $f : R = S^n \to S$ with $f(r) = rx$ for all $r \in R$ has components $d_1, d_2, \ldots, d_n$ in $\mathrm{End}_R(S)$, and so $x = \sum_{j=1}^n d_j x_j$. It is clear that this representation is unique, since $\sum_{j=1}^n d_j x_j = 0$ implies that $f = 0$.

Now since $S \cong D^n$ as a left D-module, Proposition 3.1.12 (a) shows that

$$\mathrm{BiEnd}_R(S) = \mathrm{End}_D(S) \cong \mathrm{End}_D(D^n) \cong \mathrm{M}_n(\mathrm{End}_D(D)) \ .$$

Finally, $\mathrm{End}_D(D)$ is a division ring since D is a simple left D-module. $\square$

In summary, in the proof of Theorem 3.3.1 we have shown that if S is any simple left ideal of a simple ring R, then $R \cong \mathrm{M}_n(D^{op})$, where D is the division ring $\mathrm{End}_R(S)$.

Theorem 3.3.2 (Artin–Wedderburn) *For any ring R, the following conditions are equivalent:*

(1) *R is left Artinian and $\mathrm{J}(R) = (0)$;*

(2) *$_RR$ is a semisimple module;*

(3) *R is isomorphic to a finite direct sum of rings of $n \times n$ matrices over division rings.*

Proof. (1) $\Rightarrow$ (2) We know in general that if $_RM$ is Artinian, then $M/\mathrm{J}(M)$ is a semisimple module.

(2) $\Rightarrow$ (3) If $_RR$ is a semisimple module, let $R = R_1 \oplus R_2 \oplus \cdots \oplus R_m$ be the decomposition of R into its homogeneous components. Each of these is a two-sided ideal, which is a finite direct sum of isomorphic minimal left ideals, and so it is a simple Artinian ring. Condition (3) then follows from Theorem 3.3.1.

(3) $\Rightarrow$ (1) It is clear that R is left Artinian. The Jacobson radical of each matrix ring is zero, since it is simple, and so the Jacobson radical of their direct sum is zero. $\square$

Definition 3.3.3 *A ring which satisfies the conditions of Theorem 3.3.2 is said to be* semisimple Artinian.

Recall that in a left Artinian ring the notions of primitive ideal and prime ideal coincide, so in this case $\mathrm{J}(R) = (0)$ if and only if R is a semiprime ring. We can combine Theorem 2.3.6 with the Artin–Wedderburn theorem to obtain the following corollary. We note that the last condition in the Artin–Wedderburn theorem is left–right symmetric. This means that conditions (3) through (5) of the corollary could just as well be stated for right R-modules.

Corollary 3.3.4 *The following conditions are equivalent for a ring R with identity:*

(1) *R is semisimple Artinian;*

(2) *R is left Artinian and semiprime;*

(3) *every left R-module is completely reducible;*

(4) *every left R-module is projective;*

(5) *every left R-module is injective.*

The preceding corollary allows us to restate Maschke's theorem (Theorem 2.3.8). If G is a finite group and F is a field whose characteristic is not a divisor of the order of G, then the group ring FG is a semisimple Artinian ring. This result is the foundation for Emmy Noether's module theoretic approach to group representations. She also is credited with the proof that every left R-module is semisimple if and only if R is left Artinian and semiprime.

In an Artinian ring R we can factor out the Jacobson radical $J(R)$ to obtain a semisimple Artinian ring $R/J(R)$, whose structure we now know. This also allows us to obtain further information about the ring.

Theorem 3.3.5 (Hopkins) *Any left Artinian ring is left Noetherian.*

Proof. Let R be a left Artinian ring, and let $J = J(R)$. Then J is nilpotent, and so we have a descending chain $R \supset J \supset J^2 \supset \cdots \supset J^n = (0)$ which terminates at (0), for some n. Each factor J^k/J^{k+1} is a module over R/J, so it is semisimple since R/J is semisimple Artinian. Thus J^k/J^{k+1} is isomorphic to a direct sum of simple modules, and since J^k/J^{k+1} is Artinian, this must be a finite direct sum. Therefore each term J^k/J^{k+1} has a composition series, so it follows that $_RR$ has a composition series, and R is a left Noetherian ring. $\square$

We are now ready to consider the structure of primitive rings. Example 1.5.1 shows that if V is an infinite dimensional vector space over the field F, then $\text{End}_F(V)$ is a (left) primitive ring that is not simple Artinian. If F is a field of characteristic zero, then by Corollary 2.8.4 the Weyl algebra $A_n(F)$ is a simple Noetherian ring. Since $A_n(F)$ is not Artinian, it provides another example of a primitive ring that is not simple Artinian.

Lemma 3.3.6 *Let $_RM$ be a simple module, and let $D = \text{End}_R(M)$. If $_DV$ is any finite dimensional subspace of $_DM$, then*

$$V = \{m \in M \mid \text{Ann}_R(V)m = 0\}.$$

Proof. The proof is by induction on $\dim(_DV)$, and there is nothing to prove if $_DV$ has dimension zero. Assume that $\dim(_DV) = n$ and that the result holds for all subspaces of smaller dimension. If u is a nonzero element of V, then we can extend $\{u\}$ to a basis for V, allowing us to write $V = Du \oplus W$ for a subspace W of dimension $n - 1$. Note that $\mathrm{Ann}_R(u) = \mathrm{Ann}_R(Du)$ since $dr = rd$ for all $r \in R$ and $d \in D$, and so $\mathrm{Ann}_R(V) = \mathrm{Ann}_R(u) \cap \mathrm{Ann}_R(W)$.

Let $m \in M$ with $\mathrm{Ann}_R(V)m = 0$. We must show that $m \in V$. Since $u \notin W$ and W satisfies the induction hypothesis, we must have $\mathrm{Ann}_R(W)u \neq 0$, and therefore $\mathrm{Ann}_R(W)u = M$ since M is simple. This allows us to define a mapping $\theta : M \to M$ as follows. If $x \in M$, then $x = au$ for some $a \in \mathrm{Ann}_R(W)$, so we define $\theta(x) = \theta(au) = am$. This is well-defined, since if $au = bu$ for $a, b \in \mathrm{Ann}_R(W)$, then we have $(a-b)u = 0$, which implies that $a-b \in \mathrm{Ann}_R(V)$, and thus by assumption $(a-b)m = 0$ and so $\theta(au) = \theta(bu)$. Furthermore, $\theta \in D$, since

$$\theta(rx) = \theta((ra)u) = (ra)m = r(am) = r\theta(x)$$

for all $r \in R$ and $a \in \mathrm{Ann}_R(W)$.

Finally, the induction hypothesis shows that $m - \theta(u) \in W$ because we have

$$a(m - \theta(u)) = am - a\theta(u) = am - \theta(au) = 0$$

for all $a \in \mathrm{Ann}_R(W)$. Thus $m \in V$ since $m = \theta(u) + (m - \theta(u)) \in Du \oplus W$. $\square$

Definition 3.3.7 *Let V be a left vector space over a division ring D. A subring R is called a* dense subring *of $\mathrm{End}_D(V)$ if for each $n > 0$, each linearly independent subset $\{u_1, u_2, \ldots, u_n\}$ of V, and each arbitrary subset $\{v_1, v_2, \ldots, v_n\}$ of V there exists an element $\theta \in R$ such that $\theta(u_i) = v_i$ for all $i = 1, \ldots, n$.*

Theorem 3.3.8 (Jacobson density theorem) *Any (left) primitive ring is isomorphic to a dense ring of linear transformations of a vector space over a division ring.*

Proof. Let R be a left primitive ring. By definition there exists a faithful simple left R-module V, and its R-endomorphism ring $\mathrm{End}_R(V)$ is a division ring D by Schur's lemma. Since V is faithful, the ring R is embedded as a subring of $\mathrm{BiEnd}_R(V)$, and we will show that R is a dense subring of $\mathrm{BiEnd}_R(V)$, which is the ring $\mathrm{End}_D(V)$ of D-linear transformations of $_DV$.

Let $_DW$ be a finite dimensional subspace of V. We will show by induction on the dimension of W that for each $\alpha \in \mathrm{End}_D(V)$ there exists $r \in R$ such that $(\alpha - r)W = (0)$. The result is immediate if W has dimension zero, so we may assume that $\dim(W) = n$ and the induction hypothesis holds for all subspaces of dimension less than n. If $\{v_1, \ldots, v_n\}$ is a basis for W, let U be the subspace spanned by $\{v_1, \ldots, v_{n-1}\}$. Then $\dim(U) < n$ and the induction hypothesis holds for U, so if $\alpha \in \text{End}_D(V)$, then there exists $s \in R$ with $\alpha(v_j) = sv_j$ for $j < n$. By Lemma 3.3.6, $\text{Ann}(U)v_n \subseteq U$ would imply $v_n \in U$, and so we must have $\text{Ann}(U)v_n \neq (0)$. Since $\text{Ann}(U)v_n$ is an R-submodule of V and V is simple, it follows that $\text{Ann}(U)v_n = V$. Thus there exists $a \in \text{Ann}(U)$ such that $av_n = \alpha(v_n) - sv_n$. For $r = a + s$ and $1 \leq j < n$ we have

$$rv_j = (a+s)v_j = av_j + sv_j = 0 + sv_j = \alpha(v_j) \,,$$

while

$$rv_n = (a+s)v_n = av_n + sv_n = \alpha(v_n) \,.$$

Therefore $(\alpha - r)W = (0)$, and we are done. $\square$

In the Jacobson density theorem, if the vector space is finite dimensional then instead of being a dense subring, the image of the embedding is actually equal to the ring of all linear transformations. Thus the Jacobson density theorem can be used to give a proof of Theorem 3.3.1.

We next consider the endomorphism rings of completely reducible modules.

Example 3.3.1 (Semisimple Artinian endomorphism rings)

We will show that if $_RM$ is completely reducible and finitely generated, then $\text{End}_R(M)$ is a semisimple Artinian ring. If M is completely reducible and finitely generated, then it is isomorphic to a finite direct sum of simple submodules. By grouping together the isomorphic simples, we can write M as a direct sum of its homogeneous components. Thus we have $M = M_1 \oplus \cdots \oplus M_n$, where the homogeneous component M_j is isomorphic to a finite direct sum $S_j^{m(j)}$ of copies of a simple module S_j. If $j \neq k$, then S_j is not isomorphic to S_k, so $\text{Hom}_R(S_j, S_k) = (0)$, and we must have $\text{Hom}_R(M_j, M_k) = (0)$. This implies that $\text{End}_R(M)$ is isomorphic to the direct sum of the endomorphism rings $\text{End}_R(M_j)$, and each of these is an $m(j) \times m(j)$ matrix over the division ring $\text{End}_R(S_j)$.

To describe the endomorphism ring of an arbitrary completely reducible module, we need the definition of a von Neumann regular ring.

Definition 3.3.9 *A ring R is called* von Neumann regular *if for each $a \in R$ there exists $b \in R$ such that $aba = a$.*

Proposition 3.3.10 *If $_RM$ is completely reducible, then $\text{End}_R(M)$ is von Neumann regular.*

Proof. Let $f \in \mathrm{End}_R(M)$. Since M is completely reducible, $\ker(f)$ has a complement N with $M = N \oplus \ker(f)$. Then f restricted to N is one-to-one, and so $f(N)$ has a complement K with $M = f(N) \oplus K$, again since M is completely reducible. Since $f(N) \cong N$, it is easy to define a splitting map $g : M \to N$ with $gf(x) = x$ for all $x \in N$. Given $m \in M$, we can write $m = x + y$, with $x \in N$ and $y \in \ker(f)$, and then

$$fgf(m) = fgf(x) + fgf(y) = f(x) + 0 = f(x) + f(y) = f(m) .$$

Thus $fgf = f$, and so $\mathrm{End}_R(M)$ is von Neumann regular. $\square$

As a special case of the previous proposition, if V is any vector space over a division ring D, then the ring of endomorphisms $\mathrm{End}_D(V)$ is von Neumann regular. This should serve the reader as the prototypical example of a von Neumann regular ring.

Proposition 3.3.11 *The following conditions are equivalent for a ring R:*

(1) *R is von Neumann regular;*

(2) *each principal left ideal of R is generated by an idempotent element;*

(3) *each finitely generated left ideal of R is generated by an idempotent element.*

Proof. (1) $\Rightarrow$ (2) Assume that R is von Neumann regular. Given a principal left ideal Ra, choose $b \in R$ with $aba = a$. Then $ba = (ba)^2$, so ba is idempotent. Since $a = aba \in Rba$, we have $Ra \subseteq Rba$, and it is clear that $Rba \subseteq Ra$, so $Ra = Rba$.

(2) $\Rightarrow$ (3) Using condition (2) and an inductive argument, it suffices to prove that the sum of two principal left ideals is again principal. Let Re_1 and Re_2 be principal left ideals, generated by the idempotents e_1 and e_2. By assumption there is an idempotent f with $Re_2(1 - e_1) = Rf$, and then there exist $b, c \in R$ with $e_2(1-e_1) = bf$ and $f = ce_2(1-e_1)$. Note that $e_2 = e_2e_1 + bf$ and $fe_1 = 0$. We claim that $Re_1 + Re_2 = R(e_1 + f)$. This is a consequence of the following computations:

$$
\begin{aligned}
e_1 &= (1 - f)(e_1 + f) , \\
e_2 &= e_2e_1 + bf = (e_2 - e_2f + bf)(e_1 + f) , \\
e_1 + f &= (1 - ce_2)e_1 + ce_2 .
\end{aligned}
$$

(3) $\Rightarrow$ (1) Assume that each principal left ideal of R is generated by an idempotent element. Given $a \in R$, there exists an idempotent element $e \in R$ with $Ra = Re$, and then there exists $b \in R$ with $e = ba$. For any $x \in Re$ there exists $y \in Re$ such that $x = ye$, and then $xe = (ye)e = ye^2 = ye = x$. In particular, $a(ba) = ae = a$. $\square$

Proposition 3.3.12 *If R is a von Neumann regular ring, then* $\mathrm{J}(R) = (0)$.

Proof. Let $a \in \mathrm{J}(R)$. By assumption there exists $b \in R$ with $aba = a$, and so $1 - ba$ is invertible since $ba \in \mathrm{J}(R)$. But then $a(1 - ba) = a - aba = 0$ implies that $a = 0$. $\square$

EXERCISES: SECTION 3.3

1. Let F be a field, and let $f(x) \in F[x]$. Give necessary and sufficient conditions on $f(x)$ to determine that $F[x]/(f(x))$ is a semisimple Artinian ring.

2. Let $F = \mathbf{Z}_2$, and let $G = Z_3 = \{1, x, x^2\}$, where the elements of G are multiplied like polynomials, and the exponents are reduced modulo 3. Find the fields in the decomposition of the group ring FG given by the Artin–Wedderburn theorem.

3. Repeat Exercise 2 for the field $F = \mathbf{Z}_2$ and a cyclic group G of order 5.

4. Repeat Exercise 2 for the field $F = \mathbf{Z}_5$ and a cyclic group G of order 2.

5. Let $F = \mathbf{Z}_2$ and let G be a cyclic group of order 4. Determine the lattice of ideals of FG.

6. Let M be a left R-module. Then $\mathrm{Z}(M) = \{m \in M \mid \mathrm{Ann}(m)$ is essential in $R\}$ is called the *singular submodule* of M. If $\mathrm{Z}(M) = M$, then M is called *singular*, and if $\mathrm{Z}(M) = (0)$, then M is called *nonsingular*.
Prove that $\mathrm{Z}(M)$ is a submodule of M.

7. Prove that any von Neumann regular ring is nonsingular. Thus the endomorphism ring of a completely reducible module is both left and right nonsingular.

8. Prove that if M_0 is an essential submodule of $_RM$, then $\mathrm{Z}(M/M_0) = M/M_0$, that is, M/M_0 is singular.

9. Prove that a commutative ring is nonsingular if and only if it is semiprime.

10. Prove that the following conditions are equivalent for the ring R:
    (1) R is semisimple Artinian;
    (2) R is left Noetherian and von Neumann regular;
    (3) every left R-module is nonsingular.

3.4 *Orders in simple Artinian rings

The fact that an integral domain has a quotient field is used repeatedly in the theory of commutative rings. Since a simple Artinian ring can be viewed as the noncommutative analog of a field, it is natural to consider noncommutative rings that have a simple Artinian ring of quotients.

Recall that an element $c \in R$ is *regular* if $ca = 0$ or $ac = 0$ implies $a = 0$, for all $a \in R$. We will use $\mathcal{C}(0)$ to denote the set of regular elements of R. The ring Q in the next definition is often called the *classical ring of left quotients* of R. This terminology can be extended to rings of quotients in which the denominators come from a multiplicatively closed set other than $\mathcal{C}(0)$. As our immediate interest is only in the more elementary case in which Q is a simple Artinian ring, we will use the following terminology, which has traditionally been used to state Goldie's theorem (Theorem 3.4.10).

Definition 3.4.1 *The ring R is said to be a* left order *in the ring Q if*

(i) *R is a subring of Q;*

(ii) *every regular element of R is invertible in Q;*

(iii) *for each element $q \in Q$ there exists a regular element $c \in R$ such that $cq \in R$.*

Example 3.4.1 ($\mathrm{M}_2(\mathbf{Z})$ is a left order in $\mathrm{M}_2(\mathbf{Q})$)

The subring $R = \mathrm{M}_2(\mathbf{Z})$ of matrices with integer entries is a left order in the ring $Q = \mathrm{M}_2(\mathbf{Q})$ of 2×2 matrices with rational entries. To show this, let c be any regular matrix in R. Then c must be invertible in Q, since if $\det(c) = 0$, then $ca = 0$ for the adjoint a of c. (See Definition A.3.6 in the appendix for the definition of the adjoint of a matrix.) Given any 2×2 matrix $q \in Q$, we can find a common denominator for the rational entries of q. If c is the scalar matrix determined by the common denominator, then $cq \in R$. Thus R is a left order in Q.

We begin by determining some conditions that must necessarily be satisfied, assuming that R is a left order in the ring Q. Let $\mathcal{C}(0)$ be the set of regular elements of R. If $q \in Q$, then there exists $c \in \mathcal{C}(0)$ with $cq \in R$, so $q = c^{-1}a$ for some $a \in R$. Thus each element in Q has the form $c^{-1}a$, for some $a, c \in R$, with $c \in \mathcal{C}(0)$.

Recall from Section 3.2 that a nonzero submodule N of a module $_RM$ is *essential* if and only if for each nonzero $m \in M$ there exists $s \in R$ with $sm \neq 0$ and $sm \in N$. If q is a nonzero element of Q, and $c \in \mathcal{C}(0)$ with $cq \in R$, then $cq \neq 0$ since c is invertible in Q. This shows that R is essential in Q, when viewed as an R-submodule of Q.

If $c \in \mathcal{C}(0)$ and $a \in R$, then c is invertible in Q, and so the product ac^{-1} belongs to Q. Therefore $ac^{-1} = c_1^{-1}a_1$ for some elements $a_1, c_1 \in R$ such that $c_1 \in \mathcal{C}(0)$. This shows that the following condition is satisfied: if $a, c \in R$ and $c \in \mathcal{C}(0)$, then there exist $a_1, c_1 \in R$ such that $c_1 \in \mathcal{C}(0)$ and $c_1a = a_1c$. This leads to the following definition.

Definition 3.4.2 *We say that the ring R satisfies the* left *Ore condition if for all $a, c \in R$ such that c is regular, there exist $a_1, c_1 \in R$ such that c_1 is regular and $c_1 a = a_1 c$.*

The next lemma shows that it is possible to find common denominators. As a consequence, we then show that a left order in a simple ring must be a prime ring.

Lemma 3.4.3 *Let R be a left order in the ring Q. If $q_1, q_2, \ldots, q_n \in Q$, then there exists a regular element $c \in R$ such that $cq_j \in R$ for $1 \leq j \leq n$.*

Proof. The proof is by induction, and the case $n = 1$ is clear from the definition. Assume that c is regular and $cq_j \in R$ for $1 \leq j \leq n$. For any $q_{n+1} \in Q$ we have $cq_{n+1} \in Q$, and so there exists a regular element $d \in R$ with $d(cq_{n+1}) \in R$. Thus $(dc)q_j \in R$ for $1 \leq j \leq n + 1$, and dc is regular since the product of two regular elements is again regular. $\square$

Proposition 3.4.4 *If R is a left order in a simple ring Q, then R is a prime ring.*

Proof. To show that R is prime, we will use the condition (4) of Proposition 3.1.3: R is a prime ring if and only if $aRb = (0)$ implies either $a = 0$ or $b = 0$, for $a, b \in R$.

Assume that $aRb = (0)$ and $b \neq 0$. Then QbQ is a nonzero ideal of Q, and so $QbQ = Q$ since Q is simple. For some n we must have $1 = \sum_{j=1}^{n} q_j b p_j$ for elements $q_j, p_j \in Q$. By Lemma 3.4.3, there exists $c \in \mathcal{C}(0)$ such that $cq_j \in R$ for all j, and so $c = \sum_{j=1}^{n} cq_j b p_j \in RbQ$. But then $ac = 0$ since $aRb = (0)$, and so $a = 0$ since c is a regular element. $\square$

Unlike the commutative case, in a noncommutative prime ring it is still quite possible that the product of two nonzero elements is zero. We next show that the left annihilator of an element of a left order in a simple Artinian ring cannot be too large. We first need some additional notation.

For a subset X of R it is convenient to use the following notation for the *left annihilator* and *right annihilator*, respectively, of X.

$$\ell(X) = \{a \in R \mid ax = 0, \ \forall x \in X\} \qquad r(X) = \{a \in R \mid xa = 0, \ \forall x \in X\}$$

If there is any possibility of confusion as to the ring, we will use the notation $\ell_R(X)$ or $r_R(X)$.

Let X be any subset of R. We first note that $\ell_R(X)$ is a left ideal of R, and $r_R(X)$ is a right ideal of R. To show this, let $a_1, a_2 \in \ell(X)$. Then

$(a_1 + a_2)x = a_1x + a_2x = 0$, and $(sa_1)x = s(a_1x) = 0$, for all $s \in R$ and all $x \in X$. A similar argument shows that $r_R(X)$ is a right ideal of R.

If $X_1 \subseteq X_2 \subseteq R$ and $a \in \ell(X_2)$, then certainly $a \in \ell(X_1)$, and so we see that $\ell(X_2) \subseteq \ell(X_1)$. It is also evident from the definition that for any subset X of R we have

$$X \subseteq r(\ell(X)) \quad \text{and} \quad X \subseteq \ell(r(X)) \,.$$

Since $X \subseteq r\ell(X)$, we can take left annihilators, to obtain $\ell(r\ell(X)) \subseteq \ell(X)$. On the other hand, $\ell(X) \subseteq \ell r(\ell(X))$. A similar result holds on the other side, so for any subset $X \subseteq R$ we have

$$\ell(X) = \ell r \ell(X) \quad \text{and} \quad r(X) = r \ell r(X) \,.$$

Definition 3.4.5 *The left ideal A of R is called an* annihilator left ideal *if $A = \ell(X)$ for some subset X of R. Similarly, a right ideal A of R is called an* annihilator right ideal *if $A = r(X)$ for some subset X of R.*

Definition 3.4.6 *The ring R is said to be* left nonsingular *if $\ell(a)$ is not essential in R, for any nonzero element $a \in R$.*

We can now give some additional properties of orders in simple Artinian rings. If Q is a simple Artinian ring, then it satisfies both the ascending chain condition and the descending chain condition on left ideals. These conditions do not transfer to a subring R, even if R is a left order in Q. To see this, recall the commutative case: any integral domain has a quotient field, but the integral domain need not have either chain condition on ideals. It turns out that we must focus on chains of annihilator left ideals, and on direct sums of left ideals.

Theorem 3.4.7 *Let R be a left order in a simple Artinian ring Q.*

(a) *The ring R is a left nonsingular prime ring.*

(b) *The ring R satisfies both the ascending chain condition and the descending chain condition on annihilator left ideals.*

(c) *The ring R does not contain an infinite direct sum of nonzero left ideals.*

Proof. Assume that R is a left order in a simple Artinian ring Q.

(a) We have already shown in Proposition 3.4.4 that R is a prime ring. Suppose that A is an essential left ideal of R, with $Ab = 0$ for some element $b \in R$. Then A is an essential submodule of Q, since R is an essential submodule of Q, and it follows that QA is an essential left ideal of Q. If Q is a simple Artinian ring, then any proper nonzero left ideal of Q splits off, and thus no essential left ideal of Q can be proper. We conclude that $QA = Q$, so for some n there exist elements $q_j \in Q$ and $a_j \in A$, for $1 \le j \le n$, with $1 = \sum_{j=1}^{n} q_j a_j$. By Lemma 3.4.3 there exists $c \in \mathcal{C}(0)$ with $cq_j \in R$ for all j, and so $c = \sum_{j=1}^{n} cq_j a_j \in QA$. But then $cb = 0$ since $Ab = (0)$, and so $b = 0$ since c is a regular element.

(b) We will show that either chain condition on annihilator left ideals is inherited by any subring. Suppose that S is a subring of Q, and $A_1 \subseteq A_2 \subseteq \cdots$ is an ascending chain of annihilator left ideals of S, with $A_1 = \ell_S(X_1)$, $A_2 = \ell_S(X_2)$, etc., for subsets $X_1, X_2, \ldots$ of S. The first observation is that we can assume without loss of generality that $X_1 \supseteq X_2 \supseteq \cdots$, since we can replace X_i with $r_S(\ell_S(X_i))$.

There is a corresponding chain of annihilator left ideals in Q, given by $\ell_Q(X_1) \subseteq \ell_Q(X_2) \subseteq \cdots$. If A_i is a proper subset of A_{i+1}, then there exists $a \in S$ with $a \in \ell_S(X_{i+1})$ but $a \notin \ell_S(X_i)$. For this same element we have $a \in \ell_Q(X_{i+1})$ but $a \notin \ell_Q(X_i)$, and so $\ell_Q(X_i)$ is a proper subset of $\ell_Q(X_{i+1})$.

Now it is clear that any strictly ascending chain of annihilator left ideals of S gives rise to a strictly ascending chain of annihilator left ideals of Q. If every such chain in Q must terminate after finitely many steps, then the same condition must hold in S. The argument for descending chains is similar, and so either chain condition on annihilator left ideals is inherited by any subring.

(c) The ring Q is isomorphic to a direct sum of n minimal left ideals, for some positive integer n. Suppose that $\{A_i\}_{i=1}^{n+1}$ is a collection of left ideals of R. We claim that $\sum_{i=1}^{n+1} A_i$ cannot be a direct sum. For $1 \le i \le n+1$, let $x_i \in A_i$, and consider the sum $\sum_{i=1}^{n+1} Qx_i$. This sum contains more terms than the direct sum decomposition of Q, and so it cannot be a direct sum. Thus there exists a nontrivial relation $\sum_{i=1}^{n+1} q_i x_i = 0$, for elements $\{q_i\}_{i=1}^{n}$ in Q. Since R is a left order in Q, there exists $c \in \mathcal{C}(0)$ with $cq_i \in R$ for all i. If $q_j x_j \ne 0$, then $cq_j x_j \ne 0$, and so $\sum_{i=1}^{n+1} (cq_i) x_i = 0$ is a nontrivial relation in R, showing that $\sum_{i=1}^{n+1} A_i$ is not a direct sum. $\square$

The next theorem constructs a ring of quotients Q for any ring R that satisfies the left Ore condition. In constructing Q we use as a model the construction of the quotient field of an integral domain. We will use equivalence classes of ordered pairs in which we think of an ordered pair (c, a) as corresponding to the element $c^{-1}a$ that we ultimately hope to construct in Q. A crucial question is how to define an equivalence relation on two pairs (c, a) and (d, b) to express the fact that the corresponding elements $c^{-1}a$ and $d^{-1}b$ may be equal.

In the quotient ring Q, if $c^{-1}a = d^{-1}b$, then $a = cd^{-1}b$. The element cd^{-1} can be rewritten as $d_1^{-1}c_1$, for some c_1, d_1, and thus $d_1 a = c_1 b$. We note that c_1 is invertible in Q since $c_1 = d_1 cd^{-1}$. With this motivation, we will say that (c, a) and (d, b) represent equivalent fractions if there exist regular elements c_1, d_1 with $d_1 c = c_1 d$ and $d_1 a = c_1 b$.

To add $c^{-1}a$ and $d^{-1}b$, we write $cd^{-1} = d_1^{-1}c_1$, which also yields $dc^{-1} = c_1^{-1}d_1$, and then

$$
\begin{aligned}
c^{-1}a + d^{-1}b &= d^{-1}dc^{-1}a + d^{-1}c_1^{-1}c_1 b \\
&= d^{-1}c_1^{-1}d_1 a + d^{-1}c_1^{-1}c_1 b \\
&= (c_1 d)^{-1}(d_1 a + c_1 b) \ .
\end{aligned}
$$

To multiply $c^{-1}a$ and $d^{-1}b$, we write $ad^{-1} = d_1^{-1}a_1$, and then

$$
c^{-1}a \cdot d^{-1}b = c^{-1}d_1^{-1}a_1 b = (d_1 c)^{-1}(a_1 b) \ .
$$

Theorem 3.4.8 *Let R be a ring which satisfies the left Ore condition. Then there exists an extension ring Q of R such that R is a left order in Q.*

Proof. We merely present an outline of the proof. The details are given in the proof of Theorem A.5.5, in the appendix.

Let $\mathcal{C}(0)$ denote the set of regular elements of R. We first note the following result. For $c, d \in \mathcal{C}(0)$, the Ore condition gives the existence of $d_1 \in \mathcal{C}(0)$ and $c_1 \in R$ with $d_1 c = c_1 d$. It can be shown that in fact $c_1 \in \mathcal{C}(0)$.

We introduce a relation on $\mathcal{C}(0) \times R$, by defining $(c, a) \sim (d, b)$ if there exist $c_1, d_1 \in \mathcal{C}(0)$ with $d_1 c = c_1 d$ and $d_1 a = c_1 b$. It is clear that $\sim$ is reflexive and symmetric. To show that $\sim$ is transitive, suppose that $(c, a) \sim (d, b)$ and $(d, b) \sim (s, r)$ for $c, d, s \in \mathcal{C}(0)$ and $a, b, r \in R$. By the definition of $\sim$, there exist $c_1, d_1, d_2, s_2 \in \mathcal{C}(0)$ with $d_1 c = c_1 d$, $d_1 a = c_1 b$, $s_2 d = d_2 s$, and $s_2 b = d_2 r$. From the Ore condition we obtain $c_1', s_2' \in \mathcal{C}(0)$ with $c_1' s_2 = s_2' c_1$. Then $s_2' d_1$ and $c_1' d_2$ are regular, and it can be shown that $(s_2' d_1)c = (c_1' d_2)s$ and $(s_2' d_1)a = (c_1' d_2)r$. Therefore $(c, a) \sim (s, r)$.

The equivalence class of (c, a) will be denoted by $[c, a]$, and the collection of all equivalence classes of the form $[c, a]$ will be denoted by Q. It is useful to note that if d is regular, then $[dc, da] = [c, a]$.

We define addition and multiplication of equivalence classes in $(\mathcal{C}(0) \times R)/ \sim$ as follows:

$$
[c, a] + [d, b] = [c_1 d, \ d_1 a + c_1 b]
$$

where $c_1, d_1 \in \mathcal{C}(0)$ with $d_1 c = c_1 d$, and

$$
[c, a][d, b] = [d_1 c, \ a_1 b]
$$

where $d_1 \in \mathcal{C}(0)$ and $a_1 \in R$ with $d_1 a = a_1 d$. A major part of the proof is to show that these operations respect the equivalence relation $\sim$. It must be shown that the sum is independent of the choice of $c_1, d_1 \in \mathcal{C}(0)$. Then it must be shown that addition is independent of the choice of representatives for $[c, a]$ and $[d, b]$.

To multiply $[c, a]$ by $[d, b]$, we choose $d_1 \in \mathcal{C}(0)$ and $a_1 \in R$ with $d_1 a = a_1 d$, and then let $[c, a] \cdot [d, b] = [d_1 c, a_1 b]$. It must be shown that this operation is independent of the choice of a_1 and d_1. Next, it is necessary to show that the formula for multiplication is independent of the choice of representatives for the equivalence classes.

It is clear that the commutative law holds for addition. Using the definition of addition to find $[c, a] + [1, 0]$, we can choose $c_1 = c$ and $d_1 = 1$, and then $[c, a] + [1, 0] = [c \cdot 1, 1 \cdot a + c \cdot 0] = [c, a]$, and so the element $[1, 0]$ serves as an additive identity for Q. Similar computations show that the additive inverse of $[c, a]$ is $[c, -a]$ and that $[1, 1]$ is a multiplicative identity for Q.

In verifying the associative law for addition, it is helpful to use the fact that it is possible to find common denominators. The verification of the associative and distributive laws is left as an exercise.

We can identify an element $a \in R$ with the equivalence class $[1, a]$. If $(1, a) \sim (1, b)$ for $b \in R$, then there exist $c, d \in \mathcal{C}(0)$ with $d \cdot 1 = c \cdot 1$ and $d \cdot a = c \cdot b$, which forces $a = b$. We can formalize this identification by defining the function $\eta : R \to Q$ by $\eta(a) = [1, a]$, for all $a \in R$, and showing that η is a ring homomorphism.

Given a regular element c in R, the corresponding element $[1, c]$ in Q is invertible, with $[1, c]^{-1} = [c, 1]$. Given any element $[c, a] \in Q$, we have $[1, c][c, a] = [1, a]$, and this completes the proof that R (or, more properly, $\eta(R)$) is a left order in Q. $\square$

In the proof of the next lemma, we need to make use of the following fact. The ascending chain condition on annihilator left ideals of R is equivalent to the maximum condition on annihilator left ideals, which states that every collection of annihilator left ideals contains a maximal element. It is clear that the maximum condition implies the ascending chain condition, since any ascending chain must have a maximal element. Conversely, if the ascending chain condition holds, then in any set of annihilator left ideals we can start with any left ideal A_1. If A_1 is not maximal, then there exists a larger left ideal A_2 in the set. Continuing inductively, we obtain an ascending chain of annihilator left ideals, and it must terminate in a maximal element after finitely many steps.

Lemma 3.4.9 *Let R be a prime ring which satisfies the ascending chain condition on annihilator left ideals.*

(a) *The ring R contains no nonzero nil one-sided ideals.*

(b) *The ring R is left nonsingular.*

Proof. (a) Let I be a nonzero one-sided ideal of R, and let $x \in I$ be an element such that $\ell(x)$ is maximal among left annihilators of nonzero elements of I. We will show that there exists $a \in R$ such that ax and xa are not nilpotent, so I cannot be a nil left ideal or a nil right ideal.

Since R is prime, there exists $a \in R$ such that $xax \neq 0$. Then $xax \in I$, and $\ell(x) \subseteq \ell(xax)$, so $\ell(x) = \ell(xax)$ since x was chosen such that $\ell(x)$ is maximal among left annihilators of nonzero elements of I. Thus $ax \neq 0$ implies $a(xax) \neq 0$, and then $(axa)x \neq 0$ implies $(axa)(xax) \neq 0$, etc. This implies that neither ax nor xa can be nilpotent.

(b) Suppose that a is a nonzero element of R such that $\ell(a)$ is an essential left ideal of R. We first note that $\ell(sa)$ is also essential in R, for any $s \in R$. To see this, let $0 \neq b \in R$. If $b(sa) \neq 0$, then $bs \neq 0$, and so there exists $t \in R$ with $t(bs) \neq 0$ and $t(bs) \in \ell(a)$, since $\ell(a)$ is essential. Then we have $tb \neq 0$ and $tb \in \ell(sa)$, which shows that $\ell(sa)$ is essential.

Because R is prime and a is nonzero, each term in the descending chain

$$Ra \supseteq (Ra)^2 \supseteq (Ra)^3 \supseteq \cdots$$

is nonzero, and we have a corresponding ascending chain

$$\ell(Ra) \subseteq \ell((Ra)^2) \subseteq \ell((Ra)^3) \subseteq \cdots$$

of annihilator ideals. If we can show that this is a strictly ascending chain, then this contradiction to the assumptions on R will show that $a = 0$, completing the proof.

For any $n > 0$ we have $(Ra)^{n+1} \neq (0)$. Thus the set $Ra \setminus \ell((Ra)^n)$ is nonempty, so it contains an element x such that $\ell(x)$ is maximal among left annihilators of elements in $Ra \setminus \ell((Ra)^n)$. If y is any element of Ra, then $\ell(y)$ is essential, and so $\ell(y) \cap Rx \neq (0)$. This implies that $\ell(xy)$ properly contains $\ell(x)$, since there is an element $b \in R$ with $bx \neq 0$ and $bx \in \ell(y)$. Now we note that $xy \in Ra$, and so we must have $xy \in \ell((Ra)^n)$, since $\ell(x)$ was chosen to be a maximal annihilator of the set $Ra \setminus \ell((Ra)^n)$. We conclude that $xy(Ra)^n = (0)$, and since this is true for any $y \in Ra$, it shows that $x \in \ell((Ra)^{n+1})$. Thus $\ell((Ra)^n)$ is strictly contained in $\ell((Ra)^{n+1})$. $\square$

Theorem 3.4.10 (Goldie) *The ring R is a left and right order in a simple Artinian ring if and only if R is a prime ring that contains no infinite direct sum of left or right ideals and satisfies the ascending chain condition on annihilator left ideals and on annihilator right ideals.*

Proof. The 'only if' part of the result is Theorem 3.4.7. The crucial part of the other half of the proof is to show that if R is a prime ring that contains no infinite direct sum of left ideals and satisfies the ascending chain condition on annihilator left ideals, then every essential left ideal of R contains a regular element. This argument can be repeated on the right side.

Assume that R is a prime ring that satisfies the given conditions, and let A be an essential left ideal of R. Then Lemma 3.4.9 implies that A is not a nil left ideal, and so among the elements of A that are not nilpotent we can choose one whose left annihilator is maximal, say x_1. We note that the maximality of $\ell(x_1)$ implies that $\ell(x_1) = \ell(x_1^2)$. If $\ell(x_1) \neq (0)$, then $\ell(x_1) \cap A \neq (0)$, and so we can choose a non-nilpotent element x_2 of $\ell(x_1) \cap A$ such that $\ell(x_2)$ is as large as possible. If there exist $a_1, a_2 \in R$ with $a_1 x_1 = a_2 x_2$, then $a_1 x_1^2 = a_2 x_2 x_1 = 0$, and so $a_1 x_1 = 0$. This shows that $Rx_1 \cap Rx_2 = (0)$ and $\ell(x_1 + x_2) = \ell(x_1) \cap \ell(x_2)$. If $\ell(x_1 + x_2) \neq (0)$, we choose a non-nilpotent element x_3 of $\ell(x_1 + x_2) \cap I$ such that $\ell(x_3)$ is as large as possible.

By assumption, we cannot have an infinite direct sum $Rx_1 \oplus Rx_2 \oplus \cdots$, and so we obtain elements $x_1, \ldots, x_n \in A$ such that $\ell(x_1 + \cdots + x_n) = (0)$. If we let $c = x_1 + \cdots + x_n$, then $\ell(c) = (0)$. To show that c is a regular element, it is sufficient to show that Rc is an essential left ideal, since $cx = 0$ implies $Rcx = (0)$, and this in turn implies that $x = 0$, because R is a left nonsingular ring by Lemma 3.4.9.

To show that Rc is essential, suppose that $Rb \cap Rc = (0)$, for some nonzero $b \in R$. Consider the left ideal

$$Rb + Rbc + Rbc^2 + \cdots .$$

Suppose that $\sum_{i=0}^{n} a_i bc^i = 0$ for $a_0, \ldots, a_n \in R$. Then $a_0 = 0$ and $\sum_{i=1}^{n} a_i bc^i = 0$ since $Rb \cap Rc = (0)$. Because $\ell(c) = (0)$, we have $\sum_{i=1}^{n} a_i bc^{i-1} = 0$, and then we can repeat the argument. We conclude that $a_i = 0$ for $i = 0, \ldots, n$, and thus we have a direct sum $\bigoplus_{i=0}^{\infty} Rbc^i$, which contradicts the assumption that R contains no infinite direct sum of left ideals.

Let I be any left ideal of Q. Then by Proposition 3.2.3 there exists a left ideal J maximal with respect to $I \cap J = (0)$, and $I + J$ is essential in Q. Then $I + J$ is essential as a left R-submodule of Q, and so $(I + J) \cap R$ is an essential left ideal of R. Therefore $(I + J) \cap R$ contains a regular element c of R, and thus $I + J = Q$, since c is invertible in Q. This shows that Q is completely reducible as a left Q-module, since every left ideal of Q is a direct summand of Q. It follows from Corollary 2.3.3 and the Artin–Wedderburn theorem (Theorem 3.3.2) that Q is isomorphic to a direct product of matrix rings over division rings. If $Q = I \oplus J$ for nonzero two-sided ideals I and J, then $IJ = (0)$, and this is a contradiction, since $I \cap R \neq (0)$ and $J \cap R \neq (0)$ but $(I \cap R)(J \cap R) = (0)$. We conclude that Q is a simple Artinian ring. $\quad\square$

Corollary 3.4.11 *Any left Noetherian prime ring is a left order in a simple Artinian ring.*

Proof. Let R be a left Noetherian prime ring. An infinite direct sum of left ideals does not satisfy the ascending chain condition, so the hypotheses of Goldie's theorem are satisfied for left ideals. The proof of Goldie's theorem shows that to obtain a left order we only need the conditions on left ideals. $\quad\square$

Corollary 3.4.12 *Any left Noetherian domain is a left order in a division ring.*

Proof. Let R be a left Noetherian domain, and let R be a left order in the ring Q. If A is a nonzero left ideal of Q, then $A \cap R$ is nonzero, and so it contains an element $c \in \mathcal{C}(0)$, since every nonzero element of a domain is regular. This forces $A = Q$, because c is invertible in Q, and so Q is a division ring since it has no proper nontrivial left ideals.   □

The Weyl algebra $A_n(F)$ is a Noetherian domain by Corollary 2.8.4. It follows from Corollary 3.4.12 that $A_n(F)$ has a skew field of quotients.

Goldie's theorem can be proved in greater generality. In fact, the ring R is a left order in a semisimple Artinian ring if and only if R is a semiprime ring that contains no infinite direct sum of left ideals and satisfies the ascending chain condition on annihilator left ideals. The proof of Theorem 3.4.7 carries over without change to the more general case, but a significant amount of additional work must be done to prove Lemma 3.4.9 and Theorem 3.4.10 for semiprime rings.

EXERCISES: SECTION 3.4

1. Show that the ring R of lower triangular 2×2 matrices with entries in $\mathbf{Z}$ is a left order in the ring Q of lower triangular 2×2 matrices over $\mathbf{Q}$.

2. Let R be a left order in the ring Q. Show that any subring S of Q with $R \subseteq S \subseteq Q$ is also a left order in Q.

3. Prove that a commutative ring is nonsingular if and only if it is semiprime.

4. Show that a ring R satisfies the descending chain condition for annihilator left ideals if and only if it satisfies the ascending chain condition for annihilator right ideals.

5. Prove the associative laws for addition and multiplication in the ring Q defined in Theorem 3.4.8.

6. Prove the distributive laws in the ring Q defined in Theorem 3.4.8.

7. Let R be a left order in the ring Q. Prove that if R is left Artinian, then $Q = R$.

8. Let R be a left order in the ring Q. Prove that if R is left Noetherian and Q is finitely generated as a left R-module, then $Q = R$.

9. Let R be a left order in the ring Q. Prove that if R is a prime ring, then $M_n(R)$ is a left order in $M_n(Q)$.

10. Let R be a semiprime ring, and let $P_1, \ldots, P_n$ be minimal prime ideals of R with $\bigcap_{j=1}^n P_j = (0)$. If each ring R/P_j is a left order in a simple Artinian ring Q_j, prove that R is a left order in the semisimple Artinian ring $\prod_{j=1}^n Q_j$.

Chapter 4

REPRESENTATIONS OF FINITE GROUPS

Although nonabelian groups were originally studied as certain sets of permutations, the notion of an abstract group was soon introduced in order to cover a much wider variety of examples. Cayley's theorem, which states that every group is isomorphic to a group of permutations, shows that the abstract definition does not really include additional classes of groups. Cayley's theorem also shows that any group can be 'represented' in a concrete fashion as a set of permutations.

A representation of a group may be useful even if it is not a one-to-one function. This is illustrated by the proofs of Cauchy's theorem and the Sylow theorems, which represent a group G as sets of permutations on various sets associated with the given group. Although permutation representations have been used very effectively, other representations have proved to be more important. Even before the study of permutation groups, group characters were used in working with finite abelian groups. The initial ideas can be traced to work of Gauss, but the definition of a character of an abelian group as a homomorphism from the group into the multiplicative group of nonzero complex numbers was not given until 1897 (by Dedekind).

The most useful representations of a finite group appear to be as groups of matrices. Major contributions to the initial development were made by Frobenius, at the end of the nineteenth century. The matrix groups were assumed to be subgroups of $GL_n(\mathbf{C})$, and in 1897 Frobenius formulated the modern definition of a group character as the trace of a matrix representation. The second edition of Burnside's book *Theory of Groups of Finite Order* appeared in 1911, and was the first book to present a systematic account of group representations. Numerous results on abstract groups were proved using group characters, including Burnside's famous $p^a q^b$ theorem, which states that a finite group is solvable whenever its order has only two prime divisors.

The next stage in the development of representation theory occurred in the work of Emmy Noether in 1929. By this time there was additional knowledge about ideals in noncommutative rings, as well as about modules, direct sum decompositions, chain conditions, and semisimplicity. She viewed a matrix representation of a group as a vector space together with a group action via a group of linear transformations. This led naturally to considering the representation to be a module over the group ring, and to viewing irreducible representations as simple modules over the group ring.

In this chapter we utilize Noether's point of view to introduce the representation theory of finite groups. The reader will see an important application of the general results covered in Chapter 2 and Chapter 3.

4.1   Introduction to group representations

Our goal in finding concrete ways to 'represent' a group is to do so in an efficient manner. For most purposes, groups of matrices provide more effective representations than do groups of permutations. It again follows from Cayley's theorem that every finite group is isomorphic to a group of matrices, since any permutation can be described by a permutation matrix. Unfortunately, the proof of Cayley's theorem does not provide a useful algorithm for obtaining representations. The proof shows that a group of order n is isomorphic to a subgroup of the *symmetric group* $\mathcal{S}_n$, the group of all permutations on n elements. If we start with a group such as $\mathcal{S}_3$, Cayley's theorem produces a set of permutations on six elements, but it is difficult to make use of the subgroup structure, since it is embedded in a group of order 720. On the other hand, it is possible to find various matrix representations of $\mathcal{S}_3$ that are much more manageable. For example, $\mathcal{S}_3$ is isomorphic to $GL_2(\mathbf{Z}_2)$, the group of all invertible 2×2 matrices over the field $\mathbf{Z}_2$ with two elements. (There are six invertible 2×2 matrices over $\mathbf{Z}_2$, so the group $GL_2(\mathbf{Z}_2)$ is a nonabelian group of order 6 and hence must be isomorphic to $\mathcal{S}_3$.)

Since $\mathcal{S}_n$ contains a cyclic subgroup of order n, it is also possible to represent any cyclic group as a group of $n \times n$ matrices. Since the entries of a permutation matrix are simply 0s and 1s, the representation does not depend on the field from which the entries are chosen. On the other hand, if we use the field **C** of complex numbers, then we can represent a cyclic group more efficiently by simply considering the subgroup generated by any primitive nth root of unity. This illustrates the fact that representing a group as a set of $n \times n$ matrices depends on the field F which defines the general linear group $\mathrm{GL}_n(F)$.

Example 4.1.1 (The dihedral group)

In this example we consider the dihedral group D_n, described by generators a and b of order n and 2, respectively, together with the relation $ba = a^{-1}b$. Since D_n can also be defined as the group of rigid motions of a regular n-gon lying in the plane $\mathbf{R}^2$, we can use the matrices

$$A = \begin{bmatrix} \cos(2\pi/n) & -\sin(2\pi/n) \\ \sin(2\pi/n) & \cos(2\pi/n) \end{bmatrix},$$

which represents a counterclockwise rotation through $2\pi/n$ radians, and

$$B = \begin{bmatrix} -1 & 0 \\ 0 & 1 \end{bmatrix},$$

which represents a 'flip' about the y-axis in the plane. A quick computation shows that we do indeed have $BA = A^{-1}B$.

We note that this representation involves a subgroup of $\mathrm{GL}_2(\mathbf{R})$. If p is an odd prime, then the dihedral group D_p also has a representation over the field $\mathbf{Z}_p$. We can choose for the generators A and B the matrices

$$A = \begin{bmatrix} 1 & 1 \\ 0 & 1 \end{bmatrix} \quad \text{and} \quad B = \begin{bmatrix} -1 & 0 \\ 0 & 1 \end{bmatrix}.$$

In this case A has order p, and

$$BA = \begin{bmatrix} -1 & 0 \\ 0 & 1 \end{bmatrix} \begin{bmatrix} 1 & 1 \\ 0 & 1 \end{bmatrix} = \begin{bmatrix} -1 & -1 \\ 0 & 1 \end{bmatrix} = \begin{bmatrix} 1 & -1 \\ 0 & 1 \end{bmatrix} \begin{bmatrix} -1 & 0 \\ 0 & 1 \end{bmatrix}$$
$$= A^{-1}B.$$

Example 4.1.2 (The group of quaternion units)

As shown in Example 1.1.6, the quaternion units ± 1, $\pm \mathbf{i}$, $\pm \mathbf{j}$, $\pm \mathbf{k}$ form a nonabelian group of order 8. (Up to isomorphism there are only two nonabelian groups of order 8; the other nonabelian group of order 8 is the dihedral group D_4.) The definition in Example 1.1.6 yields a representation in $\text{GL}_2(\mathbf{C})$, given by setting

$$\phi(1) = \left[\begin{array}{cc} 1 & 0 \\ 0 & 1 \end{array} \right], \quad \phi(\mathbf{i}) = \left[\begin{array}{cc} i & 0 \\ 0 & -i \end{array} \right],$$

$$\phi(\mathbf{j}) = \left[\begin{array}{cc} 0 & 1 \\ -1 & 0 \end{array} \right], \quad \phi(\mathbf{k}) = \left[\begin{array}{cc} 0 & i \\ i & 0 \end{array} \right].$$

We are now ready to give the formal definition of a representation of a group. The set of invertible $n \times n$ matrices with entries in the field F is denoted by $\text{GL}_n(F)$. If V is any n-dimensional vector space over F, then any linear transformation from V into itself determines an $n \times n$ matrix (for a given choice of a basis for V). Since invertible matrices correspond to nonsingular linear transformations, the set $\text{GL}_n(F)$ corresponds to the set of nonsingular linear transformations of the n-dimensional vector space F^n.

If V is a vector space over the field F, then we will use the notation $\text{GL}(V)$ for the *general linear group* of all nonsingular F-linear transformations from V into V.

Definition 4.1.1 *Let G be a finite group. A* linear representation *of G is a group homomorphism $\rho : G \to \text{GL}(V)$, for some finite dimensional vector space V over a field F.*

In this situation the formal term *F-linear representation* can be used to emphasize the role of the field F. Usually, ρ is simply called a *representation* of G.

If $\rho : G \to \text{GL}(V)$ is a representation of G, and V is an n-dimensional vector space over F, then choosing a basis for V allows us to assign a matrix to $\rho(g)$, for each $g \in G$. There is an associated group homomorphism $\sigma : G \to \text{GL}_n(F)$. A different choice of basis will determine another representation $\tau : G \to \text{GL}_n(F)$. If T is the matrix for the change of basis, then for each $g \in G$ we have $\tau(g) = T\sigma(g)T^{-1}$. This motivates the following general definition.

Definition 4.1.2 *Let G be a finite group, and let V and W be finite dimensional vector spaces over the field F. The representations $\sigma : G \to \text{GL}(V)$ and $\tau : G \to \text{GL}(W)$ are said to be* equivalent *if there exists an isomorphism $T : V \to W$ such that $\tau(g) = T\sigma(g)T^{-1}$, for all $g \in G$.*

Our next task is to recast the definition of a representation in terms of the module theory that we have studied in Chapter 2. We will then have at our disposal the full power of the theorems we have proved for modules.

We recall the definition of an algebra, from Definition 1.5.7. An algebra over a field F is a ring A that is also a vector space over F, in which $c(ab) = (ca)b = a(cb)$ for all $a, b \in A$ and $c \in F$. In this case the center of A must contain a copy of F, where we identify the element $c \in F$ with $c \cdot 1 \in A$. If M is a left A-module, then the multiplication $A \times M \to M$ which defines the structure of $_AM$ also makes M into a vector space over F. If $f \in \text{End}_A(M)$, then $cf(x) = f(cx)$ for all $c \in F$ and all $x \in M$. Thus $\text{End}_A(M)$ is actually a subring of $\text{End}_F(M)$.

For a field F and a finite group G, the group ring FG was constructed in Example 1.1.8. It is a vector space over F, with basis G and componentwise addition. For elements $a = \sum_{g \in G} a_g g$ and $b = \sum_{g \in G} b_g g$ in FG we have

$$\left(\sum_{g \in G} a_g g \right) \left(\sum_{g \in G} b_g g \right) = \sum_{g \in G} c_g g \,, \quad \text{where } c_g = \sum_{hk=g} a_h b_k \,.$$

With this definition of multiplication, any element of F commutes with the elements of FG, and so FG is an algebra over F. From this point on we will refer to FG as a *group algebra*.

Proposition 4.1.3 *Let G be a finite group, and let F be a field. There is a one-to-one correspondence between F-linear representations of G and finitely generated left FG-modules.*

Proof. Let V be a finite dimensional vector space over F, and let $\rho : G \to \text{GL}(V)$ be a representation of G. Example 1.2.2 shows that any group homomorphism from G into the group of units of a ring R extends uniquely to a ring homomorphism from FG into R. Since $\text{GL}(V)$ is the group of units of the ring $\text{End}_F(V)$, there is a unique ring homomorphism $\hat{\rho} : FG \to \text{End}_F(V)$ such that $\hat{\rho}(g) = \rho(g)$ for all $g \in G$. Because FG is an F-algebra, this is precisely the condition needed to induce on V the structure of a left FG-module, generated over FG by the basis elements of V.

Conversely, if V is a finitely generated left FG-module, then since FG is a finite dimensional F-algebra, it follows that V is a finite dimensional vector space over F. The left FG-module structure determines a corresponding ring homomorphism $\rho : FG \to \text{End}_F(V)$. (See Proposition 2.1.2 for additional details.) Since the elements of G are invertible when regarded as elements of the group algebra FG, it follows that $\rho(g)$ is a unit in $\text{End}_F(V)$, for each $g \in G$. Thus the restriction of ρ to G defines a group homomorphism from G into $\text{GL}(V)$. $\square$

Example 4.1.3 (The regular representation of G)

The most basic representation of G is defined by the group algebra FG itself. In G we choose an order for the elements, say

$g_1, g_2, \ldots, g_n$, with $g_1 = 1$. With respect to this basis, left multiplication by any element $g \in G$ determines a linear transformation of FG, and we denote its matrix (relative to the given basis) by $\lambda(g)$. The group homomorphism $\lambda : G \to \mathrm{GL}_n(F)$ can be extended uniquely to a ring homomorphism $\widehat{\lambda} : FG \to \mathrm{M}_n(F)$, and this extension to an element $a \in FG$ is just the matrix of the linear transformation on FG determined by left multiplication by a. In fact, the (i, j)-entry of $\widehat{\lambda}(a)$ is the coefficient of g_i in the product ag_j. This is called the *regular representation* of G.

Example 4.1.4 (The trivial representation of G)

For any vector space V over F we can consider the trivial homomorphism from G into $\mathrm{GL}(V)$, which sends each element of G to the identity of $\mathrm{GL}(V)$. This defines a 'trivial' action of FG on V, in which $gv = v$ for all $g \in G$ and all $v \in V$. Then

$$\left(\sum_{g \in G} c_g g \right) v = \sum_{g \in G} c_g v \, ,$$

for any element $\sum_{g \in G} c_g g \in FG$.

In particular, we can make the field F into a left FG-module by using the trivial action, and the resulting module is called the *trivial FG-module*.

In the new context that we have established, it is useful to interpret various definitions for modules in terms of the underlying representations.

Proposition 4.1.4 *Let G be a finite group, let F be a field, and let V be a finite dimensional vector space over F. Assume that $\rho : G \to \mathrm{GL}(V)$ is an F-linear representation of G.*

(a) A subset U of V is an FG-submodule of V if and only if it is a $\rho(G)$-invariant subspace of V.

(b) If $\sigma : G \to \mathrm{GL}(W)$ is another F-linear representation of G, where W is a finite dimensional vector space over F, then ρ is equivalent to σ if and only if the corresponding left FG-modules are isomorphic.

Proof. (a) By definition, U is a $\rho(G)$-invariant subspace of V if and only if U is a subspace of V such that $[\rho(g)](u) \in U$, for all $g \in G$ and all $u \in U$. In terms of the FG-module structure of V, this shows that if U is a $\rho(G)$-invariant subspace of V, then $g \cdot u \in U$, for all $g \in G$. Using the fact that U is a subspace of V, we have

$$\left(\sum_{g \in G} a_g g \right) u = \sum_{g \in G} a_g g u \in U \, ,$$

for any element $a = \sum_{g \in G} a_g g$ of FG.

Conversely, if U is an FG-submodule of V, then since FG is an F-algebra, it follows that U must be an F-subspace of V. For each $g \in G$, we must have $g \cdot u \in U$, for all $u \in U$. This shows that $[\rho(g)](u) \in U$, for all $u \in U$, and so U is a $\rho(g)$-invariant subspace of V.

(b) Suppose that the representations ρ and σ are equivalent. Then there exists an isomorphism $T : V \to W$ such that $\sigma(g) = T\rho(g)T^{-1}$ for all $g \in G$. For any $v \in V$ we have $[\sigma(g)T](v) = [T\rho(g)](v)$, and so $gT(v) = T(gv)$, for the multiplications induced on V by ρ and on W by σ, respectively. This extends by linearity, to show that $aT(v) = T(av)$, for all $a \in FG$ and all $v \in V$. Thus T is an FG-isomorphism.

Conversely, if $f : V \to W$ is an isomorphism of the module structures defined by ρ and σ, respectively, then $gf(v) = f(gv)$, for all $g \in G$ and all $v \in V$. This shows that $[\sigma(g)](f(v)) = f([\rho(g)](v))$, for all $g \in G$ and all $v \in V$, so $\sigma(g)f = f\rho(g)$ for all $g \in G$. The assumption that f is an FG-isomorphism implies that f is a vector space isomorphism, and so the representations ρ and σ are equivalent. $\square$

We have shown in Maschke's theorem (Theorem 2.3.8) that if the order of the group G is nonzero in F, then over the group algebra FG, every left module is completely reducible. It follows from the Artin–Wedderburn theorem (Theorem 3.3.2) and Corollary 3.3.4 that FG is a semisimple Artinian ring, and so FG is isomorphic to a finite direct sum of matrix rings over division rings.

We will show that the situation is even nicer if the field F is algebraically closed. Thus one of the nicest situations would be to study representations over an algebraically closed field of characteristic zero. In the next sections we will limit the discussion to the field $\mathbf{C}$ of complex numbers, so we will be studying what is called 'ordinary' representation theory. In the remainder of this section we will touch on some aspects of the general theory. Since an Artinian ring is semisimple if and only if its Jacobson radical is zero, the next proposition shows that if char(F) is a divisor of $|G|$ then FG is not semisimple Artinian. This shows that Maschke's theorem also has a converse.

Proposition 4.1.5 *If G is a group of order n, and the characteristic of the field F is a divisor of n, then the Jacobson radical of the group algebra FG is nonzero.*

Proof. Let $s = \sum_{x \in G} x$ be the sum in FG of the elements of G. Then for any $g \in G$ we have

$$gs = g\left(\sum_{x \in G} x\right) = \sum_{x \in G} gx = \sum_{x \in G} x = s$$

since multiplication by g simply permutes the elements of G. Similarly, for any $g \in G$ we have $sg = s$. Furthermore, for any element $a = \sum_{g \in G} a_g g \in FG$, we have

$$as = \left(\textstyle\sum_{g\in G} a_g g\right) s = \left(\textstyle\sum_{g\in G} a_g\right) s \ .$$

Thus the principal left ideal generated in FG by s is the one-dimensional subspace Fs, and Fs is actually a two-sided ideal of FG.

We next note that

$$s^2 = \left(\textstyle\sum_{g\in G} g\right) s = \left(\textstyle\sum_{g\in G} 1\right) s = ns \ .$$

Thus if n is divisible by char(F), then $(Fs)^2 = (0)$. Proposition 3.2.12 states that the Jacobson radical contains any nil ideal, and so the nonzero ideal Fs is contained in the Jacobson radical of FG. $\square$

For $s = \sum_{g\in G} g$, the ideal Fs defined in Proposition 4.1.5 is called the *trace ideal* of the group algebra FG. If $|G| = n$ is invertible in F, then $e = \dfrac{1}{n} \cdot s$ is an idempotent generator for Fs, since $e^2 = \dfrac{1}{n^2} \cdot s^2 = \dfrac{n}{n^2} \cdot s = e$. In this case e is a central idempotent of FG, and Proposition 2.2.7 shows that the trace ideal is a direct summand of FG.

There is a natural ring homomorphism $\epsilon : FG \to F$ defined by setting

$$\epsilon\left(\textstyle\sum_{g\in G} c_g g\right) = \textstyle\sum_{g\in G} c_g \ ,$$

for all $\sum_{g\in G} c_g g \in FG$. The function ϵ is called the *augmentation map*, and can be thought of as simply substituting the element $1 \in F$ for each of the group elements of G. Viewed in this way it is clear that ϵ preserves multiplication as well as addition. The kernel of ϵ is an ideal of FG, and so we make the following definition.

Definition 4.1.6 *The kernel of the augmentation map* $\epsilon : FG \to F$ *is called the* augmentation ideal *of FG, and is denoted by $\mathcal{A}(FG)$. Thus*

$$\mathcal{A}(FG) = \left\{\textstyle\sum_{g\in G} c_g g \in FG \mid \textstyle\sum_{g\in G} c_g = 0\right\} \ .$$

The augmentation map $\epsilon : FG \to F$ is an F-linear transformation. If $|G| = n$, then FG has dimension n over F, and so $\mathcal{A}(FG)$ must have dimension $n-1$ as a subspace of FG, since the image of ϵ has dimension 1. For any $g \in G$, we certainly have $g - 1 \in \mathcal{A}(FG)$. The set of these elements, ranging over all $g \in G$ except $g = 1$, forms a linearly independent set of $n - 1$ vectors, so we conclude that it must form a basis for $\mathcal{A}(FG)$, as a vector space over F.

Lemma 4.1.7 *Let H be a normal subgroup of G, and let $\overline{G} = G/H$. The kernel of the ring homomorphism $\widehat{\pi} : FG \to F\overline{G}$ induced by the natural projection $\pi : G \to \overline{G}$ is generated as a left ideal by $\mathcal{A}(FH)$.*

Proof. If $h \in H$, then $\widehat{\pi}(h - 1) = 0$, and so the $\widehat{\pi}$ maps the generators of $\mathcal{A}(FH)$ to zero. Since $\widehat{\pi}$ is an algebra homomorphism, it is F-linear, and thus $\mathcal{A}(FH) \subseteq \ker(\widehat{\pi})$.

On the other hand, let $a = \sum_{g \in G} a_g g$ be an element of FG. If H has index n, we can choose a complete set $\{g_1, g_2, \ldots, g_n\}$ of representatives of the cosets of H in G, so that $G = \bigcup_{i=1}^{n} g_i H$. Then we have

$$a = \sum_{g \in G} a_g g = \sum_{i=1}^{n} g_i \left(\sum_{x \in H} a_{g_i x} x \right) .$$

Since $\widehat{\pi}(x) = 1$ for all $x \in H$, we have

$$\widehat{\pi}(a) = \sum_{i=1}^{n} \widehat{\pi}(g_i) \left(\sum_{x \in H} a_{g_i x} \right) .$$

If $a \in \ker(\widehat{\pi})$, then since the set $\{\widehat{\pi}(g_i)\}_{i=1}^{n}$ is linearly independent, we must have $\sum_{x \in H} a_{g_i x} = 0$ for all i. Thus $\sum_{x \in H} a_{g_i x} x \in \mathcal{A}(FH)$ for all i, and so a belongs to the left ideal generated by $\mathcal{A}(FH)$. $\square$

At one end of the spectrum is the situation in which the characteristic of the field is not a divisor of the order of the group. In this case we can use Maschke's theorem to determine the structure of the group algebra. At the other end of the spectrum is the situation in which the field has characteristic p and the order of the group is p^n. Our next goal is to obtain some information about this case.

We need to review several results from group theory. Let p be any prime number, and let G be a group of order p^n, for a positive integer n. A theorem of Burnside ([4], Theorem 7.2.8; [11], Theorem 2.11.4) states that the center of G is nontrivial. Cauchy's theorem ([4], Theorem 7.2.10; [11], Theorem 2.8.2) states that in any group whose order is divisible by p there is an element of order p, and so the center of G contains an element g of order p. We note that the cyclic subgroup $\langle g \rangle$ generated by g is normal in G, because g belongs to the center of G.

Proposition 4.1.8 *Let F be a field of characteristic $p > 0$, and let G be a group of order $m = p^n$, for some positive integer n. Then the following conditions hold for the group algebra FG.*

(a) *For the augmentation ideal $\mathcal{A}(FG)$, we have $(\mathcal{A}(FG))^m = (0)$.*

(b) *The augmentation ideal coincides with the Jacobson radical $\mathrm{J}(FG)$.*

(c) *The group algebra FG has a unique maximal ideal.*

Proof. (a) The proof uses induction on n. If $n = 1$, then G is cyclic, and so the group algebra FG is commutative. Suppose that g is a generator for G. The elements $g - 1, g^2 - 1, \ldots, g^{p-1} - 1$ form a basis for $\mathcal{A}(FG)$, and each of these elements has $x = g - 1$ as a factor, so $\mathcal{A}(FG)$ is a principal ideal, generated by x. Furthermore, $x^p = (g - 1)^p = g^p - 1^p = 0$ since char$(F) = p$. It follows that $(\mathcal{A}(FG))^p = (FGx)^p = (0)$.

Now assume that $|G| = p^k$, and that the result holds for all p-groups of smaller order. The center of G is nontrivial and contains a subgroup H of order p, which is normal in G. We can form the group algebra $F\overline{G}$, with $\overline{G} = G/H$. The natural projection $\pi : G \to \overline{G}$ induces the ring homomorphism $\widehat{\pi}$ from FG onto $F\overline{G}$. If $a \in \mathcal{A}(FG)$, then it is clear that $\widehat{\pi}(a) \in \mathcal{A}(F\overline{G})$, since the sum of the coefficients of a is left unchanged by $\widehat{\pi}$. Because $|\overline{G}| = p^{k-1}$, the induction hypothesis implies that $(\mathcal{A}(F\overline{G}))^{p^{k-1}} = (0)$. It follows that $(\mathcal{A}(FG))^{p^{k-1}} \subseteq K$, where $K = \ker(\widehat{\pi})$. Thus it is sufficient to show that $K^p = (0)$.

It follows from Lemma 4.1.7 that K is generated as a left ideal by $\mathcal{A}(FH)$, and by the induction hypothesis we have $(\mathcal{A}(FH))^p = (0)$. Since H is in the center of G, the set $\mathcal{A}(FH)$ is in the center of FG, and so we conclude that $K^p = (0)$.

(b), (c) Since FG is Artinian, its Jacobson radical is equal to the intersection of its maximal ideals. Since the augmentation ideal is nilpotent, it must be contained in every maximal ideal. On the other hand, $\mathcal{A}(FG)$ itself is maximal, since $FG/\mathcal{A}(FG)$ is isomorphic to the field F. Thus FG has a unique maximal ideal $\mathcal{A}(FG)$, which is equal to its Jacobson radical. $\square$

Assume that the hypotheses of Proposition 4.1.8 hold. Since $J(FG)$ is the unique maximal ideal of FG, the factor ring $FG/J(FG)$ is a simple Artinian ring. It follows that FG has only one simple left module (up to isomorphism). In this case, the trivial representation of G is the only irreducible representation.

Before studying the structure of the group algebra FG over an arbitrary field F and any finite group G, we need to study modules over FG. To define such a module, we need a vector space V over F, and a scalar multiplication defined on V. We only need to define a multiplication gv for elements $g \in G$ and $v \in V$, since we can extend this to all of FG by linearity. To be more specific, the module structure of V over FG is determined by a ring homomorphism from FG into $\text{End}_F(V)$. To define such a ring homomorphism, we only need to define a group homomorphism from G into $\text{GL}(V)$. This amounts to saying that we must check the associative law $(g_1 g_2)v = g_1(g_2 v)$, for all elements $g_1, g_2 \in G$ and $v \in V$.

We have already noted in Example 1.2.3 that FG is isomorphic to its opposite ring $(FG)^{op}$, by mapping each basis element $g \in G$ to g^{-1}. If V is a left FG-module, then it has a natural structure as a right $(FG)^{op}$-module, and so this induces a right FG-module structure. We will make use of these operations in the following definitions. In particular, the module structure in the next definition is a special case of Proposition 2.6.7 (b). To use that proposition, we need a bimodule structure on V, and this is given on the left by using the multiplication by elements of F, while on the right we make V into an FG-module by using its structure over $(FG)^{op}$, so we define $v \cdot g = g^{-1}v$, for all $g \in G$ and $v \in V$.

Definition 4.1.9 *Let V be a left FG-module. The* dual module V^* *of V is defined as*
$$V^* = \operatorname{Hom}_F(V, F) \, ,$$
where the module structure of V^* is defined by setting $[g \cdot \lambda](v) = \lambda(g^{-1}v)$, for all $g \in G$, $\lambda \in \operatorname{Hom}_F(V, F)$, and $v \in V$.

If V and W are left FG-modules, then both are vector spaces over F, and so we can form the tensor product $V \otimes_F W$. As in Proposition 2.6.7, we could use the left module structure of V to give the tensor product a left module structure. For reasons that will become apparent later in this chapter, we introduce a different structure that depends on the module structures of both V and W. The same motivation applies to the FG-module structure that we define on $\operatorname{Hom}_F(V, W)$.

Proposition 4.1.10 *Let F a field, let G be a finite group, and let V and W be left FG-modules.*

(a) *The tensor product $V \otimes_F W$ is a left FG-module under the multiplication $g \cdot (v \otimes w) = gv \otimes gw$, for elements $g \in G$ and $v \otimes w \in V \otimes_F W$.*

(b) *The set $\operatorname{Hom}_F(V, W)$ of linear transformations from V to W is a left FG-module, under the multiplication $g \cdot f$ defined by $[g \cdot f](v) = gf(g^{-1}v)$, for elements $g \in G$, $f \in \operatorname{Hom}_F(V, W)$, and $v \in V$.*

(c) *If V and W are finitely generated FG-modules, then the vector spaces $\operatorname{Hom}_F(V, W)$ and $V^* \otimes_F W$ are isomorphic as left FG-modules.*

Proof. (a) For each $g \in G$ define a function $\beta_g : V \times W \to V \otimes_F W$ by $\beta_g(v, w) = gv \otimes gw$, for all $(v, w) \in V \times W$. Then
$$
\begin{aligned}
\beta_g(v_1 + v_2, w) &= g(v_1 + v_2) \otimes gw = gv_1 \otimes gw + gv_2 \otimes gw \\
&= \beta_g(v_1, w) + \beta_g(v_2, w) \, ,
\end{aligned}
$$
and a similar argument shows that $\beta_g(v, w_1 + w_2) = \beta_g(v, w_1) + \beta_g(v, w_2)$. Furthermore, if $a \in F$, then
$$
\begin{aligned}
\beta_g(va, w) &= g(va) \otimes gw = (gv)a \otimes gw = gv \otimes a(gw) \\
&= gv \otimes g(aw) = \beta_g(v, aw) \, .
\end{aligned}
$$

Thus β_g is a bilinear mapping, and so we obtain a corresponding group homomorphism $\overline{\beta}_g : V \otimes_F W \to V \otimes_F W$, an extension by linearity. The additive homomorphism $\overline{\beta}_g$ is easily checked to be F-linear, and Proposition 2.6.4 implies that it is an automorphism. For $g, h \in G$ we have

$$
\begin{aligned}
\overline{\beta}_{gh}(v \otimes w) &= (gh)v \otimes (gh)w = g(hv) \otimes g(hw) \\
&= \overline{\beta}_g(hv \otimes hw) = \overline{\beta}_g \left(\overline{\beta}_h(v \otimes w) \right) ,
\end{aligned}
$$

so $\overline{\beta}_{gh} = \overline{\beta}_g \overline{\beta}_h$, and we have defined a group homomorphism from G into $\mathrm{GL}(V \otimes_F W)$. The desired FG-module structure on $V \otimes_F W$ is defined by the corresponding ring homomorphism from FG into $\mathrm{End}_F(V \otimes_F W)$.

(b) On the set $\mathrm{Hom}_F(V, W)$ of F-linear transformations from V to W, we define an action of elements $g \in G$ on linear transformations $f \in \mathrm{Hom}_F(V, W)$ by letting $g \cdot f(v) = g(f(g^{-1}v))$, for all $v \in V$. It follows immediately from the definition that $gf \in \mathrm{Hom}_F(V, W)$. This action by G is associative since

$$
\begin{aligned}
[(g_1 g_2) \cdot f](v) &= g_1 g_2 (f((g_1 g_2)^{-1}v)) = g_1 g_2 (f((g_2^{-1} g_1^{-1})v)) \\
&= g_1 (g_2 (f(g_2^{-1}(g_1^{-1}v)))) = g_1 ([g_2 \cdot f](g_1^{-1}v)) \\
&= [g_1 \cdot (g_2 \cdot f)](v) ,
\end{aligned}
$$

for all $g_1, g_2 \in G$, $f \in \mathrm{Hom}_F(V, W)$, and all $v \in V$. By repeatedly using the distributive laws, we can extend this action to all of FG.

(c) Define $\theta : V^* \otimes_F W \to \mathrm{Hom}_F(V, W)$ by $[\theta(\lambda \otimes w)](v) = \lambda(v)w$, for all $\lambda \in V^*$, $w \in W$, and $v \in V$. If $\lambda \in V^*$, then either $\lambda = 0$ or λ is onto, since the codomain of λ is a field. Thus if $\theta(\lambda \otimes w)$ is the zero function, then $\lambda(v)w = 0$ for all v, and so either $\lambda = 0$ or $w = 0$. Because V and W are finitely generated over FG, they are finite dimensional vector spaces, and applying the above argument to the basis vectors shows that θ is one-to-one. Since both $V^* \otimes_F W$ and $\mathrm{Hom}_F(V, W)$ have dimension $\dim(V) \cdot \dim(W)$, it follows that θ is a vector space isomorphism.

To show that θ respects multiplication by elements of G, let $g \in G$. For all $\lambda \in V^*$, $w \in W$, and $v \in V$, we have $[g \cdot (\theta(\lambda \otimes w))](v) = g(\lambda(g^{-1}v)w)$ and $[\theta(g \cdot (\lambda \otimes w))](v) = [\theta(g\lambda \otimes gw)](v) = \lambda(g^{-1}v)gw = g(\lambda(g^{-1}v)w)$. The last equality holds since the element $\lambda(g^{-1}v)$ belongs to F and therefore commutes with g. This shows that $g \cdot (\theta(\lambda \otimes w)) = \theta(g \cdot (\lambda \otimes w))$. $\square$

Example 4.1.5 (The second dual V^{**})

Let V and W be left FG-modules, and suppose that f is an element of $\mathrm{Hom}_{FG}(V, W)$. Then we can define an FG-homomorphism $f^* : W^* \to V^*$ as follows. For $\lambda \in W^*$, let $f^*(\lambda) = \lambda f$. To show that this is a left FG-homomorphism, it suffices to show that $f^*(g \cdot \lambda) = gf^*(\lambda)$. If $v \in V$, then

$$[f^*(g \cdot \lambda)](v) = [(g \cdot \lambda)](f(v)) = \lambda(g^{-1}f(v)) = \lambda(f(g^{-1}v)),$$

and

$$[gf^*(\lambda)](v) = g[(f^*(\lambda))](v) = g(\lambda(f(v))) = \lambda(f(g^{-1}v)).$$

We also note that if $f \in \operatorname{Hom}_{FG}(V, W)$ and $h \in \operatorname{Hom}_{FG}(U, V)$, then $(fh)^* = h^*f^*$.

If V is a finite dimensional vector space, then a well-known result from elementary linear algebra states that $V \cong V^{**}$ (as vector spaces). Given $v \in V$, we define a mapping $\theta_v : V^* \to F$ by $\theta_v(\lambda) = \lambda(v)$, for all $\lambda \in V^*$. (It must be checked that θ_v is in $\operatorname{Hom}_F(V^*, F) = V^{**}$.) Then the isomorphism is given by the function $\Phi_V : V \to V^{**}$ defined by $\Phi_V(v) = \theta_v$, for all $v \in V$. The function Φ_V is actually an FG-isomorphism, as can be seen by checking that $\Phi_V(gv) = g\Phi_V(v)$, for all $g \in G$ and all $v \in V$. Since $\Phi_V(gv) = \theta_{gv}$ and $g\Phi_V(v) = g \cdot \theta_v$, the proof is completed by making the following calculation:

$$g \cdot \theta_v(\lambda) = \theta_v(g^{-1}\lambda) = [g^{-1}\lambda](v) = \lambda(gv) = \theta_{gv}(\lambda).$$

If V and W are finitely generated FG-modules, and $f : V \to W$ is an FG-homomorphism, then we can apply the results in the first paragraph of the example to get an associated FG-homomorphism $f^{**} : V^{**} \to W^{**}$. For the isomorphisms $\Phi_V : V \to V^{**}$ and $\Phi_W : W \to W^{**}$, we will show that $\Phi_W f = f^{**}\Phi_V$. This will give us the following diagram.

$$
\begin{array}{ccc}
V & \xrightarrow{\ f\ } & W \\
{\scriptstyle \Phi_V}\big\downarrow & & \big\downarrow{\scriptstyle \Phi_W} \\
V^{**} & \xrightarrow[\ f^{**}\ ]{} & W^{**}
\end{array}
$$

For any $v \in V$, on the one hand we have $\Phi_W f(v) = \theta_{f(v)}$, and evaluating this at any $\lambda \in W^*$ gives $\theta_{f(v)}(\lambda) = \lambda(f(v)) = \lambda f(v)$. On the other hand, $f^{**}\Phi_V(v) = f^{**}(\theta_v) = \theta_v f^*$, and evaluating this at any $\lambda \in W^*$ gives $\theta_v f^*(\lambda) = \theta_v(\lambda f) = (\lambda f)(v) = \lambda f(v)$.

Theorem 4.1.11 *Let F be a field, and let G be a finite group. The group algebra FG is isomorphic to its dual FG^*, as left FG-modules.*

Proof. We define a module isomorphism $f : FG \to FG^*$ as follows. For an element $a = \sum_{g \in G} a_g g$ in FG, the image $f(a)$ must belong to FG^*. For any element $b = \sum_{g \in G} b_g g$ in FG, we define

$$[f(a)](b) = \sum_{g \in G} a_g b_g \ .$$

We first show that $f(a) \in FG^* = \operatorname{Hom}_F(FG, F)$. For convenience, let $\beta : FG \times FG \to F$ be defined by $\beta(a, b) = \sum_{g \in G} a_g b_g$, where $a = \sum_{g \in G} a_g g$ and $b = \sum_{g \in G} b_g g$. Showing that $f(a)$ belongs to FG^* reduces to verifying the conditions $\beta(a, b_1 + b_2) = \beta(a, b_1) + \beta(a, b_2)$ and $\beta(a, cb) = c\beta(a, b)$ for all $a, b, b_2, b_2 \in FG$ and all $c \in F$. These conditions follow easily from the definition of β.

We next show that f is one-to-one. If $a = \sum_{g \in G} a_g g$ is a nonzero element of FG, then we must have $a_g \neq 0$ for some coefficient a_g. Then $[f(a)](g) = a_g$, and so $f(a)$ is nonzero. Because FG and FG^* have the same dimension as vector spaces over F, the linear transformation f must be an isomorphism.

Finally, to show that f is an FG-homomorphism, it is sufficient to check that $f(ga) = g \cdot f(a)$ for all $g \in G$ and all $a \in FG$. If $b \in FG$, then

$$[f(ga)](b) = \sum_{gx = y} a_x b_y \ .$$

On the other hand,

$$[g \cdot f(a)](b) = [f(a)](g^{-1}b) = \sum_{x = g^{-1}y} a_x b_y \ .$$

These are equal since $gx = y$ if and only if $x - g^{-1}y$, for all $x, y \in G$.   □

Before proving the next theorem, we need to recall from Definition 2.3.5 that a module $_RQ$ is said to be injective if for each one-to-one R-homomorphism $i : {}_RN \to {}_RM$ and each R-homomorphism $f : M \to Q$ there exists an R-homomorphism $\widehat{f} : M \to Q$ such that $\widehat{f}i = f$. The ring R is said to be *left self-injective* if the module $_RR$ is injective.

Theorem 4.1.12 *Let F be a field, and let G be a finite group. Then the group algebra FG is self-injective.*

Proof. We need to show that FG is injective as a left FG-module, so we will use Baer's criterion (Theorem 2.3.9). Let A be any left ideal of FG, with inclusion mapping $i : A \to FG$, and suppose that $f : A \to FG$ is a left FG-homomorphism. We have the following diagram, in which we must attempt to extend f to FG.

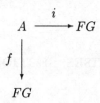

Now consider the dual of each module in the diagram, together with the dual of each of the homomorphisms in the diagram. (For a look at what happens to homomorphisms, refer to Example 4.1.5.) Since the directions of the homomorphisms are reversed, we have the following diagram.

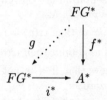

We need to show that i^* is an onto homomorphism. It follows from the diagram in Example 4.1.5 that i^{**} is a one-to-one mapping since i is one-to-one. If p is any homomorphism with domain A^* such that $pi^* = 0$, then $i^{**}p^* = (pi^*)^* = 0$, and so $p^* = 0$ since i^{**} is one-to-one. Thus $p = 0$, showing that i^* is onto. Since FG^* is isomorphic to FG, it is a projective module, so there exists an FG-homomorphism $g : FG^* \to FG^*$ such that $i^*g = f^*$. If we take duals once more we obtain the following diagram.

$$
\begin{array}{ccc}
A^{**} & \xrightarrow{\;i^{**}\;} & FG^{**} \\
{\scriptstyle f^{**}}\downarrow & {\scriptstyle g^*}\swarrow & \\
FG^{**} & &
\end{array}
$$

In Example 4.1.5 we saw that the second dual is isomorphic to the original module. Using the notation in that example, the required extension of f is $\Phi_{FG}^{-1}g^*\Phi_{FG}$, since

$$
\begin{aligned}
\Phi_{FG}^{-1}g^*\Phi_{FG}i &= \Phi_{FG}^{-1}g^*i^{**}\Phi_A = \Phi_{FG}^{-1}f^{**}\Phi_A \\
&= \Phi_{FG}^{-1}\Phi_{FG}f = f \, .
\end{aligned}
$$

This completes the proof that FG is injective as a left FG-module. $\square$

EXERCISES: SECTION 4.1

Let G be a finite group.

1. (a) Show that the trace ideal Fs of FG is the only nonzero left ideal of FG on which G acts trivially.

   (b) Assume that $n = |G|$ is not divisible by char(F). Show that the augmentation ideal is a complement of the trace ideal. That is, show that $FG = Fs \oplus \mathcal{A}(FG)$.

2. Let Λ be the regular representation of G given in Example 4.1.3. For $a \in FG$, let tr(a) be the trace of the matrix Λ_a.

   (a) Show that if $a = \sum_{g \in G} a_g g$, then tr$(a) = na_1$, where $n = |G|$.

   (b) Show that if $a \in FG$ is nilpotent, then tr$(a) = 0$.

   (c) Use (a) and (b) to show that if n is not divisible by char(F), then the Jacobson radical J(FG) is zero.

   Note: This gives another proof of Maschke's theorem.

3. Let V be a finitely generated projective FG-module. Prove that V^* is injective.

4. Let V be a finitely generated injective FG-module. Prove that V^* is projective.

5. Let V be a finitely generated projective FG-module. Prove that V^* is projective.

6. Let V be a finitely generated FG-module. Prove that V is injective if and only if it is projective.

7. Let H be a subgroup of G, and let U be a left FH-module. The FG-module $FG \otimes_{FH} U$ is called the *induced* FG-module of U, denoted by Ind$_H^G(U)$.

   Show that dim$_F($Ind$_H^G(U)) = [G : H]$ dim$_F(U)$, where $[G : H]$ denotes the index of H in G.

   Hint: Use Proposition 2.6.5.

8. Let H be a subgroup of G, and let V be a left FG-module. Then V can be regarded as a left FH-module, called the *restriction* of V to H, and denoted by Res$_H^G(V)$.

   Prove that Hom$_{FH}(U, Res_H^G(U))$ and Hom$_{FG}($Ind$_H^G(U), V)$ are isomorphic as vector spaces.

   Hint: Use Corollary 2.6.10.

   Note: This result is called the Frobenius reciprocity theorem.

4.2 Introduction to group characters

At this point we will specialize to the case in which the field F is the field $\mathbf{C}$ of complex numbers. This is often called 'ordinary' representation theory. Since we assume that the group G is finite, it is a consequence of Maschke's theorem that the group algebra $\mathbf{C}G$ is semisimple Artinian, so it is isomorphic to a direct sum of full matrix rings over division rings. Because $\mathbf{C}$ is an algebraically closed field, we can obtain additional information about the matrix rings in question. We need an extension of Schur's lemma (Lemma 2.1.10).

Proposition 4.2.1 (Burnside) *Let F be an algebraically closed field, and let A be a finite dimensional F-algebra. If S is a simple left A-module, then* $\text{End}_A(S) \cong F$.

Proof. Since $_AS$ is a cyclic module and A is a finite dimensional vector space over F, it follows that S is a finite dimensional vector space over F. If $f \in \text{End}_A(S)$, then f is a linear transformation, and so f has a nonzero eigenvalue $c \in F$, because F is algebraically closed. Thus there exists a nonzero element $x \in S$ with $f(x) = cx$, which shows that $f - c \cdot 1_S$ has a nontrivial kernel, where 1_S is the identity mapping on S. Since $\text{End}_A(S)$ is a division ring, we must have $f - c \cdot 1_S = 0$, and thus the mapping f is given by scalar multiplication by c. We conclude that $\text{End}_A(S)$ is isomorphic to F. □

Theorem 4.2.2 *Let G be a finite group. Then the group algebra $\mathbf{C}G$ is isomorphic to a direct sum of full matrix rings over $\mathbf{C}$.*

Proof. The Artin–Wedderburn theorem (Theorem 3.3.2) states that any semisimple Artinian ring is isomorphic to a direct sum of matrix rings over division rings. The theorem was proved by reducing to the case of a simple Artinian ring. Then in the proof of Theorem 3.3.1, the relevant division ring was constructed as the endomorphism ring of a minimal left ideal. When this proof is applied to the group algebra $\mathbf{C}G$, Proposition 4.2.1 shows that each of the division algebras is isomorphic to $\mathbf{C}$, and so $\mathbf{C}G$ is isomorphic to a direct sum of matrix rings over $\mathbf{C}$. □

We have already noted that $\mathbf{C}$ is a simple $\mathbf{C}G$-module, under the trivial action, and that this module is isomorphic to the trace ideal $\mathbf{C}s$ of $\mathbf{C}G$, where $s = \sum_{g \in G} g$. As a consequence of the Artin–Wedderburn theorem, every $\mathbf{C}G$-module is projective, so every simple $\mathbf{C}G$-module is isomorphic to a direct summand of $\mathbf{C}G$. If S is any simple left $\mathbf{C}G$-module, it follows from Theorem 4.2.2 that S is isomorphic to a minimal left ideal in some full matrix ring $M_n(\mathbf{C})$, for some positive integer n. Then the module S has dimension n, as a vector space over $\mathbf{C}$.

We now focus on the center of $\mathbf{C}G$. Since $\mathbf{C}G$ is a direct sum of full matrix rings over $\mathbf{C}$, the center of each matrix ring is isomorphic to $\mathbf{C}$, and so the center of $\mathbf{C}G$ is isomorphic to a direct sum of copies of $\mathbf{C}$. To find a basis for the center, we need to use the conjugacy classes of G. Recall that elements $g_1, g_2 \in G$ are *conjugate* in G if there exists an element $h \in G$ with $g_2 = hg_1h^{-1}$. This defines an equivalence relation on G, for which the equivalence classes are called the conjugacy classes of G. (See [4], Section 7.2, or [11], Section 2.11). Note that an element $g \in G$ belongs to the center of G if and only if its conjugacy class is g itself. For any $h \in G$, the function $\phi : G \to G$ defined by $\phi(g) = hgh^{-1}$ for all $g \in G$ is an isomorphism, and so it simply permutes the elements of each conjugacy class. This observation shows that if $\mathcal{K}$ is a conjugacy class of G, then $\sum_{g \in \mathcal{K}} g$ belongs to the center $Z(\mathbf{C}G)$ of $\mathbf{C}G$, since

$$h \left(\sum_{g \in \mathcal{K}} g \right) h^{-1} = \sum_{g \in \mathcal{K}} hgh^{-1} = \sum_{g \in \mathcal{K}} g \, .$$

We will see that in fact these elements form a basis for $Z(\mathbf{C}G)$.

Theorem 4.2.3 *Let G be a finite group.*

(a) *There exist positive integers $r, n_1, n_2, \ldots, n_r$ such that*

$$\mathbf{C}G \cong M_{n_1}(\mathbf{C}) \oplus M_{n_2}(\mathbf{C}) \oplus \cdots \oplus M_{n_r}(\mathbf{C}) \, .$$

(b) *There are r distinct isomorphism classes of simple left $\mathbf{C}G$-modules, and each simple module in the ith isomorphism class has dimension n_i as a vector space over $\mathbf{C}$.*

(c) $\sum_{i=1}^{r} n_i^2 = |G|$.

(d) *The number r corresponds to the number of conjugacy classes of G.*

Proof. (a) This is a restatement of Theorem 4.2.2.

(b), (c) These follow immediately from part (a) and the fact that $\mathbf{C}G$ has dimension $|G|$ as a vector space over $\mathbf{C}$.

(d) Assume that G has t distinct conjugacy classes $\mathcal{K}_1, \mathcal{K}_2, \ldots, \mathcal{K}_t$. For $i = 1, \ldots, t$, let $c_i = \sum_{g \in \mathcal{K}_i} g$. It follows from the definition of $\mathbf{C}G$ that the elements $c_1, \ldots, c_t$ are linearly independent. We will show that they form a basis for $Z(\mathbf{C}G)$.

If $a = \sum_{g \in G} a_g g$ belongs to $Z(\mathbf{C}G)$, then $hah^{-1} = a$, for any $h \in G$. Equating the corresponding coefficients, we see that $a_{hgh^{-1}} = a_g$, for all $g \in G$. It follows that the coefficients of a are constant on each conjugacy class, and we can let $a_i = a_g$ for all $g \in \mathcal{K}_i$, so that a can be rewritten as $a = \sum_{i=1}^{t} a_i c_i$. Thus $c_1, \ldots, c_t$ span $Z(\mathbf{C}G)$, and so they form a basis for $Z(\mathbf{C}G)$. We conclude that $t = r$, completing the proof. $\square$

Corollary 4.2.4 *The group G is abelian if and only if every simple left CG-module is one-dimensional.*

Proof. The group G is abelian if and only if CG is a commutative ring. Since CG is isomorphic to a direct sum of matrix rings, it is commutative if and only if in every component $M_{n_i}(C)$ of CG we have $n_i = 1$. Thus $n_i = 1$ for all i if and only if every minimal left ideal of CG is one-dimensional. Since every simple left CG-module is isomorphic to a left ideal of CG, we have $n_i = 1$ for all i if and only if every simple left CG-module is one-dimensional. $\square$

If V is a left CG-module, then each element $g \in G$ determines an invertible C-linear transformation of V, defined by left multiplication by g. If V is finitely generated over CG, then it is a finite dimensional vector space over C, and so we can choose a basis and find the matrix A corresponding to left multiplication by g. The matrix A is also the matrix of $\rho(g)$, where $\rho : G \to \mathrm{GL}(V)$ is the representation of G that defines the module structure of V. The trace of the matrix A will be denoted by $\chi_V(g)$. It is important to note that $\chi_V(g)$ is independent of the choice of basis for V, since a change of basis yields a new matrix TAT^{-1}, for some matrix T, and $\mathrm{tr}((TA)T^{-1}) = \mathrm{tr}(T^{-1}(TA)) = \mathrm{tr}(A)$ by Proposition A.3.10 of the appendix.

Definition 4.2.5 *Let G be a finite group, and let V be a finitely generated left CG-module, with representation $\rho : G \to \mathrm{GL}(V)$.*

(a) *The* character *of V is the function $\chi_V : G \to C$ defined by $\chi_V(g) = \mathrm{tr}(\rho(g))$, for all $g \in G$. The character χ_V is called the character afforded by the module V, or the character afforded by the representation ρ.*

(b) *The character defined by the module $V = CG$ is called the* regular *character of G. The character defined by the trivial module $V = F$ is called the* principal *character of G.*

(c) *The dimension of V, as a vector space over C, is called the* degree *of χ_V. If the degree of χ_V is 1, then χ_V is called a* linear *character.*

If $\rho : G \to \mathrm{GL}(V)$ and $\sigma : G \to \mathrm{GL}(W)$ are equivalent representations, then by definition there exists an isomorphism $T : V \to W$ with $\sigma(g) = T\rho(g)T^{-1}$, for all $g \in G$, and so $\mathrm{tr}(\sigma(g)) = \mathrm{tr}(\rho(g))$, for all $g \in G$. Since equivalent representations afford the same character, it follows from Proposition 4.1.4 that isomorphic CG-modules afford the same character.

We next observe that characters of G are constant on the conjugacy classes of G. If $g_1, g_2 \in G$ and there exists $h \in G$ with $g_2 = hg_1h^{-1}$, then for any representation we have $\rho(g_2) = \rho(h)\rho(g_2)\rho(h)^{-1}$, and it follows that $\mathrm{tr}(\rho(g_1)) = \mathrm{tr}(\rho(g_2))$.

For a character χ_V afforded by the representation ρ, we have $\chi_V(1) = \text{tr}(\rho(1))$. Since $\rho(1)$ is the identity matrix, it follows that $\chi_V(1)$ is equal to the degree of χ_V.

As our first examples, we consider the regular character and the principal character.

Example 4.2.1 (The regular character of G)

We will compute the character afforded by $\mathbf{C}G$ itself. If we use the standard basis for $\mathbf{C}G$, consisting of the elements of G, then left multiplication by an element of G produces a permutation matrix. Multiplying by the identity element of G leaves the basis elements fixed, while multiplying by any other element leaves no basis element fixed. Thus for the regular character afforded by $V = \mathbf{C}G$, we have $\chi_V(1) = |G|$, and $\chi_V(g) = 0$ for all other elements $g \in G$.

Example 4.2.2 (The principal character of G)

The character afforded by the trivial module $V = F$ has degree 1, and since $ga = a$ for all $g \in G$ and all $a \in F$, we have $\chi_V(g) = 1$, for all $g \in g$.

Example 4.2.3 (Linear characters correspond to homomorphisms)

Suppose that χ_V is a linear character, afforded by a one-dimensional representation ρ. Then for $g \in G$ there exists $c \in \mathbf{C}$ with $gv = cv$ for all $v \in V$, and thus $\chi_V(g) - c$. For any $g' \in G$, we have $\chi_V(g')v = g'v$, and so $\chi_V(gg') = (gg')v = g(g'v) = \chi_V(g)(\chi_V(g')v)$, which shows that χ_V is a group homomorphism from G into the multiplicative group $\mathbf{C}^\times$.

Conversely, any group homomorphism $\phi : G \to \mathbf{C}^\times$ determines a representation that affords the character $\chi_V = \phi$. Thus there is a one-to-one correspondence between linear characters of G and group homomorphisms from G into $\mathbf{C}^\times$.

Example 4.2.4 (Linear characters of G are those of G/G')

Let G' denote the derived subgroup of G, generated by all elements of G of the form $aba^{-1}b^{-1}$, for $a, b \in G$. (See [4], Section 7.6.) The essential property of G' is that homomorphisms from G into an abelian group all factor through G/G'. It follows that there is a one-to-one correspondence between group homomorphisms from G into

$\mathbf{C}^{\times}$ and group homomorphisms from G/G' into $\mathbf{C}^{\times}$. This results in the desired one-to-one correspondence between linear characters of G and linear characters of G/G'.

The next proposition provides some elementary facts about characters.

Proposition 4.2.6 *Let V be a finitely generated left $\mathbf{C}G$-module, and let $\rho : G \to \mathrm{GL}(V)$ be the corresponding representation of G. If g is an element of G of order m, then the following conditions hold.*

(a) *The matrix $\rho(g)$ is diagonalizable.*

(b) *The value $\chi_V(g)$ is equal to the sum (with multiplicities) of the eigenvalues of $\rho(g)$.*

(c) *The value $\chi_V(g)$ is a sum of mth roots of unity.*

(d) *The value $\chi_V(g^{-1})$ is the complex conjugate of $\chi_V(g)$.*

(e) *We have $|\chi_V(g)| \leq \chi_V(1)$.*

Proof. (a), (b) Since g has order m, the matrix $\rho(g)^m$ is equal to the identity matrix, and so $\rho(g)$ satisfies the polynomial $X^m - 1$. Thus the minimal polynomial of $\rho(g)$ is a factor of $X^m - 1$, and over the field of complex numbers it must factor as a product of distinct linear factors. This implies that $\rho(g)$ is diagonalizable, and so the trace of $\rho(g)$ is the sum of its eigenvalues (each repeated with the appropriate multiplicity).

(c) By the above remarks, each of the roots of the minimal polynomial is an mth root of unity.

(d) Since the matrix $\rho(g^{-1})$ is the inverse of $\rho(g)$, the eigenvalues of $\rho(g^{-1})$ are the inverses of the eigenvalues of $\rho(g)$, and corresponding eigenvalues occur with the same multiplicities. Since the inverse of a root of unity is its complex conjugate, part (d) follows immediately.

(e) Recall that $\chi_V(1)$ gives the degree of the representation. Since $\chi_V(g)$ is equal to the sum of the eigenvalues by part (b), and each of the eigenvalues has magnitude 1 since it is a root of unity, it follows from the triangle inequality for complex numbers that $|\chi_V(g)| \leq \chi_V(1)$. $\square$

Proposition 4.2.7 *Let V be a finitely generated left $\mathbf{C}G$-module, and let $\rho : G \to \mathrm{GL}(V)$ be the corresponding representation of G. Then*

$$\ker(\rho) = \{g \in G \mid \chi_V(g) = \chi_V(1)\}.$$

Proof. Let $g \in G$ have order m, and assume that V is an n-dimensional vector space over $\mathbf{C}$. We have already observed that $\chi_V(1) = n$, and that

$\chi_V(g)$ is a sum of n mth roots of unity. Thus $\chi_V(g) = n$ if and only if each eigenvalue of $\rho(g)$ is equal to 1, and this occurs if and only if $\rho(g)$ is the identity matrix. $\square$

The preceding proposition shows that the kernel of a representation can be determined just from its character. We will refer to $\ker(\rho)$ as the kernel of the character χ_V. Note that every normal subgroup of G is the kernel of some representation, so, in theory, knowledge of the characters of G should give valuable information about the normal subgroups of G.

We can define various operations on the characters of G. If χ_V is a character of G, and $c \in \mathbf{C}$, then $c\chi_V$ is a function from G to $\mathbf{C}$. We can also define sums and products of characters, which turn out to represent new characters, as will be shown in Proposition 4.2.9.

Definition 4.2.8 *Let χ_U and χ_V be characters of the finite group G. We introduce an addition $\chi_U + \chi_V$ and a multiplication $\chi_U \cdot \chi_V$ defined as follows:*

$$[\chi_U + \chi_V](g) = \chi_U(g) + \chi_V(g) \quad and \quad [\chi_U \cdot \chi_V](g) = \chi_U(g)\chi_V(g),$$

for all $g \in G$.

Proposition 4.2.9 *Let U and V be finitely generated left $\mathbf{C}G$-modules, and let $W = \mathrm{Hom}_\mathbf{C}(U, V)$. Then the following conditions hold.*

(a) $\chi_U + \chi_V = \chi_{U \oplus V}$.

(b) $\chi_U \cdot \chi_V = \chi_{U \otimes V}$.

(c) $\chi_{V^*} = \overline{\chi_V}$.

(d) $\chi_W = \overline{\chi_U} \cdot \chi_V$.

Proof. Assume that U and V are defined by the representations ρ and σ, respectively, and let $g \in G$.

(a) If we choose bases for $U \oplus (0)$ and $(0) \oplus V$, their union is a basis for $U \oplus V$, and left multiplication by g yields the matrix $A = \begin{bmatrix} \rho(g) & 0 \\ 0 & \sigma(g) \end{bmatrix}$. Thus $\chi_{U \oplus V} = \mathrm{tr}(A) = \mathrm{tr}(\rho(g)) + \mathrm{tr}(\sigma(g)) = \chi_U + \chi_V$.

(b) We have shown that $\rho(g)$ and $\sigma(g)$ are diagonalizable. Assume that $\{u_1, \ldots, u_m\}$ is a basis for U consisting of eigenvectors with respective eigenvalues $r_1, \ldots, r_m$, and assume that $\{v_1, \ldots, v_n\}$ is a basis for V consisting of eigenvectors with respective eigenvalues $s_1, \ldots, s_n$. Then $\{u_i \otimes v_j\}_{i,j}$ is a basis for $U \otimes_F V$, consisting of eigenvectors for the action of g on $U \otimes_F V$, since

$$g(u_i \otimes v_j) = gu_i \otimes gv_j = r_i u_i \otimes s_j v_j = r_i s_j (u_i \otimes v_j)$$

for all i, j. It follows that

$$\chi_{U \otimes V}(g) = \sum_{i,j} r_i s_j = \left(\sum_{i=1}^{m} r_i\right) \left(\sum_{j=1}^{n} s_j\right) = \chi_U(g) \cdot \chi_V(g) .$$

(c) Choose a basis for U as in part (b) of the proof. Let $\{\lambda_1, \ldots, \lambda_m\}$ be the dual basis of U^*, defined by $\lambda_i(u_j) = 1$, if $i = j$, and 0 otherwise. As a consequence of the module structure defined on U^*, for each i we have

$$[g\lambda_i](u_j) = \lambda_i(g^{-1}u_j) = \lambda_i(\overline{r_j}u_j) = \lambda_i(\overline{r_j}u_j) .$$

This value is $\overline{r_j}$ if $i = j$, and 0 otherwise, so $g\lambda_i = \overline{r_i}\lambda_i$ for each i. Thus the dual basis consists of eigenvectors for the transformation of U^* defined by g, so the value of $\chi_{U^*}(g)$ is $\overline{r_1} + \cdots + \overline{r_m} = \overline{\chi_U(g)}$.

(d) This follows from parts (b) and (c), since $\operatorname{Hom}_F(U, V) \cong U^* \otimes_F V$ by Proposition 4.1.10 (c). $\square$

Definition 4.2.10 *Let G be finite group, and let V be a finitely generated left $\mathbf{C}G$-module, with representation $\rho : G \to \operatorname{GL}(V)$. The character χ_V is called* irreducible *if V is a simple left $\mathbf{C}G$-module.*

It follows from part (a) of Proposition 4.2.9 that a character is irreducible if and only if it cannot be written as the sum of two nontrivial characters. Since $\mathbf{C}G$ is a semisimple Artinian ring, every finitely generated $\mathbf{C}G$-module is isomorphic to a finite direct sum of simple modules, and so every character of G is a sum of finitely many irreducible characters. We are now able to show that a finitely generated $\mathbf{C}G$-module is determined (up to isomorphism) by the character it affords.

Proposition 4.2.11 *Let G be a finite group.*

(a) *As functions from G to $\mathbf{C}$, the irreducible characters of G are linearly independent.*

(b) *Two finitely generated left $\mathbf{C}G$-modules are isomorphic if and only if their characters are equal.*

Proof. (a) As in Theorem 4.2.3, $\mathbf{C}G \cong \operatorname{M}_{n_1}(\mathbf{C}) \oplus \operatorname{M}_{n_2}(\mathbf{C}) \oplus \cdots \oplus \operatorname{M}_{n_r}(\mathbf{C})$. Let S_i be a simple left $\mathbf{C}G$-module associated with $\operatorname{M}_{n_i}(\mathbf{C})$, let χ_i be the character afforded by S_i, and let e_i be the central idempotent element of $\mathbf{C}G$ that corresponds to the identity element of $\operatorname{M}_{n_i}(\mathbf{C})$. The character χ_i is defined on G, the set of basis vectors for $\mathbf{C}G$, so there is a unique linear transformation $\widehat{\chi_i}$ that extends χ_i to all of $\mathbf{C}G$. If $c_1, \ldots, c_r \in \mathbf{C}$ and $\sum_{i=1}^{r} c_i \chi_i$ is identically zero, then $\sum_{i=1}^{r} c_i \widehat{\chi_i} = 0$ as well. Because $\widehat{\chi_i}(e_i) = n_i$ and $\widehat{\chi_i}(e_j) = 0$ for $j \neq i$, it follows that $n_i c_i = 0$ for $i = 1, \ldots, r$. Thus $c_i = 0$ for all i, and the characters χ_i are linearly independent.

(b) Suppose that V and W are finitely generated $\mathbf{C}G$-modules with $\chi_V = \chi_W$. Each module is isomorphic to a direct sum of finitely many simple modules, so each character is a linear combination of the characters χ_i. Since the irreducible characters are linearly independent by part (a), the simple components of V and W occur with the same multiplicities, and thus $V \cong W$. $\square$

Example 4.2.5 (Irreducible characters of a finite abelian group)

If G is a finite abelian group, then Corollary 4.2.4 shows that the irreducible characters of G correspond to the linear characters of G. By Example 4.2.3, these correspond to the group homomorphisms from G into $\mathbf{C}^\times$. It follows from Theorem 2.7.10, which describes modules over a principal ideal domain, that G is isomorphic to a direct sum of cyclic groups of prime power order. The irreducible characters of G are completely determined by their values on the canonical generators of this direct sum.

Example 4.2.6 (Irreducible characters of the Klein 4-group)

Let G be the Klein 4-group, described as $G = \{1, a, b, ab\}$, with relations $a^2 = b^2 = 1$ and $ba = ab$. From the previous example, the irreducible representations of G correspond to the four possible group homomorphisms from G into $\mathbf{C}^\times$. These are determined by their values on the generators a and b, and since a and b have order 2, the only possible values are 1 and -1. This produces the following list.

$$
\begin{array}{llll}
\chi_1(1) = 1 & \chi_1(a) = 1 & \chi_1(b) = 1 & \chi_1(ab) = 1 \\
\chi_2(1) = 1 & \chi_2(a) = 1 & \chi_2(b) = -1 & \chi_2(ab) = -1 \\
\chi_3(1) = 1 & \chi_3(a) = -1 & \chi_3(b) = 1 & \chi_3(ab) = -1 \\
\chi_4(1) = 1 & \chi_4(a) = -1 & \chi_4(b) = -1 & \chi_4(ab) = 1
\end{array}
$$

In the next section we will develop a standard form in which the irreducible characters can be listed.

Let V be a left $\mathbf{C}G$-module. We define the G-*fixed* submodule of V to be

$$ V^G = \{v \in V \mid gv = v \text{ for all } g \in G\} \,. $$

The subset V^G is a subspace of V, and by definition it is G-invariant, so it is a $\mathbf{C}G$-submodule of V. Let $e = \dfrac{1}{|G|} \sum_{g \in G} g$ be the idempotent generator of the trace ideal of $\mathbf{C}G$ (see the discussion following Proposition 4.1.5). We claim that

$$V^G = \{v \in V \mid ev = v\} = eV \ .$$

To show this, recall that $ge = e = eg$, for all $g \in G$, so if $ev = v$, then $gv = g(ev) = (ge)v = ev = v$, for all $g \in G$. On the other hand, if $v \in V^G$, then

$$ev = \left(\frac{1}{|G|} \sum_{g \in G} g \right) v = \frac{1}{|G|} \sum_{g \in G} gv = \frac{1}{|G|} \sum_{g \in G} v = v \ .$$

Proposition 2.2.7 implies that the G-fixed submodule V^G is a direct summand of V, corresponding to the central idempotent e.

Proposition 4.2.12 *If V is a finitely generated left $\mathbf{C}G$-module, then*

$$\sum_{g \in G} \chi_V(g) = |G| \cdot \dim_{\mathbf{C}}(V^G) \ .$$

Proof. Let $\rho : G \to \mathrm{GL}(V)$ be the representation defined by V, and let $\widehat{\rho} : \mathbf{C}G \to \mathrm{End}_F(V)$ be the extension of ρ from G to $\mathbf{C}G$. Let $e = \frac{1}{|G|} \sum_{g \in G} g$ be the central idempotent that generates the trace ideal. Since e is idempotent, it follows that the matrix $\widehat{\rho}(e)$ satisfies the polynomial $X^2 = X$. Therefore the eigenvalues of $\widehat{\rho}(e)$ are 0 and 1, and V^G is the eigenspace of $\widehat{\rho}(e)$ corresponding to the eigenvalue 1 since $V^G = \{v \in V \mid ev = v\}$. The computation

$$
\begin{aligned}
\frac{1}{|G|} \sum_{g \in G} \chi_V(g) &= \frac{1}{|G|} \sum_{g \in G} \mathrm{tr}(\rho(g)) = \mathrm{tr}\left(\frac{1}{|G|} \sum_{g \in G} \rho(g) \right) \\
&= \mathrm{tr}\left(\widehat{\rho}\left(\frac{1}{|G|} \sum_{g \in G} g \right) \right) = \mathrm{tr}(\widehat{\rho}(e)) \\
&= \dim(V^G)
\end{aligned}
$$

completes the proof. $\square$

If χ_V is an irreducible character of G, then V is a simple left $\mathbf{C}G$-module, and so either $V^G = V$ or $V^G = (0)$. In the first case G acts trivially on V, and it follows that $\dim_{\mathbf{C}}(V) = 1$. Thus for the principal character of G we have $\sum_{g \in G} \chi_V(g) = |G|$. As in Example 4.2.6, for all other irreducible characters we have $\sum_{g \in G} \chi_V(g) = 0$. This is a particular case of the more general orthogonality relations that will be developed in Section 4.3.

Let G be a finite group.

1. Show that if S is a simple $\mathbf{C}G$-module, and V is a one-dimensional $\mathbf{C}G$-module, then $S \otimes V$ is a simple $\mathbf{C}G$-module.

2. Show that a $\mathbf{C}G$-module V is simple if and only if V^* is simple.

3. Show that the linear characters of G are the only characters of G that are group homomorphisms.

4. Let G_1 and G_2 be groups with irreducible characters $\psi_1, \ldots, \psi_r$ and $\varphi_i, \ldots, \varphi_s$.

   (a) By analyzing the conjugacy classes of $G_1 \times G_2$, show that the number of irreducible characters of $G_1 \times G_2$ must be rs.

   (b) Show that the function $\chi_{ij} : G_1 \times G_2 \to \mathbf{C}^\times$ defined by $\chi_{ij}(g_1, g_2) = \psi_i(g_1)\varphi_j(g_2)$, for all $(g_1, g_2) \in G_1 \times G_2$ is a character of $G_1 \times G_2$.

   Hint: Consider the tensor product of the modules that afford ψ_i and φ_j.

   (c) Show that χ_{ij} is an irreducible character of $G_1 \times G_2$.

5. Show that if N is any normal subgroup of G, then there exist irreducible characters $\chi_1, \ldots, \chi_s$ of G, afforded by the representations $\rho_1, \ldots, \rho_s$, with $N = \bigcap_{j=1}^{s} \ker(\rho_j)$.

4.3   Character tables and orthogonality relations

We can be more precise about the statement that every character is a sum of irreducible characters. As in Theorem 4.2.3, we can write $\mathbf{C}G$ as a direct sum of r full matrix rings over $\mathbf{C}$.

$$\mathbf{C}G \cong \mathrm{M}_{n_1}(\mathbf{C}) \oplus \mathrm{M}_{n_2}(\mathbf{C}) \oplus \cdots \oplus \mathrm{M}_{n_r}(\mathbf{C})$$

With this notation, we use χ_i to denote the character of a simple module in the ith isomorphism class. We assume that χ_1 is the principal character afforded by the trivial representation of G. We recall from Example 4.2.2 that $\chi_1(g) = 1$, for all $g \in G$.

If we choose modules $S_1, \ldots, S_r$ that represent the distinct isomorphism classes of simple left $\mathbf{C}G$-modules, then any finitely generated left $\mathbf{C}G$-module V is isomorphic to a direct sum of copies of these modules. If S_i occurs with multiplicity m_i (where we may have $m_i = 0$), then it follows from Proposition 4.2.9 (a) that $\chi_V = \sum_{i=1}^{r} m_i \chi_i$.

Since the characters are class functions, each irreducible character has at most r distinct values, and it is convenient to list the values of all of the irreducible characters in an $r \times r$ matrix.

Definition 4.3.1 *Let* $X_1, \ldots, X_r$ *be the irreducible characters of* G, *and let* $g_1, \ldots, g_r$ *be representatives of the conjugacy classes of* G. *The* $r \times r$ *matrix whose* (i, j)-*entry is* $X_i(g_j)$ *is called the* character table *of* G.

The character table of the group G will be formatted as follows. The top row will list the size of each conjugacy class. The next row will list representatives of the r conjugacy classes of G. The next rows of the table will list the irreducible characters (starting with the principal character), and their values on the respective conjugacy classes.

Example 4.3.1 (The character table of $\mathbf{Z}_2$)

Let $G = \mathbf{Z}_2$, the cyclic group of order 2, and assume that g is a generator for G. We have shown in Example 4.2.5 that the irreducible characters of G correspond to the group homomorphisms from G into $\mathbf{C}^\times$. Since g has order 2, it must be mapped to 1 or -1, and this determines the two irreducible characters.

It is informative to compute the irreducible characters of G from first principles, by using the simple left $\mathbf{C}G$-modules. The group algebra $\mathbf{C}G$ is isomorphic to the factor ring $\mathbf{C}[x]/(x^2 - 1)$, and using the Chinese remainder theorem we can find primitive idempotents in this ring. The corresponding elements of $\mathbf{C}G$ are $e_1 = \frac{1}{2}(1 + g)$ and $e_2 = \frac{1}{2}(1 - g)$. It is easily checked that $e_1^2 = e_1$, $e_2^2 = e_2$, and $e_1 e_2 = 0$. Thus the simple left $\mathbf{C}G$-modules are the one-dimensional modules $\mathbf{C}G e_1$ and $\mathbf{C}G e_2$. The action of G on the modules is found by computing $g e_1 = e_1$ and $g e_2 = \frac{1}{2}(g - g^2) = \frac{1}{2}(g - 1) = -e_2$. The character table is given below.

sizes	1	1
classes	1	g
X_1	1	1
X_2	1	-1

Example 4.3.2 (The character table of $\mathbf{Z}_3$)

Let $G = \mathbf{Z}_3$, and assume that g is a generator for G. It is easiest to compute the irreducible characters of G by finding all group homomorphisms from G into $\mathbf{C}^\times$. Any homomorphism must map the generator g to an element in $\mathbf{C}^\times$ of order 3. The only possible images are ω and ω^2, where ω is a primitive cube root of unity.

We can also find the simple modules of $\mathbf{C}G$, and then compute the characters afforded by these modules. Since the group algebra

CG is isomorphic to the factor ring $\mathbf{C}[x]/\left(x^3 - 1\right)$, we can use the primitive idempotents of $\mathbf{C}[x]/\left(x^3 - 1\right)$ found in Example 1.2.11. The corresponding elements of CG are $e_1 = \frac{1}{3}(1 + g + g^2)$, $e_2 = \frac{1}{3}(1 + \omega^2 g + \omega g^2)$, and $e_3 = \frac{1}{3}(1 + \omega g + \omega^2 g^2)$. The action of G on the one-dimensional modules CGe_1, CGe_2, and CGe_3 is found by computing $ge_1 = e_1$, $ge_2 = \frac{1}{3}(g+\omega^2 g^2 +\omega g^3) = \frac{1}{3}\omega(1+\omega^2 g+\omega g^3) = \omega e_2$, and $ge_3 = \frac{1}{3}(g + \omega g^2 + \omega^2 g^3) = \omega^2 e_3$.

Either method of computation leads to the following character table for $\mathbf{Z}_3$.

sizes	1	1	1
classes	1	g	g^2
χ_1	1	1	1
χ_2	1	ω	ω^2
χ_3	1	ω^2	ω

We have noted previously that the characters of G are constant on the conjugacy classes of G. Before computing more character tables, we need additional information about such functions.

Definition 4.3.2 *A function* $\alpha : G \to \mathbf{C}$ *is called a* class function *if it is constant on the conjugacy classes of* G.

The set of all class functions on G forms a vector space over $\mathbf{C}$, since any finite linear combination of class functions is again a class function.

If α and β are class functions on G, then their *inner product* is defined to be the complex number

$$(\alpha, \beta) = \frac{1}{|G|} \sum_{g \in G} \alpha(g)\overline{\beta(g)} \ .$$

The following properties are easy to check.

(i) The inner product (,) is linear in the first variable.

(ii) For any class function α, we have $(\alpha, \alpha) \geq 0$, and, furthermore, $(\alpha, \alpha) = 0$ if and only if $\alpha = 0$.

(iii) For any class functions α, β, we have $(\alpha, \beta) = \overline{(\beta, \alpha)}$.

It follows from property (iii) that the inner product (,) is additive in the second variable, and that $(\alpha, c\beta) = \overline{c}(\alpha, \beta)$ for any class functions α, β and any $c \in \mathbf{C}$. The next result shows that the inner product of two group characters over $\mathbf{C}$ is always an integer.

Theorem 4.3.3 *Let U and V be finitely generated $\mathbf{C}G$-modules. Then*

$$(\chi_U, \chi_V) = \dim(\mathrm{Hom}_{\mathbf{C}G}(U, V)) \, .$$

Proof. Consider $\mathrm{Hom}_{\mathbf{C}}(U, V)$, which has a left $\mathbf{C}G$-module structure defined by $[g \cdot f](u) = gf(g^{-1}u)$, for all $g \in G$, $f \in \mathrm{Hom}_{\mathbf{C}}(U, V)$, and $u \in U$. The G-fixed submodule $(\mathrm{Hom}_{\mathbf{C}}(U, V))^G$ consists of all $f \in \mathrm{Hom}_{\mathbf{C}}(U, V)$ with $g \cdot f = f$. This translates into the condition that $gf(g^{-1}u) = f(u)$, for all $u \in U$. Since multiplication by g^{-1} defines an automorphism of U, this can be written as $gf(x) = f(gx)$, for all $x \in U$. Thus the G-fixed submodule of $\mathrm{Hom}_{\mathbf{C}}(U, V)$ is precisely $\mathrm{Hom}_{\mathbf{C}G}(U, V)$.

By Proposition 4.2.11,

$$(\chi_V, \chi_U) = \frac{1}{|G|} \sum_{g \in G} \overline{\chi_U(g)} \chi_V(g) = \frac{1}{|G|} \sum_{g \in G} \chi_W(g) \, ,$$

for $W = \mathrm{Hom}_{\mathbf{C}G}(U, V)$. Using Proposition 4.2.12 and the fact that $W = (\mathrm{Hom}_{\mathbf{C}}(U, V))^G$, we have $(\chi_V, \chi_U) = \dim(\mathrm{Hom}_{\mathbf{C}G}(U, V))$. Since this is an integer, $(\chi_U, \chi_V) = (\overline{\chi_V}, \chi_U) = (\chi_V, \chi_U)$, completing the proof. $\square$

As the next proposition shows, the irreducible characters of G play an important role in the study of class functions.

Proposition 4.3.4 *The irreducible characters of G form a basis for the space of class functions on G.*

Proof. Assume that G has r conjugacy classes. Since the class functions are constant on these conjugacy classes, there is an obvious basis for the space, consisting of the characteristic functions of the conjugacy classes. That is, we consider the r functions that take the value 1 on a single conjugacy class and take the value 0 on the other conjugacy classes. Since we have already shown that the irreducible characters of G form a linearly independent set, they must form a basis since there are r distinct irreducible characters. $\square$

In the next theorem, the row orthogonality relations can be stated in the following form.

$$\frac{1}{|G|} \sum_{g \in G} \chi_i(g) \overline{\chi_j(g)} = \begin{cases} 1 & \text{if } i = j \\ 0 & \text{if } i \neq j \end{cases}$$

We have already observed that the special case $j = 1$ follows immediately from Proposition 4.2.12. The row orthogonality relations can be interpreted as saying that the rows of the character table are orthogonal (hence the name). The irreducible characters then form an orthonormal basis for the space of class functions.

We have also noted one of the special cases of the column orthogonality relations. The first column of the character table of G contains the values $X_i(1)$, for $1 \leq i \leq r$. Since $X_i(1) = n_i$ (in our earlier notation), Theorem 4.2.3 (c) implies that

$$\sum_{i=1}^r (X_i(1))^2 = |G| \, .$$

In this case $C_G(1) = G$, and we have a special case of part (b) of the next theorem.

Theorem 4.3.5 (Orthogonality relations) *Let G be a finite group with irreducible characters $X_1, \ldots, X_r$ over* **C**. *Let $g_1, \ldots, g_r$ be representatives of the conjugacy classes of G, and let $C_G(g_i)$ be the centralizer of g_i.*

(a) (Row orthogonality)

$$(X_i, X_j) = \left\{ \begin{array}{ll} 1 & \text{if } i = j \\ 0 & \text{if } i \neq j \end{array} \right. .$$

(b) (Column orthogonality)

$$\sum_{i=1}^r X_i(g_j)\overline{X_i(g_k)} = \left\{ \begin{array}{ll} |C_G(g_j)| & \text{if } j = k \\ 0 & \text{if } j \neq k \end{array} \right. .$$

Proof. (a) Let $S_1, \ldots, S_r$ represent the equivalence classes of simple **C**G-modules. We have shown that $(X_i, X_j) = \dim(\mathrm{Hom}_{\mathbf{C}G}(S_i, S_j))$. This dimension is 1 if $i = j$, since $\mathrm{Hom}_{\mathbf{C}G}(S_i, S_i) \cong \mathbf{C}$, and 0 if $i = j$, since $\mathrm{Hom}_{\mathbf{C}G}(S_i, S_j) = (0)$.

(b) For each j with $1 \leq j \leq r$ we define a class function ψ_j, with $\psi_j(g_k) = 1$ if $j = k$ and $\psi_j(g_k) = 0$ if $j \neq k$. It follows from Proposition 4.3.4 that $\psi_j = \sum_{i=1}^r c_i X_i$, for some coefficients $c_i \subset \mathbf{C}$. Applying part (a) and using the linearity of the inner product, we get that

$$\frac{1}{|G|} \sum_{g \in G} \psi_k(g)\overline{X_i(g)} = (\psi_k, X_i) = c_i \, .$$

In the above summation, from the definition of ψ_k we have $\psi_k(g) = 1$ if g is conjugate to g_k, and $\psi_k(g) = 0$ otherwise. On the other hand, if g is conjugate to g_k, then $X_i(g) = X_i(g_k)$, and so each nonzero term in the summation is $\frac{1}{|G|}\overline{X_i(g_k)}$. Since the number of elements in the conjugacy class of g_k is $|G|/|C_G(g_k)|$, we have $c_i = \overline{X_i(g_k)}/|C_G(g_k)|$. We conclude that

$$\frac{1}{|C_G(g_k)|} \sum_{i=1}^r X_i(g_j)\overline{X_i(g_k)} = \sum_{i=1}^r c_i X_i(g_j) = \psi_k(g_j) = \delta_{jk} \, ,$$

and this completes the proof. $\square$

Example 4.3.3 (The character table of S_3)

Let G be the symmetric group S_3, described via generators a, of order 3, and b, of order 2, subject to the relation $ba = a^{-1}b$. There are three conjugacy classes: $\{1\}$, $\{a, a^2\}$, and $\{b, ab, a^2b\}$. The elements $1, a, a^2$ correspond to even permutations, and form the normal subgroup A_3. The factor group S_3/A_3 is isomorphic to $\mathbf{Z}_2$, so it has two linear characters that correspond to linear characters χ_1, χ_2 of S_3. Since there are three conjugacy classes, there must be one additional character χ_3. This provides the following information.

sizes	1	2	3
classes	1	a	b
χ_1	1	1	1
χ_2	1	1	-1
χ_3			

In the first column, we have $1^2 + 1^2 + \chi_3(1)^2 = 6$, so it follows that $\chi_3(1) = 2$. Then we can use the column orthogonality relations to obtain $1 \cdot 1 + 1 \cdot 1 + 2 \cdot \chi_3(a) = 0$ and $1 \cdot 1 + 1 \cdot (-1) + 2 \cdot \chi_3(b) = 0$. This allows us to complete the table.

sizes	1	2	3
classes	1	a	b
χ_1	1	1	1
χ_2	1	1	-1
χ_3	2	-1	0

Example 4.3.4 (The character table of the quaternion group)

We next consider the group $G = \{\pm 1, \pm \mathbf{i} \pm \mathbf{j} \pm \mathbf{k}\}$ of quaternion units. The elements 1 and -1 belong to the center. The centralizer $C_G(\mathbf{i})$ contains ± 1 and $\pm \mathbf{i}$, so $\mathbf{i}$ has at most two conjugates. We have $\mathbf{j}\mathbf{i}\mathbf{j}^{-1} = -\mathbf{i}$, which shows that $\mathbf{i}$ and $-\mathbf{i}$ are conjugate. Similar computations show that there are five conjugacy classes: $\{1\}$, $\{-1\}$, $\{\pm \mathbf{i}\}$, $\{\pm \mathbf{j}\}$, and $\{\pm \mathbf{k}\}$. If we let N be the center $\{\pm 1\}$ of G, then G/N is isomorphic to $\mathbf{Z}_2 \times \mathbf{Z}_2$ since every element has order 2. The irreducible characters of G/N lift to G, and as linear characters they correspond to the group homomorphisms from $\mathbf{Z}_2 \times \mathbf{Z}_2$ into $\mathbf{C}$. The first four rows of the character table are constructed from the irreducible characters of $\mathbf{Z}_2 \times \mathbf{Z}_2$, as given in Example 4.2.6.

sizes	1	1	2	2	2
classes	1	-1	i	j	k
χ_1	1	1	1	1	1
χ_2	1	1	-1	1	-1
χ_3	1	1	1	-1	-1
χ_4	1	1	-1	-1	1
χ_5					

In the first column we must have $1^2 + 1^2 + 1^2 + 1^2 + (\chi_5(1))^2 = 8$, and so $\chi_5(1) = 2$. The column orthogonality relations imply that $\chi_5(-1) = -2$, while $\chi_5(\mathbf{i}) = \chi_5(\mathbf{j}) = \chi_5(\mathbf{k}) = 0$, and this allows us to complete the table.

sizes	1	1	2	2	2
classes	1	-1	i	j	k
χ_1	1	1	1	1	1
χ_2	1	1	-1	1	-1
χ_3	1	1	1	-1	-1
χ_4	1	1	-1	-1	1
χ_5	2	-2	0	0	0

We now begin a sequence of results leading up to Burnside's $p^a q^b$ theorem. A complex number c is called an *algebraic integer* if it is a root of a monic polynomial with integer coefficients. The set of algebraic integers is a subring of $\mathbf{C}$ (see Section A.6 of the appendix for details). Thus it follows from Proposition 4.2.6 (c) that any character takes its values in the ring of algebraic integers.

Proposition 4.3.6 *Let $g_i \in G$, and let $c_i = [G : C_G(g_i)]$ denote the number of conjugates of g_i in G. Then for any irreducible character χ_j of G, the value*

$$\frac{c_i \chi_j(g_i)}{\chi_j(1)}$$

is an algebraic integer.

Proof. Let χ_j be the irreducible character afforded by the simple $\mathbf{C}G$-module S_j. Let $\mathcal{K}_i$ be the conjugacy class of g_i, and let a be the sum in $\mathbf{C}G$ of all elements in $\mathcal{K}_i$.

Since a lies in the center of $\mathbf{C}G$, left multiplication by a defines an element of $\mathrm{End}_{\mathbf{C}G}(S_j)$. By Proposition 4.2.1, there exists an element $c \in \mathbf{C}$ with $as = cs$, for all $s \in S_j$. The trace of the matrix defined by multiplication by c is $c \cdot \chi_j(1)$, while on the other hand, the trace of the matrix defined by multiplication by a is $c_i \chi_j(g_i)$, and so we have $c = \dfrac{c_i \chi_j(g_i)}{\chi_j(1)}$.

Left multiplication by a also defines a CG-endomorphism of CG, since a is central. Using the regular representation of CG, as in Example 4.1.3, let Λ_a be the matrix determined by left multiplication by a. Since a is a sum of group elements, each entry of Λ_a is an integer. Thus the characteristic polynomial $p(x)$ of Λ_a, given by $\det(xI - \Lambda_a)$, is a polynomial with integer coefficients. The group algebra contains a minimal left ideal A_j isomorphic to S_j, and for any nonzero element $s \in A_j$ we have $as = cs$. This shows that c is an eigenvalue of Λ_a, and so c is an algebraic integer since it is a root of $p(x)$. $\square$

Corollary 4.3.7 (Frobenius) *For any irreducible character χ_i of G, the value $\chi_i(1)$ is a divisor of $|G|$.*

Proof. Let $g_1, \ldots, g_r$ be a set of representatives of the conjugacy classes of G, and let $c_i = [G : C_G(g_i)]$ denote the number of conjugates of g_i in G. By Theorem 4.3.5 (a), we have $(\chi_j, \chi_j) = 1$, and so $\dfrac{1}{|G|} \sum_{g \in G} \chi_j(g) \overline{\chi_j(g)} = 1$. Thus

$$
\begin{aligned}
\frac{|G|}{\chi_j(1)} &= \frac{1}{\chi_j(1)} \sum_{i=1}^{r} c_i \chi_j(g_i) \overline{\chi_j(g_i)} \\
&= \sum_{i=1}^{r} \frac{c_i \chi_j(g_i)}{\chi_j(1)} \overline{\chi_j(g_i)} .
\end{aligned}
$$

It follows from Proposition 4.3.6 and Proposition 4.2.6 (c) that $\dfrac{|G|}{\chi_j(1)}$ is an algebraic integer, and then Proposition A.6.2 of the appendix implies that it is an ordinary integer, so $\chi_j(1)$ must be a divisor of $|G|$. $\square$

Lemma 4.3.8 *Let χ_i be an irreducible character of G afforded by the representation ρ_i. If G has a conjugacy class $\mathcal{K}_j$ such that $|\mathcal{K}_j|$ and $\chi_i(1)$ are relatively prime, then for any $g \in \mathcal{K}_j$ either $\chi_i(g) = 0$ or $\rho_i(g)$ is a scalar matrix.*

Proof. Assume that the given conditions hold. Then there exist integers m, n such that $m|\mathcal{K}_j| + n\chi_i(1) = 1$. Multiplying by $\dfrac{\chi_i(g)}{\chi_i(1)}$, we obtain

$$
m|\mathcal{K}_j| \frac{\chi_i(g)}{\chi_i(1)} + n\chi_i(g) = \frac{\chi_i(g)}{\chi_i(1)} .
$$

It follows from Proposition 4.3.6 and Proposition 4.2.6 (c) that $a = \dfrac{\chi_i(g)}{\chi_i(1)}$ is an algebraic integer. Since $\chi_i(1)$ is the degree of the representation χ_i and $\rho_i(g)$ is diagonalizable, the value $\chi_i(g)$ is a sum of $\chi_i(1)$ roots of unity. Thus we can consider a to be the average of a number of roots of unity. Proposition A.6.5 of the appendix implies that $a = 0$ or $|a| = 1$, so either $\chi_i(g) = 0$ or else $|\chi_i(g)| = \chi_i(1)$. In the second case, each entry on the diagonal of $\rho_i(g)$ is a root of unity, and so $|\chi_i(g)| < \chi_i(1)$ unless the diagonal entries are all equal. Therefore $\rho_i(g)$ must be a scalar matrix. $\square$

Theorem 4.3.9 (Burnside) *Let G be a nonabelian simple group. Then $\{1\}$ is the only conjugacy class of G whose order is a prime power.*

Proof.   Suppose that $g \in G$, with $g \neq 1$, and the conjugacy class of the element g in G has order p^n for some prime p and some positive integer n. (If $n = 0$, then the center of G is nontrivial, contradicting the assumptions on G.) By Theorem 4.3.5 (b),

$$\sum_{i=1}^{r} \chi_i(g)\chi_i(1) = 0 \,,$$

where $\{\chi_i\}_{i=1}^{r}$ are the distinct irreducible characters of G. Assuming that χ_1 is the principal character of G, we have

$$1 + \sum_{i=2}^{r} \chi_i(g)\chi_i(1) = 0 \,.$$

If p is a factor of $\chi_i(1)$ for all $i > 1$ such that $\chi_i(1) \neq 0$, then $1/p$ is an algebraic integer, contradicting Proposition A.6.2 of the appendix. Thus $\chi_i(g) \neq 0$, and p is not a divisor of $\chi_i(1)$, for some $i > 1$. The representation ρ_i that affords χ_i must be nonzero, so $\rho_i(G)$ is isomorphic to G since G is assumed to be a simple group. Lemma 4.3.8 implies that $\rho_i(g)$ is a scalar matrix, so $\rho_i(G)$ contains a nontrivial element in its center. This contradicts the assumption that G is a simple group. $\square$

The preceding theorem can be stated as a nonsimplicity condition: if the finite group G has a conjugacy class of order p^k, where p is a prime number and $k > 0$, then G cannot be a simple group. The theorem was published by Burnside in 1904, and used to prove his $p^a q^b$ theorem, the main goal of this section. This theorem (Theorem 4.3.10) had been a goal of group theorists for over ten years, and its proof utilizing character theory justified the introduction of these new techniques. In fact, it took almost seventy years for group theorists to find a proof that did not require character theory.

Before proving Theorem 4.3.10, we need to review several notions from group theory. (Chapter 7 of [4] is used as a reference.) Let G be a finite group, and let p be a prime number. A subgroup P of G is called a *Sylow p-subgroup* of G if $|P| = p^n$ for some integer $n \geq 1$ such that p^n is a divisor of $|G|$ but p^{n+1} is not. If p is a prime divisor of $|G|$, then Sylow's first theorem guarantees that G contains a Sylow p-subgroup.

A chain of subgroups $G = N_0 \supseteq N_1 \supseteq \cdots \supseteq N_n$ such that

(i) N_i is a normal subgroup in N_{i-1} for $i = 1, 2, \ldots, n$,

(ii) N_{i-1}/N_i is simple for $i = 1, 2, \ldots, n$, and

(iii) $N_n = \{1\}$

is called a *composition series* for G. The factor groups N_{i-1}/N_i are called the *composition factors* determined by the series. Any finite group has a composition series, and the Jordan–Hölder theorem states that any two composition series for a finite group have the same length.

The group G is said to be *solvable* if there exists a finite chain of subgroups $G = N_0 \supseteq N_1 \supseteq \cdots \supseteq N_n$ such that

(i) N_i is a normal subgroup in N_{i-1} for $i = 1, 2, \ldots, n$,

(ii) N_{i-1}/N_i is abelian for $i = 1, 2, \ldots, n$, and

(iii) $N_n = \{1\}$.

Such groups take their significance from the fact that a polynomial $f(x)$ over $\mathbf{C}$ is solvable by radicals if and only if its Galois group is a solvable group.

Theorem 4.3.10 (Burnside) *If G is a finite group with $|G| = p^a q^b$, where p, q are prime numbers, then G is a solvable group.*

Proof. Let G_i be any composition factor arising from a composition series for G. To show that G is solvable it suffices to show that G_i is abelian. The group G_i is simple, and $|G_i|$ is a divisor of $|G|$, so we have reduced the proof to the case in which G is a simple group.

Assume that G is simple, with $|G| = p^a q^b$, and let P be a Sylow p-subgroup of G. Since any p-group has a nontrivial center, we can find a nontrivial central element g in P. Then $P \subseteq C_G(g)$ and $|P| = p^a$, so $[G : C_G(g)]$ is not divisible by p, and this contradicts Theorem 4.3.9, unless G is abelian. $\square$

EXERCISES: SECTION 4.3

Let G be a finite group.

1. Prove that the character table of G is an invertible matrix.

2. Compute the character table of the group $\mathbf{Z}_2 \times \mathbf{Z}_4$.

3. Show that the character table of the dihedral group D_4 of order 8 is the same as the character table of the group of quaternion units given in Example 4.3.4.

4. Prove that if χ_U is a linear character of $\mathbf{C}G$ and χ_V is an irreducible character of $\mathbf{C}G$, then the product $\chi_U \chi_V$ is an irreducible character of $\mathbf{C}G$.

5. Let χ_G be the character of G afforded by $\mathbf{C}G$.
   (a) Show that for all $g \in G$, $\chi_G(g) = \sum_{i=1}^{r} \chi_i(1)\chi_i(g)$.
   (b) Show that if $g \neq 1$, then $\chi_G(g) = 0$.

6. Show that the orthogonal central idempotents of $\mathbf{C}G$ can be computed using the formula $e_i = \dfrac{1}{|G|} \sum_{g \in G} \chi_i(1)\chi_i(g^{-1})g$.

7. Let N be a normal subgroup of G, and let V be a $(\mathbf{C}(G/N))$-module. Show that if φ is the character of V, then when V is viewed as a $\mathbf{C}G$-module the corresponding character is $\varphi\pi$, where $\pi : G \to G/N$ is the natural projection.

8. Find the character table of S_4.
   Hint: Apply Exercise 7, using a normal subgroup N of S_4 with $S_4/N \cong S_3$.

$$\boxed{\text{A}}$$

APPENDIX

A.1   Review of vector spaces

Definition A.1.1 *A field F is a commutative ring for which $1 \neq 0$ and each nonzero element $a \in F$ has an inverse $a^{-1} \in F$ with $a \cdot a^{-1} = 1$.*

Let F be any field, and let a, b be nonzero elements of F. Then a has an inverse, so if $ab = 0$ then $a^{-1}(ab) = a^{-1}(0) = 0$, which implies that $b = 0$, a contradiction. We conclude that in any field the product of two nonzero elements must be nonzero.

Definition A.1.2 *Let F be a field. The smallest positive integer n such that $n \cdot 1 = 0$ is called the* characteristic *of F, denoted by $\mathrm{char}(F)$. If no such positive integer exists, then F is said to have* characteristic zero.

If $\mathrm{char}(F) = n$, then it follows from the distributive law that $n \cdot a = (n \cdot 1) \cdot a = 0 \cdot a = 0$, and so adding any element to itself n times yields 0.

Proposition A.1.3 *The characteristic of a field is either 0 or p, for some prime number p.*

Proof. If a field F has characteristic n, and n is composite, say $n = mk$, then $(m \cdot 1)(k \cdot 1) = n \cdot 1 = 0$. This is a contradiction, since in any field the product of two nonzero elements is nonzero. $\square$

Definition A.1.4 *If F and E are fields such that F is a subset of E and the operations on F are those induced by E, then we say that F is a* subfield *of E and that E is an* extension field *of F.*

Definition A.1.5 *A* vector space *over the field F is a set V on which two operations are defined, called addition and scalar multiplication, and denoted by $+$ and $\cdot$ respectively. The operations must satisfy the following conditions.*

(i) *For all $a \in F$ and all $u, v \in V$, the sum $u + v$ and the scalar product $a \cdot v$ are uniquely defined and belong to V.*

(ii) *For all $a, b \in F$ and all $u, v, w \in V$, $u + (v + w) = (u + v) + w$ and $a \cdot (b \cdot v) = (a \cdot b) \cdot v$.*

(iii) *For all $u, v \in V$, $u + v = v + u$.*

(iv) *For all $a, b \in F$ and all $u, v \in V$, $a \cdot (u + v) = (a \cdot u) + (a \cdot v)$ and $(a + b) \cdot v = (a \cdot v) + (b \cdot v)$.*

(v) *The set V contains an* additive identity *element, denoted by 0, such that for all $v \in V$, $v + 0 = v$ and $0 + v = v$. For all $v \in V$, $1 \cdot v = v$.*

(vi) *For each $v \in V$, the equations $v + x = 0$ and $x + v = 0$ have a solution $x \in V$, called an* additive inverse *of v, and denoted by $-v$.*

Example A.1.1

Let E be an extension field of F. Then any element $a \in F$ also belongs to E, so for $x \in E$ we define the scalar multiplication ax to be the product in E. Comparing Definition A.1.5 with the definition of a field shows that E is indeed a vector space over F.

Note that scalar multiplication is not a binary operation on V. It must be defined by a function from $F \times V$ into V, rather than from $V \times V$ into V. We will use the notation av rather than $a \cdot v$.

If we denote the additive identity of V by 0 and the additive inverse of $v \in V$ by $-v$, then we have the following results: $0 + v = v$, $a \cdot 0 = 0$, and $(-a)v = a(-v) = -(av)$ for all $a \in F$ and $v \in V$. The proofs are similar to those for the same results in a field, and involve the distributive laws, which provide the only connection between addition and scalar multiplication.

Definition A.1.6 *Let $S = \{v_1, v_2, \ldots, v_n\}$ be a set of vectors in the vector space V over the field F. Any vector of the form $v = \sum_{i=1}^{n} a_i v_i$, for scalars $a_i \in F$, is called a* linear combination *of the vectors in S. The set S is said to* span *V if each element of V can be expressed as a linear combination of the vectors in S.*

Let $S = \{v_1, v_2, \ldots, v_n\}$ be a set of vectors in the vector space V over the field F. The vectors in S are said to be linearly dependent if one of the vectors can be expressed as a linear combination of the others. If not, then S is said to be a linearly independent set.

In the preceding definition, if S is linearly dependent and, for example, v_j can be written as a linear combination of the remaining vectors in S, then we can rewrite the resulting equation as $a_1 v_1 + \cdots + 1 \cdot v_j + \cdots + a_n v_n = 0$ for some scalars $a_i \in F$. Thus there exists a nontrivial (i.e., at least one coefficient is nonzero) relation of the form $\sum_{i=1}^{n} a_i v_i = 0$. Conversely, if such a relation exists, then at least one coefficient, say a_j, must be nonzero. Since the coefficients are from a field, we can divide through by a_j and shift v_j to the other side of the equation to obtain v_j as a linear combination of the remaining vectors. If j is the largest subscript for which $a_j \neq 0$, then we can express v_j as a linear combination of $v_1, \ldots, v_{j-1}$.

From this point of view, S is linearly independent if and only if $\sum_{i=1}^{n} a_i v_i = 0$ implies $a_1 = a_2 = \cdots = a_n = 0$. This is the condition that is usually given as the definition of linear independence.

Theorem A.1.7 Let V be a vector space, let $S = \{u_1, u_2, \ldots, u_m\}$ be a set that spans V, and let $T = \{v_1, v_2, \ldots, v_n\}$ be a linearly independent set. Then $n \leq m$, and V can be spanned by a set of m vectors that contains T.

Proof. Given $m > 0$, the proof will use induction on n. If $n = 1$, then of course $n \leq m$. Furthermore, v_1 can be written as a linear combination of vectors in S, so the set $S' = \{v_1, u_1, \ldots, u_m\}$ is linearly dependent. Using the remarks preceding the theorem, one of the vectors in S' can be written as a linear combination of other vectors in S' with lower subscripts, so deleting it from S' leaves a set with m elements that still spans V and also contains v_1.

Now assume that the result holds for any set of n linearly independent vectors, and assume that T has $n + 1$ vectors. The first n vectors of T are still linearly independent, so by the induction assumption we have $n \leq m$ and V is spanned by some set $\{v_1, \ldots, v_n, w_{n+1}, \ldots, w_m\}$. Then v_{n+1} can be written as a linear combination of $\{v_1, \ldots, v_n, w_{n+1}, \ldots, w_m\}$. If $n = m$, then this contradicts the assumption that the set T is linearly independent, so we must have $n + 1 \leq m$. Furthermore, the set $S' = \{v_1, \ldots, v_n, v_{n+1}, w_{n+1}, \ldots, w_m\}$ is linearly dependent, so as before we can express one of the vectors v in S' as a linear combination of vectors in S' with lower subscripts. But v cannot be one of the vectors in T, and so we can omit one of the vectors $\{w_{n+1}, \ldots, w_m\}$, giving the desired set. □

Corollary A.1.8 Any two finite subsets that both span V and are linearly independent must have the same number of elements.

The above corollary justifies the following definition.

Definition A.1.9 *A subset of the vector space V is called a* basis *for V if it spans V and is linearly independent. If V has a finite basis, then it is said to be* finite dimensional, *and the number of vectors in the basis is called the* dimension *of V, denoted by* $\dim(V)$.

Proposition A.1.10 *Let $S = \{v_1, v_2, \ldots, v_n\}$ be a set of vectors in the n-dimensional vector space V. Then S is a basis for V if and only if each vector in V can be written uniquely as a linear combination of vectors in S.*

Proof. Assume that S is a basis for V. Then it spans V, and so each vector of V can be expressed as a linear combination of vectors in S. Suppose that there are two such linear combinations for $v \in V$, say $v = a_1 v_1 + a_2 v_2 + \cdots + a_n v_n$ and $v = b_1 v_1 + b_2 v_2 + \cdots + b_n v_n$. Then $(a_1 - b_1)v_1 + (a_2 - b_2)v_2 + \cdots + (a_n - b_n)v_n = 0$, and either $a_i - b_i = 0$ for $1 \leq i \leq n$, or $a_i - b_i \neq 0$ for some i. The first case implies that $a_i = b_i$ for all i, and so in this case the linear combinations of vectors in S are unique. On the other hand, if $a_i - b_i \neq 0$ for some i, then dividing through by $a_i - b_i$ (remember that F is a field) shows that v_i can be written as a linear combination of the remaining vectors in S. This contradicts the assumption that S is a basis, and so the second case cannot occur.

Conversely, assume that each vector of V can be expressed uniquely as a linear combination of vectors in S. Then S certainly spans V. For each $v_i \in S$ we have only one way to express it as a linear combination of elements of S, namely $v_i = 0 \cdot v_1 + 0 \cdot v_2 + \cdots + 1 \cdot v_i + \cdots + 0 \cdot v_n$. Thus no element of S can be expressed as a linear combination of other elements in S, and so S is a linearly independent set. $\square$

Example A.1.2

Let F be any field, let n be a fixed positive integer, and let V be the set of polynomials of the form $a_0 + a_1 x + \cdots + a_{n-1}x^{n-1}$ with coefficients in the field F. We can add polynomials and multiply them by elements of F, and since the necessary properties are satisfied, we see that V is a vector space over F. Each polynomial $f(x)$ can be expressed uniquely as $f(x) = a_0 \cdot 1 + a_1 \cdot x + \cdots + a_{n-1} \cdot x^{n-1}$, and so we see that V has a basis $\{1, x, x^2, \ldots, x^{n-1}\}$, which implies that V has dimension n as a vector space over F.

The statements in the next theorem follow directly from Theorem A.1.7. It is shown more generally in Theorem A.2.5 that even in an infinite dimensional vector space any linearly independent subset is contained in a basis.

Theorem A.1.11 *Let V be an n-dimensional vector space.*

(a) *Any linearly independent set of vectors in V is part of a basis for V, and any spanning set for V contains a subset that is a basis for V.*

(b) *Any set of more than n vectors is linearly dependent, and no set of fewer than n vectors can span V.*

(c) *Any set of n vectors in V is a basis for V if it either is linearly independent or spans V.*

Given a vector space V over a field F, a subset W of V is called a *subspace* if W is a vector space over F under the operations already defined on V. This can be shown to be equivalent to the following conditions: (i) W is nonempty; (ii) if $u, v \in W$, then $u + v \in W$; and (iii) if $v \in W$ and $a \in F$, then $av \in W$. For subspaces W_1 and W_2 of V, we can construct new subspaces $W_1 \cap W_2$ and $W_1 + W_2$, where the latter subspace is defined to be

$$W_1 + W_2 = \{v \in V \mid \exists w_1 \in W_1, w_2 \in W_2 \text{ with } v = w_1 + w_2\}.$$

Finally, we recall the definition of a linear transformation.

Definition A.1.12 *Let V and W be vector spaces over the field F. A linear transformation from V to W is a function $f : V \to W$ such that*

$$f(c_1 v_1 + c_2 v_2) = c_1 f(v_1) + c_1 f(v_2)$$

for all vectors $v_1, v_2 \in V$ and all scalars $c_1, c_2 \in F$.

If f is one-to-one and onto, it is called an isomorphism *of vector spaces.*

A.2   Zorn's lemma

The axiom of choice states that the Cartesian product of any nonempty family of nonempty sets is nonempty. Zorn's lemma provides a very useful condition that is equivalent to the axiom of choice.

Definition A.2.1 *A partial order on a set S is a relation $\leq$ such that*
  (i) $x \leq x$ *for all $x \in S$,*
  (ii) *if $x, y, z \in S$ with $x \leq y$ and $y \leq z$, then $x \leq z$,*
  (iii) *if $x, y \in S$ with $x \leq y$ and $y \leq x$, then $x = y$.*

Definition A.2.2 *Let S be a set with a partial order $\leq$.*
  (a) *An element $m \in S$ is said to be* maximal *if $m \leq x$ implies $m = x$, for all $x \in S$.*
  (b) *A subset of S is called a* chain *if for any x, y in the subset either $x \leq y$ or $y \leq x$. Such a subset is also said to be* linearly ordered.

If T is any set, let $S = 2^T$ be the collection of all subsets of T. It is easy to check that S can be partially ordered using inclusion of sets. The conditions for a partial order $\leq$ are satisfied by defining $\leq$ to be either $\subseteq$ or $\supseteq$.

The set S is said to be *well-ordered* if each nonempty subset contains a least element. With this terminology, the axiom of choice is equivalent to a general well-ordering principle: if S is a nonempty set, then there exists a linear order on S which makes S a well-ordered set.

We next give the precise statement of the axiom that is usually referred to as Zorn's lemma.

Axiom A.2.3 (Zorn's lemma) *Let $(S, \leq)$ be a nonempty partially ordered set. If every chain in S has an upper bound in S, then S has a maximal element.*

The principle of induction can be extended to sets other than the natural numbers, but it requires the use of the axiom of choice, again in the form of Zorn's lemma.

Theorem A.2.4 (Principle of transfinite induction) *Let A be a subset of a well-ordered set $(S, \leq)$. Then $A = S$ if the following condition holds for all $x \in S$:*

$$s \in A \text{ for all } s < x \text{ implies } x \in A \, .$$

Proof. Assume that A is a subset of S that satisfies the given condition. If $A \neq S$, then there is a least element x in the subset $S \setminus A$, but by assumption we must have $x \in A$, a contradiction. $\square$

We will illustrate the use of Zorn's lemma by proving two theorems from linear algebra.

Theorem A.2.5 *Let F be a field, and let V be any vector space over F. Then every linearly independent subset of V is contained in a basis for V.*

Proof. If X is any linearly independent subset of V, let S be the collection of all linearly independent subsets of V that contain X, partially ordered by set theoretic inclusion $\subseteq$. For any chain $\{W_\alpha\}_{\alpha \in I}$ in S, the union $\bigcup_{\alpha \in I} W_\alpha$ is an upper bound, since any finite subset of W is contained in W_n, for some n, and then this subset must be linearly independent. Zorn's lemma implies that there exists a maximal element B in S.

We claim that B is a basis for V. For any $v \in V$, the set $B \cup \{v\}$ cannot be linearly independent, and so there is a nontrivial linear combination $av + \sum_{i=1}^n a_i x_i = 0$, for some $x_1, x_2, \ldots, x_n \in X$ and some $a, a_1, a_2, \ldots, a_n \in F$. Since the elements $x_1, \ldots, x_n$ are linearly independent, we must have $a \neq 0$, and then a^{-1} exists in F since F is a field. Thus $v = -a^{-1}(\sum_{i=1}^n a_i x_i)$, and this linear combination is easily seen to be unique since the elements $x_1, \ldots, x_n$ are linearly independent. $\square$

Theorem A.2.6 *Let V be a vector space over the field F, and let W be any subspace of V. Then there exists a subspace Y of V such that each element $v \in V$ can be written uniquely in the form $v = w + y$ for some $w \in W$ and $y \in Y$.*

Proof. Let S be the set of all subspaces $W' \subseteq V$ such that $W \cap W' = (0)$. The set S is partially ordered by inclusion $\subseteq$, and nonempty since it contains the zero subspace. For any chain $\{W_\alpha\}_{\alpha \in I}$ of subspaces in S, the union $U = \bigcup_{\alpha \in I} W_\alpha$ is a subspace. To show that U is closed, let $u_1, u_2 \in U$. Then $u_1 \in W_\alpha$ and $u_2 \in W_\beta$ for some $\alpha, \beta \in I$. Since the set $\{W_\alpha\}_{\alpha \in I}$ is a chain, either $W_\alpha \subseteq W_\beta$, or $W_\beta \subseteq W_\alpha$. In the first case, $u_1 + u_2 \in W_\beta$, and in the second case, $u_1 + u_2 \in W_\alpha$, so $u_1 + u_2 \in U$. If $u \in U$ and $a \in F$, then $u \in W_\alpha$ for some $\alpha \in I$, so $au \in W_\alpha$, and thus $au \in U$. Thus U is an upper bound for the chain $\{W_\alpha\}_{\alpha \in I}$. Therefore we can apply Zorn's lemma, and so there is a maximal element Y in S.

Suppose that there exists $v \in V$ such that v cannot be written as a sum of elements in W and Y. Let W' be the smallest subspace containing Y and v, so that W' consists of all vectors of the form $y + av$ for some $y \in Y$ and some $a \in F$. If $w \in W \cap W'$, then $w = y + av$ for some $y \in Y$ and $a \in F$. Since $W \cap Y = (0)$, we must have $a \neq 0$, and then $v = a^{-1}w - a^{-1}y$, a contradiction. But then $W \cap W' = (0)$ and $Y \subset W'$ contradicts the maximality of Y in S. We conclude that each element $v \in V$ can be expressed in the form $w + y$, for some $w \in W$ and $y \in Y$. The uniqueness of this representation follows immediately from the fact that $W \cap Y = (0)$. $\square$

A.3 Matrices over commutative rings

In this section we consider matrices with entries from a commutative ring R. An $n \times n$ matrix over R is an array of elements of R, of the form

$$[a_{ij}] = \begin{bmatrix} a_{11} & a_{12} & \dots & a_{1n} \\ a_{21} & a_{22} & \dots & a_{2n} \\ \vdots & \vdots & & \vdots \\ a_{n1} & a_{n2} & \dots & a_{nn} \end{bmatrix}.$$

Two matrices $A = [a_{ij}]$ and $B = [b_{ij}]$ are equal if $a_{ij} = b_{ij}$ for all i, j.

Definition A.3.1 *Let R be a commutative ring. We let $\mathrm{M}_n(R)$ denote the set of all $n \times n$ matrices with entries in R.*

For $[a_{ij}]$ and $[b_{ij}]$ in $\mathrm{M}_n(R)$, the sum is given $[a_{ij}] + [b_{ij}] = [a_{ij} + b_{ij}]$, and the product is given by $[c_{ij}] = [a_{ij}][b_{ij}]$, where $[c_{ij}]$ is the matrix whose (i, j)-entry is $c_{ij} = \sum_{k=1}^n a_{ik} b_{kj}$.

Proposition A.3.2 *Matrix multiplication is associative.*

Proof. Let $[a_{ij}], [b_{ij}], [c_{ij}] \in M_n(R)$. The (i, j)-entry of $[a_{ij}]([b_{ij}][c_{ij}])$ is

$$
\begin{aligned}
\sum_{k=1}^{n} a_{ik} \left(\sum_{m=1}^{n} b_{km} c_{mj} \right) &= \sum_{k=1}^{n} \sum_{m=1}^{n} a_{ik}(b_{km} c_{mj}) \\
&= \sum_{m=1}^{n} \sum_{k=1}^{n} (a_{ik} b_{km}) c_{mj} \\
&= \sum_{m=1}^{n} \left(\sum_{k=1}^{n} a_{ik} b_{km} \right) c_{mj},
\end{aligned}
$$

and so we have $[a_{ij}]([b_{ij}][c_{ij}]) = ([a_{ij}][b_{ij}])[c_{ij}]$. $\square$

Theorem A.3.3 *For any commutative ring R, the set $M_n(R)$ of all $n \times n$ matrices with entries in R is a ring under matrix addition and multiplication.*

Proof. The associative law for multiplication was verified in the previous proposition. It is easy to check the associative and commutative laws for addition. For $[a_{ij}], [b_{ij}], [c_{ij}] \in M_n(R)$, the (i, j)-entries of $[a_{ij}]([b_{ij}] + [c_{ij}])$ and $[a_{ij}][b_{ij}] + [a_{ij}][c_{ij}]$ are $\sum_{k=1}^{n} a_{ik}(b_{kj} + c_{kj})$ and $\sum_{k=1}^{n} (a_{ik}b_{kj} + a_{ik}c_{kj})$ respectively. These entries are equal since the distributive law holds in R. The second distributive law is verified is the same way.

The zero matrix, in which each entry is zero, serves as an identity element for addition. The $n \times n$ identity matrix I_n, with 1 in each entry on the main diagonal and zeros elsewhere, is an identity element for multiplication. Finally, the additive inverse of a matrix is found by taking the additive inverse of each of its entries. $\square$

We now wish to discuss determinants. In this situation it is still possible to verify many of the results that hold for matrices over a field. We will simply state several of the results. In the following definition, the sum is taken over all permutations in $\mathcal{S}_n$. If $\sigma \in \mathcal{S}_n$ is even, we define $\operatorname{sgn} \sigma = +1$, and if σ is odd we define $\operatorname{sgn} \sigma = -1$.

Definition A.3.4 *Let R be a commutative ring, and let $A = [a_{ij}] \in M_n(R)$. The* determinant *of A is defined to be*

$$
\det(A) = \sum_{\sigma \in \mathcal{S}_n} (\operatorname{sgn} \sigma) a_{1\sigma(1)} a_{2\sigma(2)} \cdots a_{n\sigma(n)}.
$$

The next proposition is essential in our later work. For proofs of results on determinants we refer the reader to the book *Linear Algebra*, by Hoffman and Kunze [12].

Proposition A.3.5 *Let R be a commutative ring, and let $A, B \in M_n(R)$. Then $\det(A)\det(B) = \det(AB)$.*

We illustrate the use of the determinant in the case of a 3×3 matrix. Suppose that A is the matrix

$$\begin{bmatrix} a_{11} & a_{12} & a_{13} \\ a_{21} & a_{22} & a_{23} \\ a_{31} & a_{32} & a_{33} \end{bmatrix},$$

with entries from a commutative ring R. To compute $\det(A)$ we proceed as follows. From the three even permutations (1), $(1,2,3)$, and $(1,3,2)$ we obtain the products $a_{11}a_{22}a_{33}$, $a_{12}a_{23}a_{31}$, and $a_{13}a_{21}a_{32}$. From the three odd permutations $(1,3)$, $(2,3)$, and $(1,2)$ we obtain the products $a_{13}a_{22}a_{31}$, $a_{11}a_{23}a_{32}$, and $a_{12}a_{21}a_{33}$. Thus

$$\det(A) = (a_{11}a_{22}a_{33} + a_{12}a_{23}a_{31} + a_{13}a_{21}a_{32})$$
$$- (a_{13}a_{22}a_{31} + a_{11}a_{23}a_{32} + a_{12}a_{21}a_{33}).$$

We can also factor $\det(A)$ in the following way:

$$\det(A) = a_{11}(a_{22}a_{33} - a_{23}a_{32}) - a_{12}(a_{21}a_{33} - a_{23}a_{31}) + a_{13}(a_{21}a_{32} - a_{22}a_{31}).$$

The factors $A_{11} = a_{22}a_{33} - a_{23}a_{32}$, $A_{12} = -(a_{21}a_{33} - a_{23}a_{31})$, and $A_{13} = a_{21}a_{32} - a_{22}a_{31}$ are called the *cofactors* of a_{11}, a_{12}, and a_{13}.

For the second row a_{21}, a_{22}, a_{23} of A we have

$$a_{21}A_{11} + a_{22}A_{12} + a_{23}A_{13}$$
$$= a_{21}(a_{22}a_{33} - a_{23}a_{32}) + a_{22}(-1)(a_{21}a_{33} - a_{23}a_{31}) + a_{23}(a_{21}a_{32} - a_{22}a_{31})$$
$$= a_{21}a_{22}a_{33} - a_{21}a_{23}a_{32} - a_{22}a_{21}a_{33} + a_{22}a_{23}a_{31} + a_{23}a_{21}a_{32} - a_{23}a_{22}a_{31}$$
$$= 0.$$

A similar argument works for the third row, so in matrix form we have

$$\begin{bmatrix} a_{11} & a_{12} & a_{13} \\ a_{21} & a_{22} & a_{23} \\ a_{31} & a_{32} & a_{33} \end{bmatrix} \begin{bmatrix} A_{11} \\ A_{12} \\ A_{13} \end{bmatrix} = \begin{bmatrix} \det(A) \\ 0 \\ 0 \end{bmatrix}.$$

Extending this idea to the other rows of A, we have

$$\begin{bmatrix} a_{11} & a_{12} & a_{13} \\ a_{21} & a_{22} & a_{23} \\ a_{31} & a_{32} & a_{33} \end{bmatrix} \begin{bmatrix} A_{11} & A_{21} & A_{31} \\ A_{12} & A_{22} & A_{32} \\ A_{13} & A_{23} & A_{33} \end{bmatrix} = \begin{bmatrix} \det(A) & 0 & 0 \\ 0 & \det(A) & 0 \\ 0 & 0 & \det(A) \end{bmatrix}.$$

Definition A.3.6 *Let R be a commutative ring, and let $A \in M_n(R)$. The (i,j)-cofactor of A, denoted by A_{ij}, is $(-1)^{i+j}$ times the determinant of the $(n-1) \times (n-1)$ matrix constructed by omitting row i and column j of A.*

The adjoint *of the matrix A, denoted by $\operatorname{adj}(A)$, is the matrix whose (i,j)-entry is the cofactor A_{ji}.*

Proposition A.3.7 *Let R be a commutative ring, and let $A \in \mathrm{M}_n(R)$. Then $A \cdot \mathrm{adj}(A) = \det(A) \cdot I$ and $\mathrm{adj}(A) \cdot A = \det(A) \cdot I$.*

If $A \in \mathrm{M}_n(R)$ is a matrix such that $\det(A)$ is an invertible element of R, then we can compute the inverse of A as $A^{-1} = (\det(A))^{-1} \cdot \mathrm{adj}(A)$. This gives us the following corollary.

Corollary A.3.8 *If R is a commutative ring, then a matrix $A \in \mathrm{M}_n(R)$ is invertible if and only if $\det(A)$ is an invertible element of R.*

Definition A.3.9 *For an $n \times n$ matrix $A = [a_{ij}]$, we define the trace of A as the sum of the diagonal entries of A. That is, $\mathrm{tr}(A) = \sum_{i=1}^{n} a_{ii}$.*

Proposition A.3.10 *Let $A = [a_{ij}]$ and $B = [b_{ij}]$ be $n \times n$ matrices.*
(a) $\mathrm{tr}(A + B) = \mathrm{tr}(A) + \mathrm{tr}(B)$
(b) $\mathrm{tr}(AB) = \mathrm{tr}(BA)$
(c) *If T is any invertible matrix, then $\mathrm{tr}(TAT^{-1}) = \mathrm{tr}(A)$.*

Proof. The proof of (a) is clear, and (c) follows from (b) since $\mathrm{tr}(TAT^{-1}) = \mathrm{tr}(T^{-1}(TA)) = \mathrm{tr}(A)$. The proof of (b) follows from the fact that

$$\sum_{i=1}^{n} \left(\sum_{j=1}^{n} a_{ij} b_{ji} \right) = \sum_{j=1}^{n} \left(\sum_{i=1}^{n} b_{ji} a_{ij} \right). \quad \square$$

A.4   Eigenvalues and characteristic polynomials

This section of the appendix provides a brief review of some results from elementary linear algebra. We will use the notation $[a_{ij}]^t = [a_{ji}]$ for the transpose of the matrix $[a_{ij}]$, so that a column vector can be written as $[c_1 \; \cdots \; c_n]^t$.

Theorem A.4.1 *Let V be an n-dimensional vector space over the field F, and let $\{v_1, \ldots, v_n\}$ be a basis for V.*
(a) *If $T : V \to V$ is a linear transformation with $T(v_i) = \sum_{j=1}^{n} a_{ij} v_j$ for $1 \leq j \leq n$, then T is determined by matrix multiplication by the associated matrix $[a_{ij}]$.*
(b) *Assigning to each linear transformation $T : V \to V$ its associated matrix (as in part (a)) defines an isomorphism from the ring $\mathrm{End}_F(V)$ of all linear transformations on V to the matrix ring $\mathrm{M}_n(F)$.*

Proof. (a) Let $v = \sum_{i=1}^{n} c_i v_i$ be an element of V, with coordinate vector $[c_1 \; \cdots \; c_n]^t$. Then $T(v) = \sum_{i=1}^{n} c_i T(v_i)$, and substituting for $T(v_i)$ yields $[a_{ij}][c_1 \; \cdots \; c_n]^t$ as the coordinate vector for $T(v)$, relative to the given basis.

(b) The correspondence given in part (a) is one-to-one since each linear transformation is uniquely determined by its values on the basis vectors. It is onto since matrix multiplication defines a linear transformation. We have a ring homomorphism since the sum and composition of two linear transformations correspond, respectively, to the sum and product of their matrices. □

Throughout this section V will denote a finite dimensional vector space over a field F and $T : V \to V$ will be a fixed linear transformation. The goal is to choose a basis for V in such a way that the matrix corresponding to T is as close as possible to a diagonal matrix. The matrix for T is diagonal, with entries $c_1, c_2, \ldots, c_n$, if and only if the chosen basis $\{v_1, \ldots, v_n\}$ has the property that $T(v_i) = c_i v_i$ for all i. This explains the significance of the following definition.

Definition A.4.2 *A nonzero vector $v \in V$ is called an* eigenvector *for T if $T(v) = cv$ for some scalar $c \in F$, which is called the* eigenvalue *corresponding to v.*

Note that the terms *characteristic value* and *eigenvalue* are used synonymously, as are the terms *characteristic vector* and *eigenvector*. In finding eigenvectors v with $T(v) = cv$, it is often easier to use the equivalent condition $(T - cI)(v) = 0$, where I is the identity mapping. Then the eigenvectors corresponding to an eigenvalue c are precisely the vectors in $\ker(T - cI)$. Furthermore, c is an eigenvalue if and only if $\ker(T - cI) \neq (0)$. This allows us to conclude that c is an eigenvalue if and only if $\det(T - cI) \neq 0$, so the eigenvalues of T are the roots x of the equation $\det(xI - T) = 0$.

Definition A.4.3 *If A is an $n \times n$ matrix, then the polynomial $f(x) = \det(xI - A)$ is called the* characteristic polynomial *of A.*

If A is an $n \times n$ matrix, a careful computation of $\det(xI - A)$ shows that the constant term of the characteristic polynomial $f(x)$ is $(-1)^n \det(A)$. Furthermore, the coefficient of x^{n-1} is $-\operatorname{tr}(A)$.

We need to consider the characteristic polynomial of a linear transformation, independently of the basis chosen for its matrix representation. The next proposition shows that this is indeed possible.

Proposition A.4.4 *Similar matrices have the same characteristic polynomial.*

Proof. If $B = P^{-1}AP$, then

$$
\begin{aligned}
\det(xI - B) &= \det(xI - P^{-1}AP) = \det(P^{-1}(xI - A)P) \\
&= \det(P^{-1}) \cdot \det(xI - A) \cdot \det(P) = \det(xI - A) \, .
\end{aligned}
$$

This completes the proof. □

Let $f(x) = a_0 + a_1 x + \cdots + a_{n-1}x^{n-1} + x^n$ be a monic polynomial in $F[x]$. We define the *companion matrix* of $f(x)$ to be the $n \times n$ matrix

$$C_f = \begin{bmatrix} 0 & 0 & 0 & 0 & \cdots & 0 & -a_0 \\ 1 & 0 & 0 & 0 & \cdots & 0 & -a_1 \\ 0 & 1 & 0 & 0 & \cdots & 0 & -a_2 \\ \vdots & \vdots & \vdots & \vdots & & \vdots & \vdots \\ 0 & 0 & 0 & 0 & \cdots & 0 & -a_{n-2} \\ 0 & 0 & 0 & 0 & \cdots & 1 & -a_{n-1} \end{bmatrix}.$$

Proposition A.4.5 *Let $f(x) \in K[x]$ be a monic polynomial of degree n. The characteristic polynomial of the companion matrix C_f is $f(x)$.*

Proof. We need to show that $\det(xI - C_f) = f(x)$. The following calculation is made as follows: beginning with the bottom row, multiply each row by x and add it to the row above.

$$\begin{vmatrix} x & 0 & 0 & 0 & \cdots & 0 & 0 & a_0 \\ -1 & x & 0 & 0 & \cdots & 0 & 0 & a_1 \\ 0 & -1 & x & 0 & \cdots & 0 & 0 & a_2 \\ \vdots & \vdots & \vdots & \vdots & & \vdots & & \vdots \\ 0 & 0 & 0 & 0 & \cdots & -1 & x & a_{n-2} \\ 0 & 0 & 0 & 0 & \cdots & 0 & -1 & x+a_{n-1} \end{vmatrix}$$

$$= \begin{vmatrix} 0 & 0 & 0 & 0 & \cdots & 0 & 0 & f(x) \\ -1 & 0 & 0 & 0 & \cdots & 0 & 0 & (f(x) - a_0)/x \\ 0 & -1 & 0 & 0 & \cdots & 0 & 0 & \vdots \\ \vdots & \vdots & \vdots & \vdots & & \vdots & & \vdots \\ 0 & 0 & 0 & 0 & \cdots & -1 & 0 & x^2 + a_{n-1}x + a_{n-2} \\ 0 & 0 & 0 & 0 & \cdots & 0 & -1 & x + a_{n-1} \end{vmatrix}$$

Evaluating by cofactors along the first row shows that $\det(xI - C_f) = f(x)$. □

The linear transformation T is called *diagonalizable* if there exists a basis for V such that the matrix for T is a diagonal matrix.

Lemma A.4.6 *If $c_1, \ldots, c_k$ are the distinct eigenvalues of T, then any corresponding eigenvectors $v_1, \ldots, v_k$ are linearly independent.*

Proof. Choose polynomials $f_i(x)$ such that $f_i(c_j) = \delta_{ij}$. If $\sum_{j=1}^k a_j v_j = 0$, then multiplying by $f_i(T)$ gives $0 = f_i(T)\left(\sum_{j=1}^k a_j v_j\right) = \sum_{j=1}^k a_j f_i(c_j)v_j = a_i v_i$, which shows that $a_i = 0$ for all i. □

Proposition A.4.7 *Let* $c_1, \ldots, c_k$ *be the distinct eigenvalues of* T, *and let* $W_i = \ker(T - c_i I)$. *The following conditions are equivalent:*

(1) T *is diagonalizable;*

(2) *the characteristic polynomial of* T *is* $f(x) = (x - c_1)^{d_1} \cdots (x - c_k)^{d_k}$, *and* $\dim(W_i) = d_i$, *for all* i;

(3) $\dim(W_1) + \cdots + \dim(W_k) = \dim(V)$.

Proof. It is easy to see that (1) $\Rightarrow$ (2). Since the degree of $f(x)$ is equal to $\dim(V)$, we have (2) $\Rightarrow$ (3). Finally, if (3) holds then we can choose a basis for each of the nonzero subspaces W_i, and then by the previous lemma the union of these bases must be a basis for V. $\square$

Theorem A.4.8 (Cayley–Hamilton) *If* $f(x)$ *is the characteristic polynomial of an* $n \times n$ *matrix* A, *then* $f(A) = 0$.

Proof. Let $f(x) = \mid xI - A \mid = a_0 + a_1 x + \cdots + x^n$, and let $B(x)$ be the adjoint of $xI - A$. The entries of $B(x)$ are cofactors of $xI - A$, and so they are polynomials of degree $\leq n - 1$, which allows us to factor out powers of x to get $B(x) = B_0 + B_1 x + \cdots + B_{n-1} x^{n-1}$. By definition of the adjoint, $(xI - A) \cdot B(x) = \mid xI - A \mid I$, or

$$(xI - A)(B_0 + B_1 x + \cdots + B_{n-1} x^{n-1}) = (a_0 + a_1 x + \cdots + x^n) \cdot I .$$

If we equate coefficients of x and then multiply by increasing powers of A we obtain

$$
\begin{array}{ll}
-AB_0 = a_0 I & \text{and then} \qquad -AB_0 = a_0 I , \\
B_0 - AB_1 = a_1 I , & AB_0 - A^2 B_1 = a_1 A , \\
\quad \vdots & \quad \vdots \\
B_{i-1} - AB_i = a_i I , & A^i B_{i-1} - A^{i+1} B_i = a_{i-1} A^i , \\
\quad \vdots & \quad \vdots \\
B_{n-1} = I , & A^n B_{n-1} = A^n .
\end{array}
$$

If we add the right hand columns, we get $0 = a_0 I + a_1 A + \cdots + A^n$, or simply $f(A) = 0$. $\square$

We say that a polynomial $f(x) \in F[x]$ *annihilates* T if the linear transformation $f(T)$ is the zero function. The Cayley–Hamilton theorem shows that T is annihilated by its characteristic polynomial. Among all nonzero polynomials $f(x) \in F[x]$ such that $f(T) = 0$, let $p(x)$ be a monic polynomial of minimal degree. If $f(x)$ is any polynomial with $f(T) = 0$, then we can use the division algorithm for polynomials to write $f(x) = p(x)q(x) + r(x)$, where $r(x) = 0$ or

$\deg(r(x)) < \deg(p(x))$. Now $r(T) = f(T) - p(T)q(T) = 0$, and this contradicts the choice of $p(x)$ unless $r(x) = 0$. We conclude that $p(x)$ is a factor of every polynomial $f(x)$ which annihilates T.

Definition A.4.9 *The monic polynomial of minimal degree which annihilates T is called the* minimal polynomial *of T.*

Proposition A.4.10 *The characteristic and minimal polynomials of a linear transformation have the same roots (except for multiplicities).*

Proof. Let $p(x)$ be the minimal polynomial of T. We need to prove that $p(c) = 0 \Leftrightarrow c$ is an eigenvalue of T. First assume that $p(c) = 0$. Then $p(x) = (x - c)q(x)$ for some $q(x) \in F[x]$, and since $\deg(q(x)) < \deg(p(x))$, we have $q(T) \neq 0$. Therefore there exists $w \in V$ with $q(T)(w) \neq 0$. Letting $v = q(T)(w)$ gives $(T - cI)(v) = (T - cI)q(T)(w) = p(T)(w) = 0$, and thus v is an eigenvector with eigenvalue c.

Conversely, assume that c is an eigenvalue, say $T(v) = cv$ for some $v \neq 0$. Then $p(c)v = p(T)v = 0$ implies $p(c) = 0$ since $v \neq 0$. $\square$

A subspace W of V is called *invariant under T* if $T(v) \in W$ for all $v \in W$.

Lemma A.4.11 *Let W be any proper invariant subspace of V. If the minimal polynomial of T is a product of linear factors, then there exists $v \notin W$ such that $(T - cI)(v) \in W$ for some eigenvalue c of T.*

Proof. Let $z \in V$ be any vector which is not in W. Since $p(T) = 0$ we certainly have $p(T)(z) \in W$. Let $g(x)$ be a monic polynomial of minimal degree with $g(T)(z) \in W$. An argument similar to that for the minimal polynomial shows that $g(x)$ is a factor of every polynomial $f(x)$ with the property that $f(T)(z) \in W$. In particular, $g(x)$ is a factor of $p(x)$, and so by assumption it is a product of linear factors. Suppose that $x - c$ is a linear factor of $g(x)$, say $g(x) = (x - c)h(x)$ for some $h(x) \in F[x]$. Then $\deg(h(x)) < \deg(g(x))$, and so $v = h(T)(z) \notin W$, but $(T - cI)(v) = (T - cI)h(T)(z) = g(T)(z) \in W$. $\square$

Theorem A.4.12 *The linear transformation T is diagonalizable if and only if the minimal polynomial for T is a product of distinct linear factors.*

Proof. If T is diagonalizable, then it follows immediately that the minimal polynomial is a product of distinct linear factors.

Conversely, assume that the minimal polynomial for T is a product of distinct linear factors. Let W be the subspace spanned by all eigenvectors of T. This subspace has a basis of eigenvectors, and so it is an invariant subspace. To prove that T is diagonalizable it suffices to show that $W = V$, so the method of proof is to assume that $W \neq V$ and show that this contradicts the assumption that $p(x)$ has no multiple roots. With this in mind, let $p(x) = (x - c)q(x)$.

Since $W \neq V$, the previous lemma shows that there exists $v \in V$ such that $v \notin W$ but $w = (T - cI)(v) \in W$. Then $q(T)(v)$ is an eigenvector for c, since

$$(T - cI)(q(T)(v)) = p(T)(v) = 0 .$$

Dividing $q(x)$ by $x - c$ gives $q(x) = (x - c)h(x) + q(c)$, for some $h(x) \in F[x]$. Therefore

$$q(c)v = q(T)(v) - h(T)(T - cI)(v) .$$

But $q(T)(v) \in W$ because it is an eigenvector, and since W is T-invariant and $(T - cI)(v) \in W$, we also have $h(T)(T - cI)(v) \in W$. We conclude that $q(c) = 0$, since $q(c)v \in W$ but $v \notin W$. This shows that $p(x)$ has a multiple root, the desired contradiction. $\square$

A.5 Noncommutative quotient rings

This section of the appendix provides the details of the proof of Theorem 3.4.8.

An element c of the ring R is *regular* if $ca = 0$ or $ac = 0$ implies $a = 0$, for all $a \in R$. We will use $\mathcal{C}(0)$ to denote the set of regular elements of R. We will use the following definition of a ring of quotients.

Definition A.5.1 *The ring R is said to be a* left order *in the ring Q if*

(i) *R is a subring of Q,*

(ii) *every regular element of R is invertible in Q,*

(iii) *for each $q \in Q$ there exists a regular element $c \in R$ such that $cq \in R$.*

If R is a left order in Q and $a, c \in R$ with $c \in \mathcal{C}(0)$, then $ac^{-1} \in Q$, so ac^{-1} can be rewritten in the standard form $ac^{-1} = c_1^{-1}a_1$, for some $c_1 \in \mathcal{C}(0)$ and some $a_1 \in R$.

Definition A.5.2 *We say that the ring R satisfies the* left Ore condition *if for all $a, c \in R$ such that c is regular, there exist $a_1, c_1 \in R$ such that c_1 is regular and $c_1 a = a_1 c$.*

Lemma A.5.3 *If R satisfies the left Ore condition, then for any $c, d \in \mathcal{C}(0)$ there exist $c_1, d_1 \in \mathcal{C}(0)$ with $d_1 c = c_1 d$.*

Proof. Given $c, d \in \mathcal{C}(0)$, the Ore condition guarantees the existence of $d_1 \in \mathcal{C}(0)$ and $c_1 \in R$ such that $d_1 c = c_1 d$. To show that $c_1 \in \mathcal{C}(0)$, let $x \in R$.

If $xc_1 = 0$, then $d_1 c = c_1 d$, and so $x(d_1 c) = x(c_1 d) = (xc_1)d = 0$. Thus $x = 0$ since $d_1 c$ is regular.

On the other hand, if $c_1 x = 0$, then the Ore condition implies the existence of $c_2 \in \mathcal{C}(0)$ and $d_2 \in R$ with $c_2 d = d_2 c$. Furthermore, since $d_1 \in \mathcal{C}(0)$, there exist $d_1' \in \mathcal{C}(0)$ and $d_2' \in R$ with $d_1'd_2 = d_2'd_1$. We next show that $d_2'c_1 = d_1'c_2$. This follows from the fact that

$$
\begin{aligned}
(d_2'c_1)d &= d_2'(c_1d) = d_2'(d_1c) = (d_2'd_1)c \\
&= (d_1'd_2)c = d_1'(d_2c) = d_1'(c_2d) \\
&= (d_1'c_2)d
\end{aligned}
$$

and d is regular. Thus $c_1x = 0$ implies $(d_1'c_2)x = (d_2'c_1)x = d_2'(c_1x) = 0$, and so $x = 0$ since $d_1'c_2$ is regular. This completes the proof that c_1 is regular. $\square$

The next lemma shows that it is possible to find common denominators. In fact, in Lemma A.5.4 we can take $a_i = 1$, for all i. Then Lemma A.5.3 implies that we can obtain regular elements c and c_i' such that $c_i'c_i = c$, for all i. Thus if the Ore condition holds we can find a common denominators for any finite set of regular elements.

Lemma A.5.4 *Let R be a ring which satisfies the left Ore condition. Let $\{a_i\}_{i=1}^n$ be a set of elements of R, and let $\{c_i\}_{i=1}^n$ be a set of regular elements of R. Then there exist a regular element c and elements $\{a_i'\}_{i=1}^n$ such that $ca_i = a_i'c_i$ for $1 \le i \le n$.*

Proof. Given $\{a_i\}_{i=1}^n \subseteq R$ and $\{c_i\}_{i=1}^n \subseteq \mathcal{C}(0)$, applying the Ore condition to a_1 and c_1 gives $d_1 \in \mathcal{C}(0)$ and $b_1 \in R$ with $d_1a_1 = b_1c_1$. Applying the Ore condition again, to d_1a_2 and c_2, we obtain $d_2 \in \mathcal{C}(0)$ and $b_2 \in R$ with $d_2(d_1a_2) = b_2c_2$. Thus we have $(d_2d_1)a_1 = (d_2b_1)c_1$ and $(d_2d_1)a_2 = b_2c_2$. Continuing inductively, we let $c = d_n \cdots d_2d_1$, and define the elements $a_n' = b_n$, $a_{n-1}' = d_nb_{n-1}, \ldots, a_1' = d_n \cdots d_2b_1$. $\square$

The next remarks provide the motivation for the definition of addition and multiplication that is given in Theorem A.5.5.

To add $c^{-1}a$ and $d^{-1}b$, we write $cd^{-1} = d_1^{-1}c_1$, which also yields $dc^{-1} = c_1^{-1}d_1$, and then

$$
\begin{aligned}
c^{-1}a + d^{-1}b &= d^{-1}dc^{-1}a + d^{-1}c_1^{-1}c_1b \\
&= d^{-1}c_1^{-1}d_1a + d^{-1}c_1^{-1}c_1b \\
&= (c_1d)^{-1}(d_1a + c_1b) \, .
\end{aligned}
$$

To multiply $c^{-1}a$ and $d^{-1}b$, we write $ad^{-1} = d_1^{-1}a_1$, and then

$$
\begin{aligned}
c^{-1}a \cdot d^{-1}b &= c^{-1}d_1^{-1}a_1b \\
&= (d_1c)^{-1}(a_1b) \, .
\end{aligned}
$$

Theorem A.5.5 *Let R be a ring which satisfies the left Ore condition. Then there exists an extension Q of R such that R is a left order in Q.*

Proof. Let $\mathcal{C}(0)$ denote the set of regular elements of R.

(i) The equivalence relation $\sim$. We introduce a relation on $\mathcal{C}(0) \times R$ as follows. For $c, d \in \mathcal{C}(0)$ and $a, b \in R$ we define $(c, a) \sim (d, b)$ if there exist $c_1, d_1 \in \mathcal{C}(0)$ with $d_1 c = c_1 d$ and $d_1 a = c_1 b$. It is clear that $\sim$ is reflexive and symmetric. To show that $\sim$ is transitive, suppose that $(c, a) \sim (d, b)$ and $(d, b) \sim (s, r)$ for $c, d, s \in \mathcal{C}(0)$ and $a, b, r \in R$. By the definition of $\sim$, there exist $c_1, d_1, d_2, s_2 \in \mathcal{C}(0)$ with $d_1 c = c_1 d$, $d_1 a = c_1 b$, $s_2 d = d_2 s$, and $s_2 b = d_2 r$. From the Ore condition we obtain $c_1', s_2' \in \mathcal{C}(0)$ with $c_1' s_2 = s_2' c_1$. Then $s_2' d_1$ and $c_1' d_2$ are regular, with

$$
\begin{aligned}
(s_2' d_1) c &= s_2'(d_1 c) = s_2'(c_1 d) = (s_2' c_1) d \\
&= (c_1' s_2) d = c_1'(s_2 d) = c_1'(d_2 s) \\
&= (c_1' d_2) s \,.
\end{aligned}
$$

A similar computation shows that $(s_2' d_1) a = (c_1' d_2) r$, and therefore $(c, a) \sim (s, r)$.

For $c \in \mathcal{C}(0)$ and $a \in R$, the equivalence class of (c, a) will be denoted by $[c, a]$, and the collection of all equivalence classes of the form $[c, a]$ will be denoted by Q. It is useful to note that if d is regular, then $[dc, da] = [c, a]$.

(ii) The definition of addition and multiplication. We define addition and multiplication of equivalence classes in $(\mathcal{C}(0) \times R)/\sim$ as follows. Given $[c, a]$ and $[d, b]$ in $(\mathcal{C}(0) \times R)/\sim$, by the Ore condition and Lemma A.5.3 there exist $c_1, d_1 \in \mathcal{C}(0)$ with $d_1 c = c_1 d$. We define addition of equivalence classes by setting

$$
[c, a] + [d, b] = [c_1 d, d_1 a + c_1 b] \,.
$$

Similarly, to define multiplication of equivalence classes, by the Ore condition there exist $d_1 \in \mathcal{C}(0)$ and $a_1 \in R$ with $d_1 a = a_1 d$, and so we can set

$$
[c, a][d, b] = [d_1 c, a_1 b] \,.
$$

We next show that these operations respect the equivalence relation $\sim$.

(iii) Addition is well-defined. We first need to show that the addition we have defined is independent of the choice of $c_1, d_1 \in \mathcal{C}(0)$. Suppose that $[c, a]$ and $[d, b]$ in $(\mathcal{C}(0) \times R)/\sim$, with $c_1, d_1 \in \mathcal{C}(0)$ such that $d_1 c = c_1 d$. Suppose that we also have $c_1', d_1' \in \mathcal{C}(0)$ with $d_1' c = c_1' d$. Applying Lemma A.5.3 to c_1 and c_1', we obtain $c_2, d_2 \in \mathcal{C}(0)$ with $d_2 c_1 = c_2 c_1'$, and so $d_2(c_1 d) = c_2(c_1' d)$. To show that $(c_1 d, d_1 a + c_1 b) \sim (c_1' d, d_1' a + c_1' b)$, we also need to show that $d_2(d_1 a + c_1 b) = c_2(d_1' a + c_1' b)$. Since $d_2(c_1 b) = (d_2 c_1) b = (c_2 c_1') b = c_2(c_1' b)$, we only need the fact that $d_2 d_1 = c_2 d_1'$. This follows from the computation

$$
\begin{aligned}
(c_2 d_1') c &= c_2(d_1' c) = c_2(c_1' d) = (c_2 c_1') d \\
&= (d_2 c_1) d = d_2(c_1 d) = d_2(d_1 c) \\
&= (d_2 d_1) c \,,
\end{aligned}
$$

because c is regular.

We next show that addition is independent of the choice of representatives for $[c,a]$ and $[d,b]$. Suppose that $(c',a') \sim (c,a)$ and $(d',b') \sim (d,b)$. From the equivalence relation there exist $c_2, c_2', d_2, d_2' \in \mathcal{C}(0)$ with $c_2'c = c_2c'$, $c_2'a = c_2a'$, $d_2'd = d_2d'$, and $d_2'b = d_2b'$. To define the relevant sums, we choose $c_1, d_1, c_1', d_1' \in \mathcal{C}(0)$ with $d_1c = c_1d$ and $d_1'c' = c_1'd'$.

We must show that $(c_1d, d_1a+c_1b) \sim (c_1'd', d_1'a'+c_1'b')$. Using Lemma A.5.3, we obtain d_3, d_3' in $\mathcal{C}(0)$ with $d_3(c_1d) = d_3'(c_1'd')$. Next, it is necessary to show that $d_3(d_1a + c_1b) = d_3'(d_1'a' + c_1'b')$, and we will do this by checking that $d_3d_1a = d_3'd_1'a'$ and $d_3c_1b = d_3'c_1'b'$.

To show that $d_3d_1a = d_3'd_1'a'$, we choose $c_4', d_4 \in \mathcal{C}(0)$ with $d_4(d_3d_1) = c_4'c_2$. Since $d_3(c_1d) = d_3'(c_1'd')$, we have $d_3(d_1c) = d_3'(d_1'c')$, and then multiplying by d_4 yields

$$
\begin{aligned}
(d_4d_3'd_1')c' &= d_4(d_3'd_1'c') = d_4(d_3d_1c) = (d_4d_3d_1)c \\
&= (c_4'c_2)c = c_4'(c_2c) = c_4'(c_2c') \\
&= (c_4'c_2)c' \, ,
\end{aligned}
$$

from which we can cancel c' to obtain $d_4d_3'd_1' = c_4'c_2$. Now

$$
\begin{aligned}
d_4(d_3d_1a) &= (d_4d_3d_1)a = (c_4'c_2)a = c_4'(c_2a) \\
&= c_4'(c_2a') = (c_4'c_2)a' = (d_4d_3'd_1')a' \\
&= d_4(d_3'd_1'a') \, ,
\end{aligned}
$$

and since d_4 is regular, we can cancel it to obtain $d_3d_1a = d_3'd_1'a'$.

Similarly, to show that $d_3c_1b = d_3'c_1'b'$, let $c_4, d_4' \in \mathcal{C}(0)$ with $c_4(d_3c_1) = d_4'd_2'$. Since $d_3(c_1d) = d_3'(c_1'd')$, multiplying by c_4 yields

$$
\begin{aligned}
c_4(d_3'c_1'd') &= c_4(d_3c_1d) = (c_4d_3c_1)d = (d_4'd_2')d \\
&= d_4'(d_2'd) = d_4'(d_2d') \, ,
\end{aligned}
$$

from which we can cancel d' to obtain $c_4d_3'c_1' = d_4'd_2$. Now

$$
\begin{aligned}
c_4(d_3c_1b) &= (c_4d_3c_1)b = (d_4'd_2')b = d_4'(d_2'b) \\
&= d_4'(d_2b') = (d_4'd_2)b' = (c_4d_3'c_1')b' \\
&= c_4(d_3'c_1'b') \, ,
\end{aligned}
$$

and since c_4 is regular, we can cancel it to obtain $d_3c_1b = d_3'c_1'b'$.

(iv) Multiplication is well-defined. To multiply $[c,a]$ by $[d,b]$, by the Ore condition there exist $d_1 \in \mathcal{C}(0)$ and $a_1 \in R$ with $d_1a = a_1d$, and so can define $[c,a] \cdot [d,b] = [d_1c, a_1b]$. We first need to show that this operation is independent of the choice of a_1 and d_1. Suppose that we also have $d_1' \in \mathcal{C}(0)$ and $a_1' \in R$ with $d_1'a = a_1'd$.

Since $d_1 c$ and $d_1' c$ are regular, by Lemma A.5.3 there exist $d_2, d_2' \in \mathcal{C}(0)$ such that $d_2(d_1 c) = d_2'(d_1' c)$. To show that $(d_1 c, a_1 b) \sim (d_1' c, a_1' b)$, we also need to show that $d_2(a_1 b) = d_2'(a_1' b)$. Since c is regular, we have $d_2 d_1 = d_2' d_1'$. Therefore

$$
\begin{aligned}
(d_2 a_1) d &= d_2(a_1 d) = d_2(d_1 a) = (d_2 d_1) a \\
&= (d_2' d_1') a = d_2'(d_1' a) = d_2'(a_1' d) \\
&= (d_2' a_1') d \,,
\end{aligned}
$$

and so $d_2 a_1 = d_2' a_1'$ since d is regular. Multiplying by b yields the desired equality $d_2(a_1 b) = d_2'(a_1' b)$.

Next, we need to show that the formula for multiplication is independent of the choice of representatives for the equivalence classes. Suppose that $(c', a') \sim (c, a)$ and $(d', b') \sim (d, b)$. From the equivalence relation there exist $c_2, c_2', d_2, d_2' \in \mathcal{C}(0)$ with $c_2' c = c_2 c'$, $c_2' a = c_2 a'$, $d_2' d = d_2 d'$, and $d_2' b = d_2 b'$. To define the relevant products, we choose $d_1, d_1' \in \mathcal{C}(0)$ and $a_1, a_1' \in R$ with $d_1 a = a_1 d$ and $d_1' a' = a_1' d'$. We must show that $(d_1 c, a_1 b) \sim (d_1' c', a_1' b')$.

By Lemma A.5.3 there exist $d_2, d_2' \in \mathcal{C}(0)$ with $d_2 d_1 c = d_2' d_1' c'$. We only need to show that $d_2 a_1 b = d_2' a_1' b'$. We can choose $d_3, d_3' \in \mathcal{C}(0)$ with $d_3(d_2 d_1) = d_3' c_2'$. Then

$$
\begin{aligned}
(d_3 d_2' d_1') c' &= d_3(d_2' d_1' c') = d_3(d_2 d_1 c) = (d_3 d_2 d_1) c \\
&= (d_3' c_2') c = d_3'(c_2' c) = d_3'(c_2 c') \\
&= (d_3' c_2) c' \,,
\end{aligned}
$$

and so $d_3 d_2' d_1' = d_3' c_2$ since c' is regular. Next, we have

$$
\begin{aligned}
d_3(d_2' d_1' a') &= (d_3 d_2' d_1') a' = (d_3' c_2) a' = d_3'(c_2 a') \\
&= d_3'(c_2' a) = (d_3' c_2') a = (d_3 d_2 d_1) a \\
&= d_3(d_2 d_1 a) \,,
\end{aligned}
$$

and so $d_2' d_1' a' = d_2 d_1 a$ since d_3 is regular. This shows that

$$
\begin{aligned}
d_2'(a_1' d') &= d_2'(d_1' a') = d_2' d_1' a' = d_2 d_1 a \\
&= d_2(d_1 a) = d_2(a_1 d) \,.
\end{aligned}
$$

The next step is to choose $d_4, d_4' \in \mathcal{C}(0)$ with $d_4(d_2 a_1) = d_4' d_2'$. Then

$$
\begin{aligned}
(d_4 d_2' a_1') d' &= d_4(d_2' a_1' d') = d_4(d_2 a_1 d) = (d_4 d_2 a_1) d \\
&= (d_4' d_2') d = d_4'(d_2' d) = d_4'(d_2 d') \\
&= (d_4' d_2) d' \,,
\end{aligned}
$$

and so $d_4 d_2' a_1' = d_4' d_2$ since d' is regular. Therefore

$$
\begin{aligned}
d_4(d_2 a_1 b) &= (d_4 d_2 a_1) b = (d_4' d_2') b = d_4'(d_2' b) \\
&= d_4'(d_2 b') = (d_4' d_2) b' = (d_4 d_2' a_1') b' \\
&= d_4(d_2' a_1' b') \,,
\end{aligned}
$$

and so $d_2(a_1 b) = d_2'(a_1' b')$ since d_4 is regular. This completes the proof that multiplication is well-defined.

(v) Q is a ring. It is clear that the commutative law holds for addition. Using the definition of addition to find $[c, a] + [1, 0]$, we can choose $c_1 = c$ and $d_1 = 1$, and then $[c, a] + [1, 0] = [c \cdot 1, 1 \cdot a + c \cdot 0] = [c, a]$, and so the element $[1, 0]$ serves as an additive identity for Q. Similar computations show that the additive inverse of $[c, a]$ is $[c, -a]$ and that $[1, 1]$ is a multiplicative identity for Q.

Lemma A.5.4 states that if $\{a_i\}_{i=1}^n$ is a set of elements in R and $\{c_i\}_{i=1}^n$ is a set of regular elements of R, then there exist a regular element c and elements $\{a_i'\}_{i=1}^n$ such that $ca_i = a_i' c_i$ for $1 \leq i \leq n$. If we take $a_i = 1$, for all i, then $c = c_i' c_i$ for each i, with c_i' regular. Thus for each i we have $[c_i, a_i] = [c_i' c_i, c_i' a_i] = [c, c_i' a_i]$. This fact is useful in verifying the associative law for addition. The verification of the associative and distributive laws is left as an exercise for the reader.

(vi) R is a left order in Q. We can identify an element $a \in R$ with the equivalence class $[1, a]$. If $(1, a) \sim (1, b)$ for $b \in R$, then there exist $c, d \in \mathcal{C}(0)$ with $d \cdot 1 = c \cdot 1$ and $d \cdot a = c \cdot b$, which forces $a = b$. We can formalize this identification by defining the function $\eta : R \to Q$ by $\eta(a) = [1, a]$, for all $a \in R$, and showing that η is a ring homomorphism.

Given a regular element c in R, the corresponding element $[1, c]$ in Q is invertible, with $[1, c]^{-1} = [c, 1]$. Given any element $[c, a] \in Q$, we have $[1, c][c, a] = [1, a]$, and this completes the proof that R (or, more properly, $\eta(R)$) is a left order in Q. □

A.6 The ring of algebraic integers

Definition A.6.1 *A complex number c is called an* algebraic integer *if it is a root of a monic polynomial with integer coefficients.*

Proposition A.6.2 *An algebraic integer that is a rational number is an ordinary integer.*

Proof. Let a/b be a rational number, where a and b are relatively prime numbers in $\mathbf{Z}$. Suppose that a/b is a root of the monic polynomial $f(x) = x^n + c_{n-1} x^{n-1} + \cdots + c_1 x + c_0$, where the coefficients of $f(x)$ are integers. Then

$$\left(\frac{a}{b}\right)^n + c_{n-1} \left(\frac{a}{b}\right)^{n-1} + \cdots + c_1 \left(\frac{a}{b}\right) + c_0 = 0$$

and multiplying through by b^n shows that a^n is divisible by b. This contradicts the assumption that a and b are relatively prime. □

Proposition A.6.3 *The following conditions are equivalent for a complex number u:*

(1) *u is is an algebraic integer;*

(2) *the subring $\mathbf{Z}[u]$ generated by u and $\mathbf{Z}$ is finitely generated as a $\mathbf{Z}$-module;*

(3) *there exists a subring R with $\mathbf{Z} \subseteq R \subseteq \mathbf{C}$ such that $\mathbf{Z}[u] \subseteq R$ and R is finitely generated as a $\mathbf{Z}$-module.*

Proof. (1) $\Rightarrow$ (2) Let $f(x) = x^n + a_{n-1}x^{n-1} + \cdots + a_1 x + a_0$ be a monic polynomial in $\mathbf{Z}[x]$ with $f(u) = 0$. Then $u^n = -a_{n-1}u^{n-1} - \cdots - a_1 u - a_0$, and an inductive argument shows that $\sum_{i=0}^{n-1} \mathbf{Z}u^i$ contains u^m for all $m \geq 0$, so $\mathbf{Z}[u]$ is generated as an $\mathbf{Z}$-module by $\{1, u, \ldots, u^{n-1}\}$.

(2) $\Rightarrow$ (3) We can simply let $R = \mathbf{Z}[u]$.

(3) $\Rightarrow$ (1) Assume that $R = \sum_{i=1}^{n} \mathbf{Z}t_i$, for $t_1, \ldots, t_n \in \mathbf{C}$. Since $ut_i \in R$, for $1 \leq i \leq n$ we have $ut_i = \sum_{j=1}^{n} a_{ij}t_j$, with $a_{ij} \in \mathbf{Z}$. The coefficients a_{ij} define a matrix $A = [a_{ij}]$, and if we let x be the column vector with entries $t_1, \ldots, t_n$, then in matrix form we have the equation $uIx = Ax$, where I is the $n \times n$ identity matrix.

Let $d = \det(uI - A)$. Multiplying the matrix equation $(uI - A)x = 0$ by the adjoint (see Proposition A.3.7) of the matrix $uI - A$, we obtain $dIx = 0$, so $dt_i = 0$ for $1 \leq i \leq n$. Since $dR = (0)$, it follows that $d = 0$. Expanding $\det(uI - A)$ yields an expression of the form $u^n + b_{n-1}u^{n-1} + \cdots + b_1 u + b_0$, which must equal zero, and this produces the necessary monic polynomial in $\mathbf{Z}[x]$ that has u as a root. $\square$

Proposition A.6.4 *The set of algebraic integers is a subring of $\mathbf{C}$.*

Proof. Let u, v be algebraic integers, with $f(u) = 0$ and $g(v) = 0$ for monic polynomials in $\mathbf{Z}[x]$ of degrees n and m, respectively. The subring $\mathbf{Z}[u, v]$ is generated as a $\mathbf{Z}$-module by $\{u^i v^j\}$, for $0 \leq i < n$ and $0 \leq j < m$. Since $u - v$ and uv belong to $R[u, v]$, it follows from Proposition A.6.3 (3) that both are algebraic integers. This implies that the set of algebraic integers is a subring of $\mathbf{C}$. $\square$

Proposition A.6.5 *Let c be a complex number that is an average of mth roots of unity. If c is an algebraic integer, then either $c = 0$ or $|c| = 1$.*

Proof. Let c be an average of mth roots of unity, say

$$c = \frac{a_1 + a_2 + \cdots + a_d}{d} \, ,$$

where $a_1, a_2, \ldots, a_d$ are roots of $x^m - 1$. Since $|a_i| = 1$ for $1 \leq i \leq d$, the triangle inequality for complex numbers shows that

$$|c| \leq \frac{|a_1|}{d} + \frac{|a_2|}{d} + \cdots + \frac{|a_d|}{d} = 1 \, .$$

Assume that c is an algebraic integer, so that c is a root of a polynomial $f(x)$ with integer coefficients. We may assume without loss of generality that $f(x)$ is irreducible over $\mathbf{Q}$. Let G be the Galois group of the polynomial $f(x)$ over $\mathbf{Q}$. (See Chapter 8 of [4] for the necessary definitions and results.) If σ is any element of G, then since σ is an automorphism of the splitting field of $f(x)$ we have

$$\sigma(c) = \frac{1}{d}\left(\sigma(a_1) + \sigma(a_2) + \cdots + \sigma(a_d)\right) .$$

Furthermore, since a_i is a root of $x^m - 1$ it follows that $\sigma(a_i)$ is a root of $x^m - 1$, for $1 \le i \le d$. Therefore $|\sigma(c)| \le 1$, as shown above for $|c|$. Finally, we define

$$b = \prod_{\sigma \in G} \sigma(c) .$$

For each automorphism $\sigma \in G$, the number $\sigma(c)$ is also a root of $f(x)$, and so $\sigma(c)$ is an algebraic integer. Since the set of algebraic integers is a subring of $\mathbf{C}$, it follows that the product b of these values is an algebraic integer. From the definition of b we see that $\sigma(b) = b$ for all $\sigma \in G$, so $b \in \mathbf{Q}$ since the subfield of elements left fixed by G is precisely $\mathbf{Q}$. It follows from Proposition A.6.2 that b is an ordinary integer. Thus if $|c| \ne 1$, then $|b| < 1$, so $b = 0$. This forces $c = 0$, completing the proof. $\square$

Bibliography

[1] Alperin, J. L., with R. B. Bell, *Groups and Representations*, Graduate Texts in Mathematics, Vol. 162, Springer-Verlag, New York, 1995.

[2] Anderson, F. W., and K. R. Fuller, *Rings and Categories of Modules*, Graduate Texts in Mathematics, Vol. 13, 2nd Ed., Springer-Verlag, New York, 1992.

[3] Auslander, M., and D. A. Buchsbaum, *Groups, Rings, Modules*, Harper and Row, New York, 1974.

[4] Beachy, J. A., and W. D. Blair, *Abstract Algebra*, 2nd Ed., Waveland Press, Prospect Heights, Ill., 1996.

[5] Brewer, J. W., and M. K. Smith, Ed., *Emmy Noether: a Tribute to Her Life and Work*, Marcel Dekker, New York–Basel, 1981.

[6] Collins, M. J., *Representations and Characters of Finite Groups*, Cambridge Studies in Advanced Mathematics, Vol. 22, Cambridge University Press, Cambridge, 1990.

[7] Coutinho, S. C., *A Primer of Algebraic D-Modules*, London Mathematical Society Student Texts, Vol. 33, Cambridge University Press, Cambridge, 1995.

[8] Eisenbud, D., *Commutative Algebra with a View toward Algebraic Geometry*, Graduate Texts in Mathematics, Vol. 150, Springer-Verlag, New York, 1995.

[9] Goodearl, K. R., and R. B. Warfield, Jr, *An Introduction to Noncommutative Noetherian Rings*, London Mathematical Society Student Texts, Vol. 16, Cambridge University Press, Cambridge, 1989.

[10] Herstein, I. N., *Noncommutative Rings*, Carus Mathematical Monographs, Vol. 15, Mathematical Association of America, Washington, D.C., 1968.

[11] Herstein, I. N., *Abstract Algebra*, 3rd Ed., Prentice-Hall, Upper Saddle River, N.J., 1996.

[12] Hoffman, K., and R. Kunze, *Linear Algebra*, 2nd Ed., Prentice-Hall, Upper Saddle River, N.J., 1971.

[13] Hungerford, T., *Algebra*, Graduate Texts in Mathematics, Vol. 73, Springer-Verlag, New York, 1974.

[14] Jacobson, N., *Structure of Rings*, Colloquium Publications, Vol. 37, American Mathematical Society, Providence, R.I., 1964.

[15] Jacobson, N., *Basic Algebra I*, 2nd Ed., W. H. Freeman & Company Publishers, San Francisco, 1985.

[16] Jacobson, N., *Basic Algebra II*, 2nd Ed., W. H. Freeman & Company Publishers, San Francisco, 1989.

[17] James, G. D., and M. W. Liebeck, *Representations and Characters of Groups*, Cambridge University Press, Cambridge, 1993.

[18] Lam, T. Y., *A First Course in Noncommutative Rings*, Graduate Texts in Mathematics, Vol. 131, Springer-Verlag, New York, 1991.

[19] Lambek, J., *Lectures on Rings and Modules*, 3rd Ed., Chelsea Publishing Co., New York, 1986.

[20] Lang, S., *Algebra*, 3rd Ed., Addison-Wesley Publishing Co., Inc., Reading, Mass., 1993.

[21] Matsumura, H., *Commutative Ring Theory*, Cambridge University Press, Cambridge, 1986.

[22] Passman, D. S., *A Course in Ring Theory*, Brooks/Cole Publishing Co., Pacific Grove, Cal., 1991.

[23] Sharp, R. Y., *Steps in Commutative Algebra*, London Mathematical Society Student Texts, Vol. 19, Cambridge University Press, Cambridge, 1990.

[24] Van der Waerden, B. L., *Algebra*, Springer-Verlag, New York, 1991.

[25] Van der Waerden, B. L., *A History of Algebra: from al-Khwarizmi to Emmy Noether*, Springer-Verlag, New York, 1985.

By topic:

Background: Herstein [11], Beachy and Blair [4]
Comprehensive: Van der Waerden [24], Lang [20], Hungerford [13], Auslander and Buchsbaum [3], Jacobson [15], [16]
Noncommutative rings: Jacobson [14], Lambek [19], Herstein [10], Anderson and Fuller [2], Goodearl and Warfield [9], Lam [18], Passman [22]
Commutative rings: Matsumura [21], Sharp [23], Eisenbud [8]
Group representations: Collins [6], James and Liebeck [17], Alperin [1]
History: Van der Waerden [25], Brewer and Smith [5]

List of Symbols

Index